Physics and Metaphysics in Descartes and in His Reception

This volume explores the relationship between physics and metaphysics in Descartes' philosophy. According to the standard account, Descartes modified the objects of metaphysics and physics and inverted the order in which these two disciplines were traditionally studied. This book challenges the standard account in which Descartes prioritizes metaphysics over physics. It does so by taking into consideration the historical reception of Descartes and the ways in which Descartes himself reacted to these receptions in his own lifetime. The book stresses the diversity of these receptions by taking into account not only Cartesianisms but also anti-Cartesianisms and by showing how they retroactively highlighted different aspects of Descartes' works and theoretical choices. The historical aspect of the volume is unique in that it not only analyzes different constructions of Descartes that emerged in the eighteenth, nineteenth, and twentieth centuries, but also reflects on how his work was first read by philosophers across Europe. Taken together, the essays in this volume offer a fresh and up-to-date contribution to this important debate in early modern philosophy.

Delphine Antoine-Mahut is Professor of Early Modern Philosophy at the École normale supérieure de Lyon. Her research focuses on the genesis and reception of Cartesian dualism between the seventeenth and the nineteenth centuries. She is especially interested in the intersections of metaphysics and physiology.

Sophie Roux is Professor of the History and Philosophy of Science at the École normale supérieure, Paris. She published extensively on natural philosophy in the Early Modern period, but also in philosophy of science (e.g., on thought experiments and on mathematization) and in history of philosophy of science (e.g., on Duhem and on Couturat).

Routledge Studies in Seventeenth-Century Philosophy

Receptions of Descartes
Cartesianism and Anti-Cartesianism in Early Modern Europe
Edited by Tad M. Schmaltz

Material Falsity and Error in Descartes' Meditations
Cecilia Wee

Leibniz's Final System
Monads, Matter, and Animals
Glenn A. Hartz

Pierre Bayle's Cartesian Metaphysics
Rediscovering Early Modern Philosophy
Todd Ryan

Insiders and Outsiders in Seventeenth-Century Philosophy
Edited by G.A.J. Rogers, Tom Sorell and Jill Kraye

Vanishing Matter and the Laws of Nature
Descartes and Beyond
Edited by Dana Jalobeanu and Peter R. Anstey

Locke and Leibniz on Substance
Edited by Paul Lodge and Tom Stoneham

Locke's Science of Knowledge
Matthew Priselac

The Idea of Principles in Early Modern Thought
Interdisciplinary Perspectives
Edited by Peter R. Anstey

Physics and Metaphysics in Descartes and in His Reception
Edited by Delphine Antoine-Mahut and Sophie Roux

Physics and Metaphysics in Descartes and in His Reception

Edited by
Delphine Antoine-Mahut
and Sophie Roux

Routledge
Taylor & Francis Group
NEW YORK AND LONDON

First published 2019
by Routledge
711 Third Avenue, New York, NY 10017

and by Routledge
2 Park Square, Milton Park, Abingdon, Oxon OX14 4RN

Routledge is an imprint of the Taylor & Francis Group, an informa business

First issued in paperback 2021

Library of Congress Cataloging-in-Publication Data
A catalog record for this book has been requested

ISBN: 978-1-138-35144-8 (hbk)
ISBN: 978-1-03-209481-6 (pbk)
ISBN: 978-0-429-43529-4 (ebk)

Typeset in Sabon
by Apex CoVantage, LLC

Contents

Abbreviations

A	Leibniz, Gottfried Wilhelm. *Sämtliche Schriften und Briefe*. Darmstadt and Berlin: Berlin Academy, 1923–.
AG	Leibniz, Gottfried Wilhelm. *Philosophical Essays*. Translated by Roger Ariew and Daniel Garber. Indianapolis: Hackett, 1989.
AT	Descartes, René. *Œuvres de Descartes*, edited by Charles Adam and Paul Tannery, 11 vols. Paris: Vrin, 1964–1976, reed. 1996.
C	Leibniz, Gottfried Wilhelm. *Opuscules et fragments inédits*, edited by Louis Couturat. Paris: Alcan, 1903.
CSM	Descartes, René. *The Philosophical Writings of Descartes*, 2 vols. Translated by John Cottingham, Robert Stoothoff, Dugald Murdoch. Cambridge: Cambridge University Press, 1984.
CSMK	Descartes, René. *The Philosophical Writings of Descartes*, vol. 3 (*Correspondence*). Translated by John Cottingham, Robert Stoothoff, Dugald Murdoch and Anthony Kenny. Cambridge: Cambridge University Press, 1991.
DHC	Bayle, Pierre. *Dictionnaire historique et critique*, 4 vols. Amsterdam, Leiden, The Hague and Utrecht, 1740. Translated by Richard Popkin in Pierre Bayle, *Historical and Critical Dictionary. Selections*. Indianapolis: Bobbs-Merrill, 1965.
GM	Leibniz, Gottfried Wilhelm. *Die mathematische Schriften von Gottfried Wilhelm Leibniz*, edited by Carl I. Gerhardt, 7 vols. Reprint Georg Olms Verlag: Hildesheim, 1962.
GP	Leibniz, Gottfried Wilhelm. *Die philosophischen Schriften von Gottfried Wilhelm Leibniz*, edited by Carl I. Gerhardt, 7 vols. Reprint G. Olms Verlag: Hildesheim, 1965.
Gr	Leibniz, Gottfried Wilhelm. *Textes inédits*, edited by Gaston Grua, 2 vols. Paris: Presses universitaires de France, 1948.
JS	Malebranche, Nicolas. *Dialogues on Metaphysics and on Religion*, edited by Nicholas Jolley, Translated by David Scott. Cambridge: Cambridge University Press, 1997.
L	Leibniz, Gottfried Wilhelm. *Philosophical Papers and Letters*. Translated by Leroy E. Loemker, 2 vols. Dordrecht and Boston: Reidel, 1969 (reed. 1985, 1 vol.).

M	Leibniz, Gottfried Wilhelm. *Monadology* (1714). In GP IV. Translated in AG 213–25.
OCM	Malebranche, Nicolas. *Œuvres complètes de Malebranche*, edited by André Robinet. Paris: Vrin, 1958–1978.
OD	Bayle, Pierre. *Œuvres diverses*, 4 vols. The Hague, 1727–1731. Reprinted and supplemented with a fifth volume by Georg Olms Verlag: Hildesheim, 1968–1982.
ST	Thomas Aquinas. *Summa theologiae*, Blackfriars edition, 61 vols. New York: McGraw-Hill, 1964–1981.
T	Leibniz, Gottfried Wilhelm. *Theodicy*. In GP VI. Translated by E.M. Huggard. La Salle, IL: Open Court, 1985.

Introduction

*Delphine Antoine-Mahut and Sophie Roux**

Physics and Metaphysics: An Old Problem

It is known that Descartes modified the object of metaphysics, by defining it, no longer as the science of being qua being, or as the science of the first of all beings, but as the science of the first known beings—namely, the soul and God.[1] It is also known that Descartes modified the object of natural philosophy, by arguing that physical phenomena had to be explainable in terms of corpuscles, which were wholly determined by the properties of a matter whose essence was extension. Lastly, it is recognized that Descartes held, starting at a certain point in his career, that physics needed a metaphysical foundation, such that he ended up proposing a tree of knowledge that was inverted in relation to the Scholastic order of disciplines: metaphysics now came first, not last.[2] In that sense, one could say that based on a modification of the objects of metaphysics and physics, Descartes inverted the traditional order in which these two disciplines were studied, while also asserting the requirement that the one be grounded by the other. Nevertheless, three reasons can be given that justify why we seek to address the issue of the relation between physics and metaphysics anew.

1. From a historical point of view, the question of the relation between physics and metaphysics was a source of polemics starting with Descartes' first publications, and gradually became central for Descartes himself.

The popularization of the physiology contained in Part Five of the *Discourse on Method* and the scientific essays, particularly the *Dioptrics*, by Henricus Regius at the University of Utrecht, drew an immediate hostile response from the Utrecht theologians, due to the implications of the main thesis of this physics, namely, the reduction of matter to extension and the denial of substantial forms. For substantial forms were a boundary object between natural philosophy, first philosophy, and theology.[3]

To the memory of Desmond M. Clarke

While Regius's reaction to Gisbertus Voetius and Marten Schoock's attacks consisted in extending the principle of parsimony that Descartes had applied to physical forms, to the human soul and then the idea of the God—given that in Regius's view, this was Descartes' original gesture—Descartes, on his part, responded with the *Meditations* and the elaboration of the tree of knowledge of the prefatory letter to the *Principles of Philosophy*, which grounds physics in metaphysics. Against Regius, Descartes thus intended to set down once and for all what the relation between physics and metaphysics should be in Cartesian philosophy: a "Cartesian" philosophy would follow this order, and correlatively, a "non" or "anti"-Cartesian philosophy would grant the trunk autonomy in relation to the roots, or even cut it from its roots.

But Regius persisted in defending the coherence of his approach, viewing it as the means of ensuring the survival and credibility of Cartesianism. He denounced Descartes' compromise with the theologians, and the resulting philosophical abstractions—innate ideas of God and the soul are, for him, only revivals of these abstractions, which should not be allowed in an epistemology that claims to be rid of occult faculties. If Descartes saw Regius's work as the breaking apart of an organic totality, Regius, on his part, felt that he was actually reinforcing the coherence of the Cartesian system by applying the same principles everywhere.[4]

2. Whatever Descartes' authority may have been in defining what was to be the interpretation of his own doctrine, his physics and metaphysics ended up *de facto* being separated. Jacques Rohault, the French Cartesian with the most long-lasting publishing success, published a *Traité de physique*, in which he deliberately avoided metaphysical issues (and especially, theological issues). While the earliest condemnations of Cartesianism concerned Descartes' physical theses, because they were incompatible with the dogma of transubstantiation, from the 1690s onwards, the condemnations concerned metaphysical theses—particularly, the fact of requiring radical doubt to establish knowledge.[5] Indeed, anti-Cartesian polemics developed in a direction following the official condemnations: in the 1670s, Antoine Rochon and Ignace-Gaston Pardies attacked Cartesian physics, while in the 1690s, it was mainly his metaphysics which was attacked by the Bishop of Avranches, Pierre-Daniel Huet.[6]

The decoupling of physics and metaphysics grew more pronounced in the following centuries. When, around 1740, Newtonianism sounded the death knell of Cartesian science,[7] one of the ways to preserve part of Descartes' work was to separate it from more metaphysical writings. François Azouvi had indeed already observed this:

> Until the 1730s, Cartesianism remained monolithic. . . . By mid-century [the eighteenth century], it was possible to pick and choose elements of

> his doctrine, as they did not all age at the same rate. New species will appear: metaphysical Cartesians who defend Newtonian physics, and (Lockean) Newtonians who maintain an unchanged admiration for the *Discourse on Method*.[8]

According to the "Discours préliminaire" of the *Encyclopédie*, what remains of Cartesian philosophy is geometry, not metaphysics or physics—both of which were considered at that point to be out of date.[9]

When we turn to the nineteenth century, the return of metaphysics is striking. The Descartes of Part Four of the *Discourse* and of the *Meditations* becomes the figurehead of a rational psychology that claims to study the soul with just as much scientificity as positivism, yet without reducing it to a natural object. In French spiritualist historiography,[10] Descartes' way of proceeding comes to be valorized in contrast to Bacon's—Bacon being the reference point for the *Idéologues*, who hold that physics is the foundation of metaphysics. But by identifying "bad" ways of reading Descartes, notably, those which split off the physiological branch from the metaphysical branch, these historiographies also open onto alternative interpretations. While Charles Renouvier points to the Cartesian medical materialist tendency—coming out of Regius, continuing with La Mettrie and then Cabanis—as a perversion, Marx makes use of the same tendency, but treats it as a genuine potential resource in Descartes' work, one that was complementary to the materialism of the English.[11]

3. Given these conditions, it is not surprising that throughout the twentieth century, commentators defended various points of view concerning the three following questions. First, what exactly is the connection between physics and metaphysics in Descartes? Second, which of these is the primary discipline? And third, should this primacy be understood as *de facto* and temporal, or *de jure* and determined by the order of reasons? We can only give some examples here of the variety of responses given to these three questions.[12]

In 1922, in his Thèse complémentaire,[13] a future historian of science, Alexandre Koyré, reproached the holder of the chair in medieval philosophy at the Sorbonne, Étienne Gilson, for having followed Louis Liard's mistaken reading of Descartes as a physicist rather than a metaphysician:

> Mr Gilson seeks to present us with a scientist Descartes, solely concerned with science and developing his metaphysics merely as a sort of preface to his physics, building it hastily with elements obtained at random, modifying traditional doctrines as little as possible. It is a kind of mosaic, which serves as his banner, and as the banner covers the wares, it enables him to peddle his physics. . . . The history of science would not have been noticeably different if Descartes had not existed, whereas the history of philosophy would have been profoundly affected.[14]

Seventy years later, the publication in the same year of the works by Daniel Garber and Desmond M. Clarke brought the issue of the relation between physics (on the one hand) and metaphysics and theology (on the other hand) back to center stage. As he indicates in the very title of his work, *Descartes' Metaphysical Physics*, Daniel Garber held that Descartes' physics is fundamentally metaphysical and cannot be separated from questions concerning God and his action in the world.[15] In *Descartes' Theory of Mind*, Desmond M. Clarke defended the opposite view: Descartes was obliged by the theologians to take into account the metaphysical implications of his physics, but failed to establish the coherence of the emergent language of the new scientific revolution and the dualist language of the Scholastic tradition.[16] Along with Catherine Wilson, Desmond M. Clarke held that Regius's empiricist reading was a legitimate possible reading of Cartesianism, and quite coherent—indeed, the most coherent reading.[17]

Physics and Metaphysics: A New Way of Treating This Old Problem

We do not claim to cut the Gordian knot of the various responses that have been given to the question of how physics and metaphysics relate, but rather, to address it based on three methodological commitments that are currently widely shared in the community of scholars working on the intellectual history of the seventeenth century.

1. Cartesianisms reveal a certain historical plurality, as reflected in the title and approach employed in the edited volume *Qu'est-ce qu'être cartésien?* as well as in Tad Schmaltz's newest book, *Early Modern Cartesianisms*.[18] More precisely, if each element in the relation we are examining (i.e., metaphysics and physics) is located in its intellectual context, this plurality then presents two distinct dimensions to us.

Cartesian metaphysics has been compared to Scholastic metaphysics, whether the aim was to better grasp Descartes' own positions or to show how themes he had passed under silence reemerged in the first systematic works that proposed syntheses of the new philosophy and Scholastic doctrine.[19] This is an important point for our volume, because it allows us to attend to the influence of other types of metaphysics in the articulation of the new physics and physiology. Thus in addition to the resurgence of Aristotelianism, we must add the influence of Renaissance thought and of Plotinus on the Cambridge Platonists—Glanvill, More, Cudworth—in the discussion on the relations between soul and body in the second part of the seventeenth century.

Cartesian physics has been related to the emergence of new sciences and, contrary to the image of a "rationalist" (that is, exclusively metaphysical, for present purposes) Descartes, scholarship has begun to acknowledge that empiricist Cartesians really existed. Geneviève Rodis-Lewis was indeed the first to use this hybrid category to describe the thought of Pierre-Sylvain Régis or Robert Desgabets,[20] but the full usage of these results for further research came much later.[21] The most collective outcome so far is the work presented by Mihnea Dobre and Tammy Nyden in *Cartesian Empiricisms*.[22]

2. If this plurality is to be understood, this requires attending not just to Cartesianisms but also to anti-Cartesianisms.

How can one not see, for instance, that neither the Cartesian "party" nor the Society of Jesus can be studied as separate realms? On the one hand, Descartes' work circulated beyond his party, such that even his adversaries were affected by it: anti-Cartesian attacks were often presented according to the rules of the game that Descartes' work had established, concerning problems opened up by the tensions it presented, using intellectual tools borrowed from it. On the other hand, the definition of a party, a school, or a movement involves, to a considerable extent, a negative dimension of differentiation from another party, school, or movement—thus the study of Cartesians of the period cannot be separated from the study of Jesuit anti-Cartesianism.

Without claiming that the study of controversies is somehow the "panacea of intellectual history,"[23] the scholar seeking to understand the plurality of Cartesianisms thus has a lot to gain by attending to the controversies that opposed Cartesians and anti-Cartesians, as well as to the internal polemics during which the Cartesians sought to define what comprised their identity.[24] Leaving aside the distinction between debates, controversies, polemics, and quarrels, let us specify, as briefly as possible, the value of this polemical understanding of Cartesianism.

In fact, the study of controversies is almost a field of its own in the history of science—in this field, micro-configurations are emphasized (a restricted editorial or experimental space, a delimited historical period, or a minute scientific object), which enables—when the project is successful—the articulation of structures and processes traditionally classified as scientific (or belonging to the history of science) and structures and processes traditionally classified as social.[25] In the history of philosophy, the study of controversies can contribute to dispelling or avoiding two illusions:

i. A retrospective illusion. The historian of philosophy will tend to believe that a patient study of Descartes' writings will enable her to define their meaning once and for all, as well as point to who were the authentic Cartesians. Yet a controversy calls this meaning into question, given that the identification of the relevant questions for determining the demarcation between Cartesians and anti-Cartesians is sometimes itself controversial. Thus, from one controversy to another, the retrospective illusion consisting in the idea that the meaning of Cartesianism can be laid down, once and for all, for this meaning is both historicized and pluralized, without thereby being relativized, as it is inscribed within the objective conditions of the controversies, which both set down its key traits and transform it.

ii. A scholastic illusion. Because the historian of philosophy takes a disinterested stance on her object, she tends to think that she can grasp Cartesianism *sub specie aeternitatis*, thus neglecting certain issues that she deems to be inessential. Controversies enable one to resist this tendency, insofar as they operate on different discursive levels, disturb the standard categories, and make it more difficult to isolate texts, as all internalist readings do.

Hence, one comes to the view that Cartesianism and anti-Cartesianism are not essences but categories that came about historically, through debates, polemics, and controversies. It is also one of the great contributions of the theories of dissimulation to have problematized this essentialist illusion afflicting the historian of philosophy.[26]

3. The study of receptions has a further effect on the interpretations we can propose of the original corpus, as interpretive conflicts do not emerge from nowhere—they originate in the possibilities opened up by Descartes' texts themselves. If in the course of Descartes' reception the discussion of such and such problem of physics can occur independently of the rest of the system, it is indeed because a certain number of initial texts called for that kind of reading. Whether one chooses, with Renouvier and Marx, the Regius/La Mettrie/Cabanis line, in order to conceive of a "Cartesian materialism,"[27] or one opts for the Clauberg/Andreae/Victor Cousin line in order to defend, on the contrary, a "spiritualist" reading, in both cases, it is because the Cartesian writings themselves harbor both of the possible readings, in a relation that itself is a worthwhile object of study.[28]

Physics and Metaphysics: The Present Volume

The project of this volume came out of the Research Program *Anthropos. Vers une physique de l'âme: la constitution d'une science de l'homme. France, Allemagne, Italie, Pays-Bas, Grande-Bretagne, XVIIe-XIXe siècle*, which concluded in 2015.[29] The first focal point of this research program, which we headed, dealt with the specificities of the reception of Descartes' natural philosophy. We organized a conference in two parts, first in the

École Normale Supérieure of Paris and second in the École Normale Supérieure of Lyon. The very fruitful exchanges that came out of this project have led to a volume organized around three major themes.

1. The first describes the main ways in which the relations between Descartes' physics and metaphysics have been understood through the ages (Mariafranca Spallanzani for the eighteenth century, Delphine Antoine-Mahut for the nineteenth century, and Delphine Bellis for the twentieth century). The volume begins with these three contributions because these historiographies reveal a great deal about the potentialities latent in the Cartesian corpus, as well as about the mediations (sometimes unconscious) through which it has reached us. In sum, they problematize the naïve idea of the historian of philosophy's "direct access" to the original corpus.
2. The next four contributions (Christoph Lüthy on the period before Descartes, Emanuela Scribano on the topic of sensations, Desmond M. Clarke on *a priori* knowledge, and Sophie Roux on causal agency), show how Descartes himself played on the relations between physics and metaphysics. The goal here is to provide some core examples of the way in which the original text opens, closes, combines, and hierarchizes materials that will then be reused in the course of its various receptions.
3. This plurality is examined, in turn, in five contributions dealing with specific geographic areas (the Low Countries for Antonella Del Prete, Italy for Pierre Girard, England for Philippe Hamou, France and Germany for Jean-Pascal Anfray, Germany with Mogens Lærke). One can see here the respective and contrasting roles played by theology, physics, and metaphysics in the traditions commonly linked to these geographic areas by the historiography that was problematized in part 1, according to the different relations studied in part 2.

Notes

* Delphine Antoine-Mahut, ENS de Lyon, IHRIM, CNRS UMR 5317 and LABEX "Comod", delphine.antoine-mahut@ens-lyon.fr; Sophie Roux, ENS de Paris, République des savoirs, Mathesis and LABEX "TransferS," sophie.roux@ens.fr. We thank Charles Wolfe for translating this introduction and Marco Storni for carefully and diligently proofreading this volume.

1 Jean-Christophe Bardout, *Malebranche et la métaphysique* (Paris: Vrin, 1999).

2 Roger Ariew, "Les *Principia* et la *Summa philosophica quadripartita*," and Laurence W.B. Brockliss, "Rapports de structure et de contenu entre les *Principia* et les cours de philosophie des collèges," in *Descartes: Principia philosophiae (1644–1994). Atti del Convegno per il 350 anniversario della pubblicazione dell'opera (Parigi, 5–6 maggio 1994; Lecce, 10–12 novembre 1994)*, eds. Jean-Robert Armogathe and Giulia Belgioioso (Napoli: Vivarium, 1996), 473–89 and 491–516.

3 Theo Verbeek, *La querelle d'Utrecht* (Paris: Les Impressions Nouvelles, 1988) and *Descartes and the Dutch: Early Reactions to Cartesian Philosophy,*

1637–1650 (Carbondale and Edwardsville: Southern Illinois University Press, 1992).

4 Delphine Antoine-Mahut, *Descartes radical: Historicité d'un canon philosophique* (Paris: Vrin, forthcoming).

5 Roger Ariew, *Descartes and the First Cartesians* (Oxford: Oxford University Press, 2015) and Tad Schmaltz, *Early Modern Cartesianisms: Dutch and French Constructions* (Oxford: Oxford University Press, 2016).

6 Sophie Roux, "A French Partition of the Empire of Natural Philosophy (1670–1690)," in *The Mechanization of Natural Philosophy*, eds. Daniel Garber and Sophie Roux (New York, Dordrecht, Boston and London: Kluwer Academic Publishers, 2013), 55–98.

7 Carlo Borghero, *Les Cartésiens face à Newton: Philosophie, science et religion dans la première moitié du XVIIIe siècle* (Turnhout: Brepols, 2011).

8 François Azouvi, *Descartes et la France: Histoire d'une passion nationale* (Paris: Arthème Fayard, 2002), 95.

9 Mariafranca Spallanzani in this volume.

10 Delphine Antoine-Mahut in this volume.

11 *The Holy Family* (in German: *Die heilige Familie, Kritik der kritischen Kritik*) was the first work jointly authored by Karl Marx and Friedrich Engels, but mainly attributed to Marx. It was first published in February 1845. Marx made use of Charles Renouvier's *Manuel de philosophie moderne* (Paris: Paulin, 1842)—the proximity between Marx and Renouvier's texts was uncovered by Olivier-R. Bloch, "Marx, Renouvier et l'histoire du matérialisme," in *La Pensée*, 191, 1977, reprinted in his *Matières à histoires* (Paris: Vrin, 1997), 384–441.

12 Delphine Bellis in this volume.

13 The "second doctorate" that existed then in the French academic system.

14 Alexandre Koyré, *Essai sur l'idée de Dieu et les preuves de son existence chez Descartes* (Paris: Ernest Leroux, 1922), 5–6.

15 Daniel Garber, *Descartes' Metaphysical Physics* (Chicago: University of Chicago Press, 1992).

16 Desmond M. Clarke, *Descartes' Theory of Mind* (Oxford: Oxford University Press, 1992).

17 Desmond M. Clarke, "The Physics and Metaphysics of the Mind: Descartes and Regius," in *Mind, Method, and Morality: Essays in Honour of Anthony Kenny*, eds. John Cottingham and Peter Hacker (Oxford: Oxford University Press, 2010), 187–207; Catherine Wilson, "Descartes and the Corporeal Mind Some Implications of the Regius Affair," in *Descartes' Natural Philosophy*, eds. Stephen Gaukroger, John Schuster and John Sutton (London and New York: Routledge, 2000), 659–79.

18 *Qu'est-ce qu'être cartésien?* ed. Delphine Kolesnik-Antoine (Antoine-Mahut) (Lyon: ENS Editions, 2013) and Tad M. Schmaltz, *Early Modern Cartesianisms* (Oxford: Oxford University Press, 2017).

19 Robert Pasnau, *Metaphysical Themes 1274–1671* (Oxford: Oxford University Press, 2011); Antonella Del Prete, "Discussioni sul metodo nel cartesianismo olandese Il caso di Johannes de Raey," in *La ragione e le sue vie: Saperi e procedure di prova in età moderna*, eds. Carlo Borghero and Claudio Buccolini (Firenze: Le Lettere, 2015), 146–67.

20 These contributions are grouped in *L'anthropologie cartésienne* (Paris: Presses universitaires de France, 1990).

21 Roger Ariew was one of the first firsts to employ the expression "Cartesian empiricism" (in an article in the *Revue roumaine de Philosophie*, 2006) to refer to the work of Desgabets, Régis, Lamy, and Bayle (François), but also Du Roure. Cf. also Thomas M. Lennon and Patricia Ann Easton, *The Cartesian*

Empiricism of François Bayle (New York: Garland, 1992), 501–16, and Monte Cook, "Desgabets as a Cartesian Empiricist," *Journal of the History of Philosophy* 46, no. 4 (2008): 501–15; Delphine Antoine-Mahut (Kolesnik-Antoine), "Les expériences physiologiques chez Henricus Regius: les pierres lydiennes du cartésianisme?" *Journal of Early Modern Studies* II (2013), "The Creative Role of Experimentation in Early Modern Science," ed. Dana Jalobeanu (April 2013): 125–45 and "Élisabeth philosophe: un cartésianisme empirique?" in *Élisabeth face à Descartes: deux philosophes?* eds. Marie-Frédérique Pellegrin and Delphine Kolesnik-Antoine (Antoine-Mahut) (Paris: Vrin, 2014), 119–38; Sophie Roux, "Was There a Cartesian Experimentalism in Seventeenth Century France?" in *Cartesian Empiricisms*, eds. Mihnea Dobre and Tammy Nyden (Dordrecht: Springer, 2014), 47–88.

22 *Cartesian Empiricisms*, eds. Mihnea Dobre and Tammy Nyden (Dordrecht: Springer, 2014).

23 "Le sésame de l'histoire intellectuelle" (Antoine Lilti, "Querelles et controverses: Les formes du désaccord intellectuel à l'époque moderne," *Mil neuf cent* 25 (2007): 13–28, here p. 13). See also Mogens Lærke's theoretical analysis of "Historical Perspectivism," in *Les Lumières de Leibniz: Controverses avec Bayle, Huet, Régis et More* (Paris: Classiques Garnier, 2015). The following paragraphs partly reprise Sophie Roux's contribution, "Pour une conception polémique du cartésianisme: Ignace-Gaston Pardies et Antoine Dilly dans la querelle de l'âme des bêtes," in *Qu'est-ce qu'être cartésien?* eds. Delphine Kolesnik-Antoine (Antoine-Mahut) (Lyon: ENS Editions, 2013), 315–37.

24 Roux, "A French Partition,"; Denis Moreau, *Deux cartésiens: La polémique entre Antoine Arnauld et Nicolas Malebranche* (Paris: Vrin, 1999).

25 Dominique Raynaud, *Sociologie des controverses scientifiques* (Paris: Presses universitaires de France, 2003); *Mil neuf cent*, 25, 2007; Jean-Louis Fabiani, "Controverses scientifiques, controverses philosophiques: Figures, positions, trajets," *Enquête* 5 (1997): 11–34.

26 See in particular Jean-Pierre Cavaillé, "L'art d'écrire des philosophes," *Critique* 631 (December 1999): 959–80; repr. in *Les Déniaisés—Irréligion et libertinage au début de l'époque moderne* (Paris: Garnier, 2014); "Libertinage et dissimulation, quelques éléments de réflexion," *Libertinage et Philosophie au XVIIe siècle* 5 (2001): 57–82.

27 For his materialist reception in Italy, see Pierre Girard in this volume.

28 Olivier R. Bloch in "Marx, Renouvier et l'histoire du matérialisme," 384–441.

29 ANR Blanc, 2012–2015, co-directed by D. Antoine-Mahut, P. Girard, and J-F. Goubet.

Part I

Historiography

1 First Philosophy, Metaphysics, and Physics

The Implications of Order in Cartesian Philosophy and in the Philosophy of Enlightenment

Mariafranca Spallanzani

I

> *I would think I knew nothing in physics if I could say how things can be but could not demonstrate that they can not be otherwise. Such demonstrations are perfectly possible once physics has been reduced to the laws of mathematics. I think I can provide them for the small area to which my knowledge extends; but I did not do them in my Essays because I did not want to present my principles there.*[1]

These few lines from the letter to Mersenne dated 11 March 1640, three years after the *Discours*, contain one of the clearest statements Descartes provides concerning his theory of truth in physics. Both as an instance of the epistemological foundations of a science yet to be constructed and as a recognition of the limits of the already existing sciences, these lines might well summarize Descartes' intellectual path from the *Discours* and the *Essais* to the *Meditationes*, that is, from natural science to first philosophy.

As a philosophical reflection on the modalities of judgment, Descartes in this letter engages in the debate at that time concerning the models of scientific knowledge and delineates the elements of his rationalist epistemology of the science of bodies through a strict correlation between physics and mathematics. As a philosopher, Descartes affirms that only a necessary physics can resist and respond to the objections put forward by skeptics, and that only a true physics can emerge victorious from both the ruins of the Aristotelian doctrine of forms and from the ambiguities of modern scientific hypotheses. The legitimization of physics must satisfy above all the demand for the justification of the laws of nature, that is, for the mathematical truths created for nature and the demand for stability for physical objects, i.e., bodies as objects of science.

These are difficult but fundamental questions for Cartesian science in its claim of necessity, truth, and reality. Can the objects of physics be known through the clear, distinct, and complete idea of extension?[2] Moreover, can bodies be truly subject to a mathematical treatment that guarantees their necessity, and can they be comprehended through the chain of theorems that follow from their definition? Lastly, are bodies true beings, in which the evidence of reason grasps the truth of the thing, in line with the necessary conditions of a real physics? Or are they no more than a seductive construction of the imagination, an extraordinary production of the intellect?[3]

Descartes' interests as a man of science are intertwined with his demands as a philosopher. According to his philosophy, the path that leads to the truth and necessity of science is not provided by or within science itself, but goes beyond science, without, however, sidestepping the imperative of order: it is defined as the way to the founding principles. The theoretical solution that allows physics to be saved from the dangers of arbitrariness and the risks of contingency, as well as from any purely hypothetical solutions, can be no other than a doctrine of the principles. This doctrine will be given the old Aristotelian name of "first philosophy" but it will be completely renewed, in an anti-Aristotelian perspective. First philosophy, which concerns "in general all the first things"[4] conceived by the soul, is defined by Descartes as the science of the first principles that underpin all other knowledge, which naturally derives from these principles according to the order of a coherent chain—or coherent chains—of deductions. The movement of deduction is guided by the sole criterion of evidence in a multiplicity of points of access to the truth. The reversibility of the deductive connections guarantees the consistency of the solutions by means of the distinction between the proofs given "by the effects" and the explanations given "by the causes" mentioned by Descartes in his response to Morin on 13 July 1638.

Descartes repeatedly expressed his doubts regarding the possibility of concretely achieving physics by means of the principles. "I still cannot see anything to persuade me to present them in the future," he confessed to Mersenne again in March 1640. Nevertheless, his solution had already been clearly outlined in the 1630s, while he was fully focused on "the great mechanics of nature." His solution was "to demonstrate the principles of physics by means of metaphysics." It will be legitimated in the *Meditationes de Prima Philosophia* thanks to the discovery of the immaterial first principles, i.e., the existence of the soul and of God, which, according to him, give physics the theoretical consistency of necessity, truth, and reality. The existence of the *Ego sum*, with the demonstration of the unconditional validity of the general rule of truth, along with the existence of infinite and perfect God, creator of the existences and the essences—the "foundations of metaphysics" mentioned in the *Discours*—act indeed as the "first principles" of physics, offering it the intelligibility, stability, and truth that allow physics to be possible, necessary, true, and real. Physics is possible in its geometric model—"my physics is no other than geometry"—from which can be inferred its necessity, comparable to the necessity of geometry; physics is true in its object, i.e., the extension, which the intellect conceives as "a determinate nature . . . that is eternal and unchanging"[5] and essentially distinct from the soul.

Descartes intended to complete his project of a "scientia perfectissima" of nature that proceeds from causes to effects[6] in the *Principia Philosophiæ*, in which a strong theory of the principles, anchored to metaphysics and presented as the condition for the passage from a mathematics of possible beings to a physics of necessary and real beings, sustains the equivalence between the object of geometry and the matter of bodies. In this work, written in 1644, the entire realm of physics, articulated according to the *a priori* principles of metaphysics, takes the shape of a true rational system of the world that is true and real: true of the truth of the clear and distinct idea of bodies, i.e., the extension,[7] and real of the reality of the mathematical laws created by God for the matter in the eternity of his immutable decrees.

Equivalent in the *Principia* to first philosophy, metaphysics thus makes Descartes' physics a philosophical physics,[8] and even a metaphysical physics.[9] Metaphysics is a reasoned knowledge of the laws of nature founded on the clear and distinct, true, and real principles that it imposes on physics as the first principles of reason that also act as causes, according to the equivalence established by Descartes *causa sive ratio*.[10] Writing the *Principia* "as a Philosopher"[11] and providing all "the true reasons,"[12] Descartes offers a unified corpus of his philosophy[13] that is tied to the *Meditationes* as regards the metaphysical subjects of the first part—"almost the same things," he confirms—but that is put forward in "another style" and "in such an order" that it could "easily be taught."[14]

"As a Philosopher," Descartes renews in the *Lettre-Préface* to his work the traditional figure of the tree of philosophy:[15] the trunk—physics—grows on the roots of metaphysics "as upon rocks" and extends into the three branches of medicine, mechanics, and morals according to the law of the unity of universal wisdom,[16] which in the end Descartes offers to men as the most noble and precious fruit of the philosophical orchard.

However, the letter-response to Mersenne written by Descartes on March 1640, which defines the philosophical conditions of the truth of physics, also contains one of the most explicit admissions of the limits of his *Essais* published with the *Discours*: this work, according to Descartes, risked missing the truth of science because of the voluntary omission of the principles. The readers of the *Discours* had repeatedly noted these limits, lamenting the obscurities, the omissions, sometimes even the "ambiguities"[17] of a work that presented important scientific results that, however, were never entirely developed, and the "key"[18] to them apparently deliberately kept hidden. The readers agreed that these results were excellent but they were put forward more as oblique suppositions than through complete "explanations" and exhaustive "clarifications." In their opinion, the obscurities, omissions, and ambiguities made the work unsatisfactory and Descartes' physics weak: it was not a system of the world, but an orderly exploration of natural phenomena.

Descartes himself, for that matter, had openly recognized the limits of his work. The principles of the *Essais* were insufficient and their demonstrations incomplete due to the elisions produced by the model of a hypothetical-deductive science. Furthermore, the fourth part of the *Discours* was obscure because of the simplification of its arguments due to the treatment of metaphysics as an essay of method and to Descartes' philosophical reticence in this text written in French for an audience of readers who were not necessarily philosophers.

Only the Latin text of the *Meditationes de Prima Philosophia*, dedicated to the Doctors of the Sorbonne, would in fact render all theoretical and argumentative complexity of Descartes' metaphysics. In this work, Descartes conceives first philosophy as the universal science, and metaphysics a specific science[19] not far removed from theology.[20] Radically overthrowing all traditional paradigms, Descartes' theory of truth abandons the classical theory of categorical adequacy, in the same way as it rejects scholastic metaphysics in both its general ontological function and its specific articulation in theology, pneumatology, and cosmology. Indeed, first philosophy is defined as the science of "the first principles of knowledge," which the universality of the *Mathesis* validates not according to a hierarchy of the *genera* of being but according to the order of the reasons. Following Aristotle but going against his model, Descartes continues to think of metaphysics in relation to first philosophy: "the most evident and certain" principles of first philosophy—the finite soul and God, "ens summe perfectum et infinitum"—are "intellectual natures,"[21] and thus metaphysical subjects.

This theory was not entirely new in Descartes' philosophy: the two subjects of metaphysics had appeared as titles in the "petit Traité" written in 1629—"the existence of God and of our souls"—, had been mentioned in the letters to Mersenne in the spring of 1630 as principles of science that had guided the philosopher's earliest scientific studies[22] and had been set as the foundation of physics in *Le Monde*. In the metaphysics of Part Four of the *Discours*, the soul and God are conceived by the intellect as immaterial subjects: as first principles according to the order of evidence and as first "reasons" according to the order of method. Finally, in the *Meditationes*, these subjects are examined and demonstrated as truths of first philosophy, are intertwined in their metaphysical primacy with the notion of matter in order to found a physics of extension, and are defended against the attacks of skeptics and atheists as absolutely certain and evident philosophical truths as well as rational arguments made available to a Christian apologetic. Descartes had every reason to be satisfied, given the unbending strength of his demonstrations and the originality of his discoveries, which were as new *de facto*—"no one before me"[23]—as they were ancient *de jure*—nothing being older than truth.[24]

And yet, if Descartes sees metaphysics as a science that is absolutely necessary to the new encyclopedia—the tree of philosophy is rooted precisely in metaphysics-Cartesian metaphysics is ultimately a kind of knowledge that is constitutionally limited to its essential function as a foundation. This science is indeed the evident science of "the principles of knowledge"—"the explanation of the principal attributes of God and of the immateriality of our souls"[25]—and is charged with indexing the simple notions of the human spirit "that one can know philosophizing with order."[26] But Cartesian metaphysics is profoundly different from the other sciences, which it ensures and authorizes without, however, offering them a specific content or putting their progressive discoveries in order. Descartes' metaphysics never presents itself as a psychology: it is just a "punctual" knowledge of the soul[27] limited to affirming its spiritual essence through an intuitive and always identical form of knowledge and to demonstrating its real distinction from the body through the procedures of induction and deduction authorized by the method. Descartes' metaphysics is not an anthropology but a form of knowledge whose aim is to "make understood" the distinction between substances: in the case of mankind, metaphysics is limited to affirming the union of the soul and the body, leaving all further elaboration to sensitive knowledge and life experience.[28] What's more, Descartes' metaphysics is not a theology but a form of knowledge bound to the method: it rather radically denies any possibility of imposing itself as a rational theology with the claim of being complete and adequate to its object, the infinity of God, given that this object infinitely exceeds the limits of reason and absolutely does not submit itself to the methodical imperatives of order and measure. Cartesian metaphysics, indeed, essentially consists in the exercise of the intellect that conceives in all evidence the principles of the other sciences and can therefore neither progress over time nor expand to provide further results. Discovered and defined "once and for all"[29] and for everyone, metaphysics disappears as such in its function as foundation. This is why Descartes affirmed that we should not occupy ourselves with it for long: he himself, he used to say, did so for a few hours each year.[30]

II

> *Chancellor Bacon was followed by the illustrious Descartes. That exceptional man, whose fortune has varied so much in less than a century, possessed all the qualities necessary to change the face of philosophy: a strong imagination, a most logical mind, knowledge drawn from himself more than from books, great courage in battling the most generally accepted prejudices, and no form of dependence which forced him to spare them.*[31]

Thus writes d'Alembert in the *Discours Préliminaire*. This introduction of the *Encyclopédie* gave d'Alembert the opportunity to draw some conclusions on Descartes' philosophy and one hundred years of Cartesian philosophy, from the splendor and glory of a great and often quite varied philosophical tradition of the seventeenth century to the vicissitudes and misadventures encountered in the philosophical and scientific culture of the Enlightenment, which had become critical toward the Cartesian heirs and Cartesian doctrines.

D'Alembert portrays Descartes in the chiaroscuro of history, at the dawn of modern culture and at the origin of the "revolution" of philosophy, a revolution that was no doubt laborious and contested in the beginning but was well established during the Enlightenment, "le siècle philosophe."[32] For the *Philosophes*, Descartes was indeed the main actor of this revolution: he was the creative genius of which Newton and Locke and the *Philosophes* themselves were the heirs. It was in fact Descartes who inaugurated by his "revolt" against the tradition the "revolution" of philosophy: Descartes is the *esprit créateur* "who searched for the reason of things," both master and disciple of natural light; Descartes is the *ami de la solitude*, audacious, solitary, scornful philosopher, master of himself and teacher of queens who was persecuted and misinterpreted by his contemporaries and destined to be recognized and honored only in posterity.

D'Alembert sees Descartes as a philosophical figure whose life was an adventure of thought, and whose thought was a philosophy of freedom, situated within the genealogy of reason in its progress toward the light. Thus, d'Alembert depicted the Cartesian epic as a gesture of freedom and a militant program of truth,[33] as a paradigm of how philosophy can inspire the exercise of enlightened reason and the endeavor of enlightened action.

D'Alembert repeatedly expresses this opinion in his articles written for the *Encyclopédie*. In his article *Expérimental*, for example: without disregarding the "errors" of Cartesian physics, he recognizes that experimental philosophy owes much to Descartes, who, together with Bacon, was among the first to open the way for the study of nature. In his article *Intégral*, he sees Descartes as the "restorer of science"; in his article *Géomètre*, he calls Descartes one of the greatest geniuses of all time; in his article *Bibliomanie*, he appropriates the beautiful and enlightened passage of the *Discours de la méthode* about books, and in the pages of his article *Forme*, he calls Descartes "the boldest and the brightest of philosophers."

Yet, when he passes from Descartes' philosophical figure to his work, d'Alembert's opinion changes radically, and the monumental image of Descartes gives way to a rather fragmentary view of his thought. D'Alembert writes: "one can view Descartes as a geometer or as a philosopher."[34]

As a great geometer, Descartes is worthy of immortal glory for the "new acquisitions in algebra," for the "rule of signs"[35] and the method of undetermined coefficients—"a very ingenious and very subtle artifice, which has been applied to a large number of searches,"[36]—and for the "tangent method, which is like the first step to be taken towards infinitesimal calculus."[37] Above all, d'Alembert attributes to Descartes the merit of "the application of algebra to geometry," which is "one of the grandest and most fortunate ideas that the human mind has ever had"[38] and which opens the road to the solution of an infinite number of other difficult questions.[39]

However, according to d'Alembert, Descartes was not as great a philosopher as he was a great geometer. While inventive in his method and brilliant in his *Dioptrique* thanks to his ingenious idea of applying geometry to physics, Descartes was less fortunate in his vortex physics, of which the Newtonian physics had sanctioned the defeat. Even though the vortex theory was the first modern rational explanation of the formation and movement of celestial bodies,[40] with time this theory turned out to be "purely hypothetical," perhaps a little more fortunate than so many others,[41] but "a philosophical conjecture"[42] like so many others. The Newtonian system of gravitation had contradicted it, as d'Alembert repeats. We have another principle—he writes—which explains the system of nature, "a principle whose present existence is fully manifested in nature: we want to speak of the gravitation of the bodies."[43]

But, although Descartes "was mistaken," d'Alembert recognizes he had conceived the first genial idea to propose a rational demonstration of the order and structure of the universe,[44] "formulating the laws of motion" from which physical reasons for natural phenomena may be deduced. "This courage merits the recognition of other philosophers because [Descartes] introduced those coming after him to the search for the true laws of nature."[45]

> Descartes, who was forced to create a completely new physics, could not have created it better; it was necessary, so to speak, to pass by way of the vortex theory in order to arrive at the true system of the world, and, if he was mistaken concerning the laws of movement, he was the first, at least, to see that they must exist.[46]

Thus, despite "the mistakes" that future science and philosophy would correct, when d'Alembert dates the birth of modern physics, he does not hesitate to include Descartes in the history of the "revolutions" of the human mind moving toward truth. Newton and Locke, creators of the new physics and of the new metaphysics, can be considered in the genealogy of Cartesian philosophy as those who achieved the Cartesian revolution. Thus, any praise of Newton and Locke must be extended to praising Descartes. The history of reason discovers secret continuities.

Lastly, as a metaphysician, d'Alembert presents Descartes in the *Discours Préliminaire* as "a creative spirit, but also as unfortunate as the philosopher and even more." "His metaphysics, as ingenious and new as his physics, suffered virtually the same fate. And it too can be justified by more or less the same reasons."[47] For d'Alembert, Descartes, "biased by the prejudice of innate ideas," was far from the truth when he identified the principles of metaphysics with "abstract ideas,"[48] from which he had expected to deduce all other knowledge and to "explain clearly everything."[49] Thus, "ignoring the science that is acquired through the senses, Descartes, being accustomed to confining himself entirely within the intellectual ideas that had no reality, went with a lot of genius from error to error."[50] Any science that would "reasonably" know our soul and our affections must begin to abandon Descartes' illegitimate presumptions. For d'Alembert, metaphysics is not the "science of abstract ideas"[51] but the natural "history of our thoughts" based on experience, "that is, on the natural progress of the operations of human intellect."[52]

No doubt, therefore, Descartes "was mistaken in admitting the existence of innate ideas," as d'Alembert, at the school of Locke and Condillac, repeatedly states. But d'Alembert also highlights that

> had [Descartes] retained that single truth taught by the Aristotelians concerning the origin of ideas through the senses, perhaps it would have been more difficult to uproot the errors that debased this truth by being alloyed with it. Descartes dared at least to show intelligent minds how to throw off the yoke of scholasticism, of opinion, of authority—in a word, of prejudices and barbarism. And by that revolt whose fruits we are reaping today, he rendered a service to philosophy perhaps more difficult to perform than all those contributed thereafter by his illustrious successors. He can be thought of as a leader of conspirators who, before anyone else, had the courage to arise against a despotic and arbitrary power and who, in preparing a resounding revolution, laid the foundations of a more just and happier government, which he himself was not able to see established.[53]

D'Alembert's judgment of Descartes' metaphysics, inspired by a rational philosophy of history, evidently follows two epistemological models: the inexorability of rational history of truth, which condemns Descartes' doctrine on the one hand, and the contingency of human history, which exalts his figure on the other. In the history of truth, Descartes' metaphysics is condemned as an erroneous theory of ideas. Within the framework of the history of the human mind, however, Descartes' philosophy appears to have been a necessary step that contributed to eradicating the prejudices of tradition: one of the first steps toward truth.

Thus, as did Malebranche, and Perrault before him, as did Fontenelle and Du Marsais, as did Voltaire in the *Quatorzième Lettre*, d'Alembert, although recalling the "errors" of Cartesian metaphysics and physics, affirms: "the weapons we use to fight him belong nonetheless to him, because we turn against him." Descartes was a strange master and a precious, special teacher who taught his heirs how to detach themselves from any teacher, any doctrine, and any authority. Descartes, therefore, despite his mistakes and the fate of his doctrines, which his epigones had transformed into a new scholastic system, remained the thinker of the past to whom d'Alembert and the Encyclopedists felt the greatest obligation.[54] At the dawn of the new age of philosophy, Descartes was the "sublime genius" who devoted his intelligence to the quest for truth and his life to the ideals of freedom and wisdom; he is the "great philosopher"[55] who had to suffer persecution and exile to defend these values.[56]

In conclusion, the philosophy of Enlightenment offers an image of Descartes as a *Philosophe* more than it provides a historical analysis of his work and thought. Moreover, during the Enlightenment, after Newton and Locke, after Boerhaave and Condillac, after "the light had shone in all the sciences" and both a philosophy of experience and an intelligent application of mathematics banned "the taste for systems," after the new scientific naturalism corrected the illegitimate ambitions of the Cartesian *Mathesis* that aspired to be *Universalis*, very little remains of Descartes' philosophy. If mathematics "constitutes the most solid and the least contested part of his glory," not much remains of his metaphysics, which the *Philosophes* define with a touch of irony as a *roman de l'âme*. There is also very little left of Descartes' ideal of wisdom as the fruit of the tree of philosophy, which the *Philosophes*, following a philosophy of experience, see as fragmented in moral actions. Not much is left of Descartes' method: the *Philosophes* highly praise its innovative and analytical character, but, following Condillac, refuse it as dependent on a "false" theory of ideas.[57] Little remains of Descartes' physics and of his philosophy of nature, and almost nothing of his vortex theory, which Newton demonstrates to be both incapable of explaining the movement of the planets and opposite to astronomic phenomena and the laws of mechanics. Very little remains also of Descartes' doctrine of light, whose indispensable foundation, that is, instantaneous propagation, is negated first by the Newtonian theory of the *Principia* and the *Opticks*, and later by the reports of Rœmer and Picard. Not much remains of the Cartesian theses on meteors, whose vortex system-based explanations are corrected by the Encyclopedists thanks to the observations of modern experimental science, developed by Newton and Dutch physicists. Little remains of Cartesian mechanics and dynamics of fluids, which Leibniz, Newton, and Huygens correct and which the Encyclopedists, d'Alembert in particular, quite simply forget to mention, preferring to discuss more recent results. Nothing remains of Descartes' "paradoxical" theory of the animal-machines, which appeared to the *Philosophes* opposed to common sense and scientific experience, nor much of his theory of man, which medicine, psychology, and the life sciences to a large extent forget or surpass.

But then again, if the *Philosophes* hold a rather critical attitude towards Descartes' system, their attention to his works is inevitably limited by the fact that the Age of Enlightenment, unlike the *Grand Siècle*,[58] largely ignores his texts even after the publication in 1701 of the *Opuscula Posthuma*, the collective publication of the *Lettres* in 1724–1725 and the Compagnie des Libraires' publication of Descartes' texts in 1723–1729.[59] The authors of the *Encyclopédie* rarely cite Descartes: Turgot, Quesnay, d'Alembert, and the Abbé Pestré mention only a few passages of the *Discours*, Formey reproduces just a couple of the articles of the *Principia*, d'Alembert and Diderot quote a few lines of the *Géométrie* and the *Dioptrique*, and Abbé Pestré provides a few observations from his letters. These texts, however, are almost never quoted first-hand; often, the passages are transcribed here and there, and the passages Abbé Pestré in particular reproduces within quotation marks in his article *Cartésianisme* are in general quotations from other texts that are culturally closer: the classic biography of Descartes by Adrien Baillet,[60] the *Mémoires* by Father Jean-Pierre Niceron,[61] the *Histoire du Ciel* by Antoine Noël Pluche,[62] and the *Essai sur l'origine des connaissances humaines* by Condillac. Whether explicitly admitted or tacitly understood, these are the sources of the articles of the authors of the *Encyclopédie.* In addition, they have a slight interest in how Descartes' corpus was received and in the history of his school. Their few concise judgments condemn the Cartesian scholasticism and substantially repeat what Leibniz expressed and concentrated in his famous anagram of Descartes' name *Cartesius-Sectarius.*

Thus, for the Encyclopedists, Descartes is *géomètre*, *philosophe*, and *métaphysicien.* The tree of human knowledge, which supports the *Encyclopédie*'s new order of sciences inspired by Bacon, has fragmented the unity of the Cartesian tree and with it the unity of Descartes' philosophy[63] into different and specific disciplines that are now evaluated according to their own scientific contents. Just one century after Descartes' death, the history of his philosophy and of Cartesianisms and anti-Cartesianisms—both necessarily plural—are in fact more the history of a manifold philosophy that over time split into a series of particular doctrines, all inspired by Descartes but not necessarily tied together in the same way as in accordance with the Cartesian model of the *Mathesis universalis* and with the conceptual image of the Cartesian tree of philosophy.

How was this possible? By the "Voltaire effect," as François Azouvi writes.[64] In 1734, in his *Lettres philosophiques*, Voltaire denounces Descartes' philosophy for its excesses. Voltaire's judgment could not be clearer: Descartes was "carried away," he writes, "by that systematic spirit that blinds even the greatest men." This severe judgment becomes the leitmotif of the Enlightenment critique against Descartes. In this *esprit de système*, Voltaire sees the true source of the philosophical excesses of the Cartesian philosophy, a splendid, arrogant, and audacious philosophy: the exact inverse of the "wisdom" and the "modesty" that characterize both Locke's philosophy and Newton's science.

Underlining the "wisdom" and the "modesty" of English philosophy, Voltaire corrects the theories of Cartesian philosophy, which were no doubt brilliant, dazzling even, but often full of errors and unable to enlighten mankind in the search for truth. If in the *Questions sur l'Encyclopédie* (1770) Voltaire is more explicit, in the *Lettres philosophiques* he limits himself to underlining the two most blatant Cartesian errors that English philosophy will later correct: the first is that Descartes wrote the imaginative *roman* of the innate idea beginning with the wrong notion of thought as the essence of the soul, while Locke set out to write the *histoire* of the soul, following "step by step" the progress of the human knowledge. The second is that Descartes wrote the *roman* of a vortex world beginning with the wrong notion of extension as the essence of bodies, while Newton discovered "the true system of the world" by refusing hypotheses and by beginning with "the certain and indisputable effects of an unknown but real principle, the attraction."

Voltaire's comparison plays on two opposing philosophical paradigms declined in different forms: deductive order versus inductive method; *a priori* versus *a posteriori*; hypotheses versus facts; *roman* versus *histoire*; presumption versus modesty.[65]

The *Philosophes* repeat these Voltairean formulas during the Enlightenment, criticizing the Cartesian system and sanctioning its definitive abandonment. Condillac helps to provide also an epistemological foundation for these criticisms; thus the "Voltaire effect," but also the "Condillac effect." In his *Traité des systèmes* written in 1749, Condillac proposes indeed a new configuration of the syntactic order of knowledge, founded on an empiricist theory of knowledge and anchored to a powerful theory of principles[66] that is also a vigorous critique of modern philosophical theories established on abstract principles or illegitimate premises. Condillac formulates this theory by analyzing not the contents of the different systems but the logical and formal conditions of their discourse, which make systematic thought unacceptable. A good system, Condillac writes, is not a synthetic unity of consequences derived from *a priori* abstract principles, but a consequently interconnected multiplicity of elements that derive from well-chosen facts and are supported by experience. Newton's system shows the ideal model of system in science, whose paradigm should be exported, according to Condillac, to other disciplines such as philosophy, mechanical arts, fine arts, ethics, and politics. By virtue of this Newtonian definition of system, in chapter VI of his *Traité* Condillac deconstructs the theoretical foundations and the entire development of Cartesian philosophy. In Condillac's opinion, Cartesian philosophy is nothing more than a set of metaphors, vague comparisons, and obscure notions. From this derives the fragility of the Cartesian system, founded as it is on abstract and confused principles: unintelligible explanations and incomplete ideas and rules, such as that of the evidence, which are overly strict. Condillac underlines the "bad" abstraction that is implied by the criterion of evidence and denounces the dangers that would arise if it were to be extended.

But this critique of the Cartesian system from the standpoint of a new logic of experience does not represent a refusal of metaphysics for Condillac; rather, this critique permits him to redefine a new metaphysics. Metaphysics, for Condillac, is not "the science of the first truths, of the first principles of things," but "the science making the greatest contribution to the mind's lucidity, precision, extension, and, thus, so necessarily preparing it for the study of every other science."[67]

This new, reasonable, "wise and modest metaphysics," founded on the analysis of ideas and not on Cartesian analysis,[68] shatters the unity of Descartes' system. By destroying its principles, Enlightenment philosophy discredits the latter's entire corpus. According to Abbé Yvon, author of the article ANALYSE in the *Encyclopédie*, Descartes, "biased by the prejudice of innate ideas," drifted far from the truth while trying to identify the foundations of philosophy with "general principles and abstract notions:"[69] an ambitious and audacious theory, certainly, but "badly founded." Thus, Pestré affirms in the article *Cartésianisme*, that Descartes, "condemning the science acquired through the senses, and closing himself up entirely in intellectual ideas, proceeded from error to error."[70] This is why Abbé Pestré believes that Descartes' works were "novels" of the soul instead of its history; though fascinating and full of interesting suggestions, they were nonetheless "novels."[71] This is why Descartes imagined hypotheses of the physical universe: they were seductive, but opposite to the reality of phenomena and the necessity of natural laws. This is why Descartes formulated the "absurd" thesis of animal-machines, a well-deduced and coherent theory, certainly, but a fierce and paradoxical doctrine. This is why, even though Descartes had theorized an excellent method,[72] in his metaphysics he proceeded contrary to the natural order of thought, turning to synthesis where analysis was required.[73] Lastly, the errors of Descartes' metaphysics and physics must be attributed to the errors of his method, which led him "to begin by defining things, and to consider these definitions to be principles capable of leading him to discover properties." For the Encyclopedists, indeed, the definition can't reach the essence of things. On the contrary, the definition must be searched for in the "genealogy of knowledge," that is, in the "development of the simple ideas that are contained in a word."[74]

Hence, as regards the decisive subject of truth, the authors of the *Encyclopédie* abandon Descartes' unity of evidence and its modulations, which control the necessary and continuous chains of the *Mathesis Universalis*. The Encyclopedists even go so far as to fragment the notion of truth itself according to the different modalities of judgment—"internal truths are inevitable and evident, while external truths are uncertain and faulty"[75]—and to disintegrate it into a multiplicity of different meanings of truth that pluralize the question of order in relation to the plurality of the principles and facts of each science.[76] The long inventory of meanings that the *Encyclopédie* provides for the word *Vérité* clearly shows its authors' opposition to the preceding century's metaphysical systems.

Above all, the Encyclopedists do not accept Descartes' absolutism concerning evidence; they multiply truth through the subtle and irreducible distinctions of a theory of correspondence, presence, convenience, and order. Truth is therefore multiple: "metaphysical or transcendental" truths, "eternal" truths, "fundamental" truths, "moral" truths, the truths of "sacred critique," and the truths "in mythology." "Metaphysical or transcendental" truths, completely detached from Descartes' metaphysics of immaterial substances, are now defined as the persistence of all things in being, the "sole and unmovable foundation" of all knowledge. "Eternal" truths, which Chevalier de Jaucourt explicitly deprives of any reference to theology and of cosmological pertinence, involve a theory of the temporal invariance of the relation of "convenience" between abstract ideas and the names that designate them, rather than Descartes' thesis of divine creation and the attributes of God. Furthermore, as concerns "fundamental" truths, the Encyclopedists abandon the radical and necessary status of Cartesian metaphysics and more modestly define them as "fertile truths" with a large number of consequences: universal gravitation in physics, neighborly love in ethics.

This is a definitive, anti-Cartesian acquisition of Enlightenment philosophy. Following Locke, the *Philosophes* counter Descartes' metaphysics of essences and the Cartesian theory of innate ideas by neutralizing the question of substance and allowing the intellect to analyze its own operations, in an anti-metaphysical destabilization of Cartesian reason introduced by Locke's "wise and modest" metaphysics of the soul and Newton's victorious science. For d'Alembert, who in the *Discours Préliminaire* writes the history of philosophy at the school of Locke and Condillac as a philosophical history of reason, the principles of metaphysics are no more than the "results of our sensations."

III

> *Reasonable metaphysics can only consist, as does experimental physics, in the careful assembling of all these facts, in reducing them to a corpus of information, in explaining some by others, and in distinguishing those which ought to hold the first rank and serve as the foundation. In brief, the principles of metaphysics, which are as simple as axioms, are the same for the philosophers as for the general run of people.* [77]

Metaphysics is therefore not defined by d'Alembert as "the science of ideas";[78] it does no more than outline the "genealogy of our knowledge." As a prudent and modest account of the "natural history of the soul," metaphysics is thus inscribed by the Encyclopedists in the theory of man, and the principles that it provides for itself and for the other sciences derive from the system of operations of the human intellect in its regulated exercise on itself, on souls and bodies. For d'Alembert, who, together with Diderot, sets out the "genealogical or encyclopedic tree" of knowledge,[79] this empiricist and "experimental" metaphysics dictates the classification of the sciences in the branch of reason. Thus, the hierarchies of this new classification are not given by ontological privileges or systems of values, nor by a confidence in an ingenuous realism, but rather by the modalities of reflection of the intellect in its essential functions of analysis, abstraction, and generalization or, conversely, of synthesis and recomposition. The speculative genesis of concepts and the relative order of sciences now entirely rest on a method of abstraction through which the mind analyzes, selects, reduces, and isolates certain properties of concrete and individual objects—"common properties by way of which we compare them, and dissimilar properties by way of which we distinguish them"—in order to offer these properties to the various rational sciences[80] in the form of simple, distinct intellectual ideas.

The *Métaphysique générale*, which d'Alembert places in the branch of philosophy, seems to enjoy an eminence all its own, given that it is the first of the rational sciences. Presented through synonyms such as *Ontologie, ou Science de l'être en general*[81] that derive from a tradition that goes back to Wolff, the *Métaphysique générale* is defined as the science of the most general notions that refer to the "general properties" of all beings, "such as existence, possibility, duration, etc."[82] This science of the most elevated abstractions that the mind can attain in its experience of itself and of things, however, runs no risk of being confused with abstract traditional metaphysics. If metaphysics is still possible for d'Alembert, who theorizes the need for a "reasonable" metaphysics and calls for a particular metaphysics for each science as an analysis of its ideas, it will have to be a "wise and modest metaphysics" that lies within the limits of the human intellect, similar to Locke's "experimental physics of the soul." Directed by the analysis that proceeds from the facts of concrete experience to the most general ideas, this metaphysics methodically follows the intellect in its most delicate operations of reflection, generalization, connection, and subordination, which are carried out on the transparent data of psychic facts and on the observations of physical phenomena. As science based on evidence, "clear and precise in its general and philosophical truths,"[83] metaphysics is a science *a posteriori*, following the rules of the *esprit systématique* that Condillac offers to the philosophy of the *Encyclopédie*, freed from the conjectures and the overindulgence of the *esprit de système*. According to d'Alembert,

> reasonable metaphysics can only consist, as does experimental physics, in the careful assembling of all these facts, in reducing them to a corpus of information, in explaining some by others, and in distinguishing those which ought to hold the first rank and serve as the foundation.[84]

Certainly, the knowledge map, which Diderot and d'Alembert draw in the *Discours Préliminaire*, cuts and divides knowledge, whereas nature is continuous, global, and total. Certainly, the *Système figuré*, which in the first volume schematizes the encyclopedic tree of knowledge, is entirely fashioned and modulated according to clear articulations and put forward in a proportionate symmetry, whereas psychic phenomena are complicated, tortuous, and often obscure. Certainly, in the *Encyclopédie* the chain of knowledge is not as continuous, the deductions not as necessary, the logic not as linear as in Descartes' philosophy. Through these complex, broken, and fragmentary chains, however, the encyclopedic tree of knowledge sanctions the humanism of the entire encyclopedic enterprise, while affirming at the same time the naturalization of culture and history. Philosophy is no longer a question of the principles of ontology but is a question mankind poses in his search for truth and happiness. Ideas are the result of experience, values are derived from the fabric of social relations, and the sciences are no more than a network created from the abstractions of the human intellect.

In the *Système figuré*, it is not a love of novelty or of originality that inspires the Encyclopedists to place the *Science de l'homme* before the *Science de la nature*, inverting Bacon's order of the sciences. This inversion is in fact required by the analytical order: the *Science de l'homme*, which is the result of the speculative genealogy of ideas, rises above all others as their principle. Thus, following the order of analysis, this science methodically sets out a body of notions according to the progressive acquisitions of an epistemological Adam who discovers his own existence "through internal sense or consciousness"[85] and discovers the existence of external objects, including his own body, through the sensations that affect him.[86] In spite of the natural union and the mutual correspondence of his own body with his own "principle of action," mankind finds his truest most authentic *myself* in the soul—"the faculty of will and thought, . . . the principle that acts, or, which is the same thing, the substance that wills and knows"[87]—which is accompanied by a system of moral values and is able to acquire, compose, and communicate ideas. In the *Système figuré*, the *Science de l'âme* is thus first in the philosophical branch of the sciences of man as the evident, rational knowledge of psychical facts, and is both the foundation of the *Logique*, which follows it in its classical threefold division of the *Art de penser*, *Art de retenir*, and *Art de communiquer*, and is the basis of the *Morale*, articulated in turn into the two sections of the *Science du bien et du mal en général* and the *Science des lois, ou Jurisprudence*.

If, then, the sciences of man are set out according to an analytically ordered sequence, by following the inverse order of synthesis and the complication of properties d'Alembert sets out the map of the *Sciences de la nature* on the results of reflection, descending from the most abstract and the most simple sciences to the most complex, particular, and concrete sciences. The most abstract, simple, and extended science is the *Métaphysique des corps ou Physique générale*[88] as the knowledge of the most general properties of matter. It is followed by the *Mathématiques*, which are in turn subdivided into *Pures*, *Mixtes*, and *Physico-mathématiques*, and then by the *Physique particulière*.

The degree of abstraction determines "the different objects" of the different sciences, that is, the properties that constitute the domain of each science, "designated by an abstract name."[89] The degree of simplicity of the object determines both the extension of the science and its degree of certitude: the *evidence* of metaphysics and mathematics; the *certainty* and the *exactitude* of physico-mathematics; the *probabilities* and the *hypotheses* of experimental physics; the *conjectures* and the *analogies* of particular physics, which is a "simple collection of facts" or an inventory of "individuals" and a list of "singular and different truths" that elude the generalities of abstraction and highlight the limits of the experimental method.[90]

And, finally, the *Système figuré* reviews the *abuses of reason*. As nature has its "deviations" and its "monsters," reason also has its own troubles, infidelities, distractions, and excesses, Diderot writes. The *Encyclopédie*, charged with enumerating the most elevated sciences and the most noble arts that bear witness to mankind's genius and its capabilities, must also not forget "all the sciences and all the arts which show only the greediness, the wickedness, and the superstition of man, and which dishonor him."[91]

The obligations of order thus lead to a call for an open denunciation, assuming the responsibility for a transparent critique. The morphology of the *Système figuré* now becomes militant philosophy.

Notes

1 Descartes to Mersenne, 11 March 1640, AT III 39, CSMK 145: "Pour la Physique, je croirais n'y rien savoir si je ne savais que dire comment les choses peuvent être, sans démontrer qu'elles ne peuvent être autrement; car l'ayant réduite aux lois des Mathématiques, c'est chose possible, et je crois le pouvoir en tout ce peu que je crois savoir, bien que je ne l'aie pas fait en mes *Essais*, à cause que je n'ay pas voulu y donner mes Principes."

2 Arnauld raised this question when he asked Descartes to define the complete and adequate ideas of extension and thought (*Objectiones Quartæ*, AT VII 200–1). Descartes answered with a theory of the complete but not adequate notion (*Quartæ Responsiones*, AT VII 226).

3 As Gassendi was led to believe, reading the Fifth Meditation on material things (*Objectiones Quintæ*, AT VII 319–21).

4 Descartes to Mersenne, 11 November 1640, AT III 235, 239.

5 *Meditationes de Prima Philosophia. Meditatio Quinta,* AT VII 64.

6 *Principia Philosophiæ*, I 24, AT VIII-1, 14. It must be admitted, however, that the *Principia* do not always remain faithful to the ideal of a perfect science proceeding "by composition" from causes to effects. Attentive to the order of the reasons and to the theoretical constraints of a text that he wants to be unitary and systematic, Descartes is nevertheless conscious of the specificity of the different subjects treated and of the didactic ease of the various arguments, so that he never stops at a unique demonstrative style. If indeed the synthetic style imposes itself as the method of the "scientia perfectissima" that more directly follows the natural order from the simple to the complex and proceeds *a priori* from the causes to the effects, he often integrates to the deductive approach the order of analysis and the results obtained *a posteriori.* This is a proof of the flexibility of the Cartesian method, which is controlled in its demonstrative movements by the priority question of the order. This is also a proof of the impossibility of a Cartesian "scientia perfectissima," which remains rather an operative ideal for the philosophical research than a completed system. Descartes himself will recognize these limits of the *Principia* in the *Lettre—Preface* to the French translation of the work (1647), expressing his confidence in the progress of the science of which future generations will bear.

7 Descartes to Gibieuf, 19 January 1642, AT III 474.

8 Michel Fichant, *Science et métaphysique dans Descartes et Leibniz* (Paris: Presses universitaires de France, 1998), 29–58, 77.

9 Dan Garber goes as far as defining Descartes' physics a "metaphysical physics" (*Descartes' Metaphysical Physics*, Chicago: University of Chicago Press, 1992).

10 *Meditationes de Prima Philosophia. Quartæ Responsiones*, AT VII 236.

11 *Les Principes de la Philosophie. Epître à la Sérenissime Princesse Elisabeth*, AT VIII-2, 21.

12 Descartes to Mersenne, 11 November 1640, AT III 233.

13 *Ibid.*

14 *Epistola ad P. Dinet*, AT VII 577: "non eodem ordine et stylo." On the principle of "variation of styles" in Descartes' philosophy, see Jean-Marie Beyssade, *Études sur Descartes: L'histoire d'un esprit* (Paris: Éditions du Seuil, 2001), in particular the article "*Scientia perfectissima:* Analyse et synthèse dans les *Principia*".

15 *Les Principes de la Philosophie. Lettre-Preface*, AT IX-2, 14: "Ainsi toute la Philosophie est comme un arbre, dont les racines sont la Métaphysique, le tronc est la Physique, et les branches qui sortent de ce tronc sont toutes les autres sciences, qui se réduisent à trois principales, à savoir la Médecine, la Mécanique et la Morale, j'entends la plus haute et la plus parfaite Morale, qui, présupposant une entière connaissance des autres sciences, est le dernier degré de la Sagesse."

16 *Les Principes de la Philosophie. Lettre-Preface*, AT VIII-1, 5.

17 De Beaune to Mersenne, 13 November 1638, *Correspondance de Mersenne*, VIII 172.

18 Rivet to Mersenne, 23 May 1638, *Correspondance de Mersenne*, VII 212.

19 Jean-Luc Marion dedicates the first chapter of his book *Sur le prisme métaphysique de Descartes* (Paris: Presses universitaires de France, 1986), 9–72 to this theoretical question and its historical implications. He underlines the role of Aristotle, but also of Suárez (*Disputationes metaphysicæ*, 1597), whose theoretical decisions—the distinction between *Prima Philosophia* and *Metaphysica*—became "determinant for all modern philosophers, especially for Descartes. Fundamental, again, was the role of Pereria, whose work *De communibus omnium rerum naturalium principiis* (1576), often reprinted, seems to have influenced Descartes not only in its vocabulary, like the syntagm *Mathesis universalis* found in the *Regulæ*, but also for the theoretical syntax of the disciplines.

20 Descartes even called metaphysics theology, following the tradition, but he reversed the primacy of metaphysics with respect to first philosophy: "Metaphysica sive Theologica" (to Regius, [January 1642], AT III 505).

21 Descartes to ***, [March 1637], AT I 353.

22 Descartes to Mersenne, 15 April 1630, AT I 144.

23 *Mediatationes de Prima Philosophia. Objectiones Septimæ. Responsiones*, AT VII 549: "a nemine ante me."

24 *Mediatationes de Prima Philosophia. Sapientissimis Clarissimisque viris Sacræ Facultatis Theologiæ Parisiensis Decano et Doctoribus Epistola*, AT VII 3: "Nihil est veritate antiquius."

25 *Les Principes de la Philosophie. Préface*, AT VIII-2, 14.

26 Descartes to Mersenne, 11 November 1640, AT III 239.

27 Here, I take up the definition given by Étienne Balibar, *Identité et difference: Introduction* (Paris: Éditions du Seuil, 1998), 38.

28 Descartes to Elisabeth, 28 June 1643, AT III 691–2: "It is in using only life and ordinary conversations . . . that we learn to conceive the union of the soul and the body."

29 *Discours de la méthode*, AT VI 13.

30 Pierre Guenancia, *Descartes, chemin faisant* (Paris: Les Belles Lettres, 2010), 273.

31 *Encyclopédie ou Dictionnaire raisonné des sciences, des arts et des métiers* . . . (Paris: Briasson, David, Le Breton, Durand, 1751–1772 - henceforth referred to as *Encyclopédie* -, rpt Forman) (addition), *Discours Préliminaire*, I, XXIV (*The Encyclopedia of Diderot & d'Alembert*. Collaborative Translation Project, Michigan Publishing, https://quod.lib.umich.edu/d/did/intro.html).

32 *Ibid*., art. *ENCYCLOPÉDIE, V, 644.

33 *Ibid*., *Discours Préliminaire*, I, XXVI.

34 *Ibid.*, *Discours Préliminaire,* I, XXIV.
35 *Ibid.*, art. *Equation,* V, 842; art. *Racine d'une équation*, XIII, 748.
36 *Ibid.*, *Discours Préliminaire,* I, XXV.
37 *Ibid.*, art. *Tangente*, XVI, 885.
38 *Ibid.*, *Discours Préliminaire,* I, XXV.
39 *Ibid.*, art. *Géométrie*, VII, 631; art. *Algèbre*, I, 261; art. *Cartésianisme*, II, 717; art. *Central*, II, 824. The Encyclopedists quote Descartes also among the greatest mathematicians to whom is ascribed the invention of the cycloid (art. *Cycloïde*, IV, 590) and the equation of the parabola called "trident" (art. *Trident*, XVI, 637). To Descartes they attribute also the distinction between algebraic and transcendental curves (art. *Courbe,* III, 378; art. *Méchanique*, X, 224).
40 *Ibid.*, art. *Moderne*, X, 601.
41 *Ibid.*, art. *Newtonianisme*, XI, 123.
42 *Ibid.*, art. *Cosmogonie*, IV, 293.
43 *Ibid.*, art. *Tourbillon*, XVI, 471; art. *Wolstrope*, XVII, 630.
44 *Ibid.*, *Discours Préliminaire*, I, XXVI; art. *Astronomie*, I, 793; *Ciel,* III, 442; art. *Nature*, XI, 40; art. *Paradis*, XI, 894; art. *Physique corpusculaire*, IV, 269.
45 *Ibid.*, art. *Expérimental*, VI, 299.
46 *Ibid.*, *Discours Préliminaire*, I, XXV: "Reconnaissons donc que Descartes, forcé de créer une Physique toute nouvelle, n'a pu la créer meilleure; qu'il a fallu, pour ainsi dire, passer par les tourbillons pour arriver au vrai système du monde; et que s'il s'est trompé sur les lois du mouvement, il a du moins deviné le premier qu'il devait y en avoir."
47 *Ibid., Discours Préliminaire*, I, XXVI: "Sa Métaphysique, aussi ingénieuse et aussi nouvelle que sa Physique, a eu le même sort à peu près."
48 *Ibid.*, *Discours Préliminaire*, I, 401.
49 *Ibid.*, *Discours Préliminaire*, I, XXVI.
50 *Ibid.*, art. *Cartésianisme*, II, 719.
51 *Ibid.*, art. *Éléments des sciences*, V, 493.
52 *Ibid.*, *Discours Préliminaire*, I, XXVI.
53 *Ibid.*, *Discours Préliminaire*, I, XXVI: "Descartes se trompa sans doute en admettant les idées innées: mais s'il eût retenu de la secte Péripatéticienne la seule vérité qu'elle enseignait sur l'origine des idées par les sens, peut-être les erreurs qui déshonoraient cette vérité par leur alliage, auraient été plus difficiles à déraciner. Descartes a osé du moins montrer aux bons esprits à secouer le joug de la scholastique, de l'opinion, de l'autorité, en un mot des préjugés et de la barbarie; et par cette révolte dont nous recueillons aujourd'hui les fruits, la Philosophie a reçu de lui un service, plus difficile peut-être à rendre que tous ceux qu'elle doit à ses illustres successeurs. On peut le regarder comme un chef de conjurés, qui a eu le courage de s'élever le premier contre une puissance despotique et arbitraire, et qui en préparant une révolution éclatante, a jeté les fondements d'un gouvernement plus juste et plus heureux qu'il n'a pu voir établi."
54 *Ibid.*, art. *Cartésianisme*, II, 717.
55 *Ibid.*, art. *Flèche, (La),* VI, 850.
56 This image of Descartes is very important for d'Alembert, who repeats it in many passages of his work. For instance, he quotes Descartes in describing the characteristics of the true *philosopher*, when, in the *Avertissement* of the *Essai sur les Eléments de Philosophie*, he intends to defend "the philosophical quality of the *Philosophe*" (emphasis in original): "a loyal citizen to his duties and attached to his own country, subject to the laws of religion and state; who prefers, following the principle of Descartes, *to regulate his desires that the world order*" (emphasis in original, *Œuvres Complètes de d'Alembert* [Paris: A. Blin, 1821–22], I, P. I, 119).

57 *Encyclopédie*, art. CARTÉSIANISME, II, 719.
58 Albert-Jean Guibert, *Bibliographie des Œuvres de René Descartes publiées au XVII*[e] *siècle* (Paris: Editions du Centre National de la Recherche Scientifique, 1976).
59 The Compagnie des Libraires published in Paris, between 1723 and 1729, an edition of Descartes' texts that is no more than a reprint of each work, with the exception of the Latin text of a few letters that had only been available in translation, and a few French versions of Latin letters. Only the six volumes of the *Lettres* (1724–1725) are set out in a consecutive series of volumes.
60 Adrien Baillet, *La Vie de Monsieur Descartes* (Paris: Chez D. Horthemels, 1691): P. II, L. VII, Ch. VII, 282–3; P. I, L. I, Ch. V, 25; P. I, L. II, Ch. IX, 131–4 (*passim*); P. II, L. VII, Ch. VIII, 289–90; P. II, L. VIII, Ch. VII, 503–8; P. II, L. VI, Ch. II, 99–100, 107–10 and 113–4.
61 Jean-Pierre Niceron, *Mémoires pour servir à l'Histoire des Hommes illustres dans la République des Lettres avec un catalogue raisonné de leurs Ouvrages* (Paris: Chez Briasson, 1728–1738), T. XXXI (1735): 175–314.
62 Noël-Antoine Pluche, *Histoire du Ciel, où l'on recherche l'origine de l'idolâtrie, et les méprises de la Philosophie sur la formation, et sur les influences des corps célestes*, 2nd ed. (Paris: Chez la Veuve Estienne, 1740), T. II, 173–86, 189–90, 215–22.
63 See Tad M. Schmaltz, *Descartes and the Enlightenment* (Montréal: McGill-Queen's University Press, 1989) and *Radical Cartesianism: The French Reception of Descartes* (Cambridge: Cambridge University Press, 2002); François Azouvi, *Descartes et la France* (Paris: Hachette, 2006) and "Être cartésien: hier ou aujourd'hui?" in *Qu'est-ce qu'être cartésien?* ed. Delphine Kolesnik-Antoine (Lyon: ENS Éditions, 2014), 387–94. I would refer to my studies on the legacy of Descartes in the eighteenth century, particularly *Immagini di Descartes nell'Encyclopédie* (Bologna: Il Mulino, 1999) and *L'arbre et le labyrinthe. Descartes selon l'ordre des Lumières* (Paris: Champion, 2009).
64 Azouvi, *Descartes et la France*, Ch. V, 95–125.
65 *Ibid.*, Ch. V, 97.
66 Étienne Bonnot de Condillac, *Traité des systèmes* (La Haye: Chez Neaulme, 1749; Paris: Fayard, 1991), 1–2: "Un système n'est autre chose que la disposition des différentes parties d'un art ou d'une science dans un ordre où elles se soutiennent toutes mutuellement, et où les dernières s'expliquent par les premières. Celles qui rendent raison des autres, s'appellent principes; et le système est d'autant plus parfait, que les principes sont en plus petit nombre: il est même à souhaiter qu'on les réduise à un seul."
67 Étienne Bonnot de Condillac, *Essai sur l'origine des connaisances humaines* (Paris: Mortier, 1746; Paris: Vrin, 2002), Introduction, III.
68 *Encyclopédie*, *Discours Préliminaire*, I, X. Condillac dedicates the entire second section of the second part of the *Essai sur l'origine des connaissances humaines* to this topic (P. II, S. II, Ch. III).
69 *Ibid.*, art. ANALYSE, I, 401.
70 *Ibid.*, art. CARTÉSIANISME, II, 719.
71 *Ibid.*, art. LOGIQUE, IX, 638.
72 *Ibid.*, art. ANALYSE, I, 401. D'Alembert illustrates these procedures in the article ÉLÉMENTS DES SCIENCES (V, 493–4) in which he proposes the traditional distinction between analysis, the method that is "more apt for finding truths and demonstrating clearly how we have found them," and synthesis, the method that is "more apt at explaining and communicating the truths that we have already found." Analysis ascends from already known consequences to the principles that are not as yet known and, by generalizations, it allows the latter to be

discovered; synthesis is the method that can be successfully applied "in those sciences whose objects lie outside us." As d'Alembert emphasizes in this text which seems to be inspired by Descartes, but which turns in reality against Descartes, "the method of analysis must combine simplicity and clarity, the most essential qualities that the elements of a science must have." It is only beginning with certain simple principles that one can arrive at certain consequences. Descartes, instead, starting with abstract ideas, was wrong, continues d'Alembert. "He proceeded contrary to the natural order of thought, turning to synthesis where analysis was required."

73 *Encyclopédie*, art. IDÉE, VIII, 490. This observation is taken from Condillac, *Essai sur l'origine des connaissances humaines*.

74 *Ibid.*, art. ÉLÉMENTS DES SCIENCES, V, 494.

75 *Ibid.*, art. VÉRITÉ, XVII, 68.

76 See Walter Tega, *Arbor scientiarum. Enciclopedie e sistemi in Francia da Diderot a Comte* (Bologna: Il Mulino, 1984); Francine Markovits, *L'ordre des échanges* (Paris: Presses universitaires de France, 1986); Florent Guénard, Francine Markovits and Mariafranca Spallanzani, eds., "L'ordre des renvois dans l'*Encyclopédie*," *Corpus* 51 (2006): 1–391; André Charrak, *Empirisme et théorie de la connaissance. Réflexion et fondement des sciences au XVIII*[e] *siècle* (Paris: Vrin, 2009).

77 *Encyclopédie, Discours Préliminaire*, I, XXVII–XXVIII: "La Métaphysique raisonnable ne peut consister, comme la Physique expérimentale, qu'à rassembler avec soin tous ces faits, à les réduire en un corps, à expliquer les uns par les autres, en distinguant ceux qui doivent tenir le premier rang et servir comme de base. En un mot les principes de la Métaphysique, aussi simples que les axiomes, sont les mêmes pour les Philosophes et pour le Peuple."

78 *Encyclopédie*, *Discours Préliminaire*, I, XXVIII.

79 *Ibid.*, I, XIV: "Il ne nous reste plus qu'à former un Arbre généalogique ou encyclopédique qui les rassemble sous un même point de vue, et qui serve à marquer leur origine et les liaisons qu'elles ont entre elles. Nous expliquerons dans un moment l'usage que nous prétendons faire de cet arbre."

80 D'Alembert, *Essai sur les Eléments de Philosophie*, Éclaircissement, V, Vol. I, P. I, 30.

81 *Encyclopédie*, *Discours Préliminaire*, I, XVII.

82 *Ibid.*, I, XVI.

83 *Ibid.*, art. ELÉMENTS DES SCIENCES, V, 492.

84 *Ibid.*, *Discours Préliminaire*, I, XXVII. See: Michel Malherbe, "L'Encyclopédie: histoire, système et tableau," in *L'Encyclopédie ou la création des disciplines*, ed. Martine Groult (Paris: CNRS Editions, 2003), 39–42.

85 *Encyclopédie*, **Explication détaillée du système des connaissances humaines*, I, XLVIII: "par conscience ou sens interne."

86 *Ibid.*, art. EXISTENCE, VI, 262.

87 *Ibid.*, *Discours Préliminaire*, I, III.

88 *Ibid.*, **Explication détaillée du système des connaissances humaines*, I, XLVII.

89 See Martine Groult, *D'Alembert et la mécanique de la vérité dans l'Encyclopédie* (Paris: Champion, 1999); Michel Malherbe, *Introduction* to the edition of the *Discours Préliminaire de l'Encyclopédie* (Paris: Vrin, 2000).

90 *Encyclopédie*, *Discours Préliminaire*, I, XIV.

91 *Ibid.*, **Explication détaillée du système des connaissances humaines*, I, XLVII—L: "La Nature a ses écarts, et la Raison ses abus. Nous avons rapporté les *monstres* aux écarts de la Nature; et c'est à l'abus de la Raison qu'il faut rapporter toutes les Sciences et tous les Arts, qui ne montrent que l'avidité, la méchanceté, la superstition de l'Homme, et qui le deshonorent."

2 To Replant and Uproot. Typology of the Cartesian Tree of Knowledge in Nineteenth-Century French Histories of Philosophy

Delphine Antoine-Mahut

Introduction

From the 1820s and, more or less markedly, until the death of Victor Cousin[1] in 1867, French spiritualism was characterized by a strong return of metaphysics, directed by a triple reaction.

1. A reaction to the sensualist and skeptic philosophy of Locke and his circle and, therefore, also to Bacon, Hobbes, Gassendi, and Condillac.
2. A reaction to the "scientist" and "positivist" ideology—particularly in its physiological dimension—as revived in the works of Broussais and Comte.[2]
3. A reaction to another type of spiritualism: the theological spiritualism represented by de Maistre or de Bonald,[3] in particular.

This triple reaction had double implications. On the one hand, the criticisms of these adversaries of rational spiritualism had to be addressed; namely the accusations of abstraction and musings far removed from the empirical data of natural philosophy. On the other hand, in response to theological spiritualism, it was necessary to make short work of the criticisms of the extravagant claims of rationality to display a certainty which only revelation could bestow. It also implied that the nascent history of philosophy should be structured upon the basis of a system enabling metaphysics to be protected from these various assaults.

Cousin found the solution in Descartes and an empirical-rational interpretation of the *cogito*.[4] In so doing, he redefined metaphysics both in opposition to the natural sciences while adopting their methods of observation, and in opposition to theology while refocusing on psychology.

My objective is to rapidly review this framework, which is still present today,[5] in order to reveal certain boundary issues and displacements that testify to the complex relationships existing between metaphysics and physics in Cartesianism for Cousin himself. I will then use two successive prisms to turn my attention to the papers submitted for the 1839 competition concerning the history of Cartesianism organized by the *Académie des Sciences Morales et Politiques*.

The first prism will be the report about these papers written by Jean-Philibert Damiron.[6] Here they will be examined from the perspective of the "official" or "dominant" philosophy. This official character was proven by the publication of Damiron's report in its entirety as an introduction to his *Essai sur l'histoire de la philosophie en France au XVIIe siècle*, in 1846.

The second prism will be the actual content of the competition entries. We will see the historiography of Cartesianism shift according to how major elements were jointly taken into consideration in order to renew and reinforce metaphysics understood as being the study of the soul. These elements were the subversive power of physics, on the one hand, and the need to account for its progress, on the other.

In all, seven papers were in competition, one of which arrived after the deadline. In addition to the prize-winning paper submitted by Jean-Baptiste Bordas Demoulin, two other entries, written by Charles Renouvier and Francisque Bouillier,[7] respectively, were recognized as being worthy of the accolade: "*très honorable*" i.e., most honorable. In all, however, four entries were published since, in 1843, Jean-André Rochoux independently published in unmodified form his subversive *Épicure opposé à Descartes*. These four texts will provide us with our privileged area of investigation into the roots and main body of Cartesian philosophy. They typologize different ways to update Cartesian philosophy in a restrictive Cousinian context.

My intention here is not only to inquire into the "dominant" conception of the relationships between metaphysics and physics in these histories of philosophy, but also to question the reasons for which the form of heterodoxy, which I wish to highlight, took so long to acquire legitimacy in Cartesian studies. If, today, we have stopped or almost stopped mentioning Renouvier, Bordas-Demoulin, Bouillier, or Rochoux[8] and if we tend to level Cousin's position toward a plain and simple refusal of the problematics linked to physics, this is undoubtedly not merely because Descartes, himself, chose to harden his positions along these lines in the polemical texts. This is also because we are still reluctant to grant all their meaning to objective facts such as the contextualization provided by the *Querelle d'Utrecht*, to the "Lettre-Préface" to the *Principes de la Philosophie*, and thus to the metaphor of the tree of knowledge. From this point of view, a detour via the nineteenth century can prove fruitful and even necessary in order to focus or refocus upon the multiple potentialities of the Cartesian text. In the eighteenth century, Mariafranca Spallanzani painted a contrasted picture of these potentialities, while the followers of Cousin later strove to caricature them and relegate them to the darkest shadows.

Is There a Porosity of Frontiers Between Physics and Metaphysics in the Writings of Victor Cousin?

The position of Cousin was characterized by the identification of philosophy with metaphysics and the progressive relegation of everything pertaining to physics to an area outside philosophy. To revisit the eighteenth century, we can refer to this extract of the *Fragments de philosophie cartésienne* (Pour faire suite aux *Fragments philosophiques*[9]):

> Let us dare to speak the truth: the 18th century in France [which has been] so rich in great men, has not produced a single one in philosophy, if at least by philosophy we mean metaphysics.

However, even if we cannot return to every text Cousin wrote on the subject, there are three other points which must be underlined: (1) the claim to responsibility for importing the methods of physics into metaphysics reconceived as psychology; (2) the redefinition by Cousin and his followers of physics as a subject which was "spiritualist without knowing it," which, for all that, did not identify itself with a hylomorphic projection that had been definitively defeated by Cartesianism; and (3) the contrasted treatment of boundary objects or subjects. Particularly those borrowed from physiology, echoes of which could be found in spiritualist doctors and which, in turn, were reflected in Cousin's rewritings of his lessons.

I will rapidly examine these three points which show that, even in Cousin's very own writings, the position of the problem of the relationship between the trunk and the roots of the tree of knowledge is not as Manichaean as it is sometimes made out to be.

1. The importation of the methods of physics into metaphysics, reconceived as psychology.

Cousin set himself the objective of requalifying metaphysics as the set of facts of internal observation and rational facts by means of an adaptation of the method of observation practiced in physics since Bacon to the objects sacrificed by this latter. This favored metaphysics at the core of which philosophy enabled the finding of what natural science had sought to dismiss: namely, ontology. Experience applied to the consciousness revealed substantial activity that was both anterior and superior to any phenomenal activity and specific to the thinking substance. This rooted force on the side of spirituality.

2. Because of this, far from physics dictating its laws to psychology, physics was spiritualist without knowing it. If modern physics is concerned exclusively with forces and laws, it can't be materialist. It becomes necessarily spiritualist when it rejects all other methods except observation and induction, which can only lead to forces and laws, and nothing is materialist in forces and laws.

Only—and this included the work of the man Cousin called "the first French metaphysician of his time," Maine de Biran[10]—the Cartesian critique of the hylomorphic projection had to remain an intangible line of conduct. Just as one did not import exteriority into the soul, one did not export one's interiority into the physical universe. In Cousin's work, thinking about the relationships between physics and metaphysics remained profoundly dualist and activity remained within the sphere of the thinking substance.

3. However, one area resisted this dualism. This area was physiology, which Maine de Biran defined as being situated i.e., "between the dynamics of the bodies and the dynamics of the spirits."[11] Thus:

> The study of medicine supposes that of the physical and natural sciences; it develops the taste and the talent for observation and, in this respect, it may be said that the study of medicine is an excellent preparation for metaphysics; but, it must be added for a well-formed mind, for when we are continually surveying phenomena of organic life, it is easy, it is natural to be surprised and carried away by the appearance, and to confound with these phenomena other phenomena which are very different; and I pray you not to forget that, in fact, in the review which I have presented to you of all the philosophical schools we have seen sensualism and empiricism as well as skepticism often proceed from schools of natural philosophers and physicians.[12]

The shifting of activity toward the area of the bodies led to a belief in the uselessness, even the falseness, of a different spiritual principle likely to engender thoughts. One possible solution consisted in recalling that, to be exact, this Leibnitzian debate was "not born at the time of Descartes" and only appeared toward 1691 or 1694.[13]

Conversely, however, the reduction of matter, and living matter in particular, to pure extension threw the door wide open to a Spinozist and Malebranchist interpretation of Cartesian physics and to its potential degeneration into pantheist animism.[14] In order to avoid proving the Leibnizian objections right, the only alternative was thus to reinject metaphysics into Cartesian physiology, while simultaneously maintaining the banishment of final causes from the remainder of physics as thematized in article 28 of the first part of the *Principes de la philosophie*.

In this extremely astute way, Cousin succeeded by inserting another prism between Descartes and Leibniz. This prism took the form of Claude Clerselier and the division into paragraphs, which he proposed for the posthumous editions of *l'Homme* and *De la description du corps humain* (renamed traité *De la formation du foetus*) in 1664 and 1677, editions in which he replaced the Cartesian lexicon of the "description" of the parts and functions with the finalist vocabulary of their "use." This subterfuge enabled Cousin to conclude that "the determination of the end of a phenomenon is necessary for knowledge, the seeking of the final causes is an integral part of science and should not be referred to a foreign science."[15]

Therefore, for Cousin, physics, and particularly physiology, could be recovered once the materialist and pantheist risk had been foiled by the revealing of the primacy of the method and principles of psychology over those of natural philosophy. In the history of ideas, metaphysical theories have always been applied to the sciences, in general, and to physiology, in particular, no matter how pertinent or not the predominant philosophy in the period under consideration might be.

Let us revisit the eighteenth-century example by Condillac below. The main argument is as follows: just as in metaphysics the ego or the soul is simply the sum of our sensations, so in physiology, life is simply the sum of functions without unity. If that is the case, then the harmony of these finctions is strange; but people jumped with both feet right over all these difficulties and medicine had its totally empirical philosophy.

At the beginning of my presentation of this section devoted to Cousin, I evoked the various returns and inflections of this conception of the relationships between physics and metaphysics among spiritualist doctors. I shall take one example here, a doctor from Montpellier[16] who is quite unknown today. Alphonse Jaumes, who, in a paper entitled *De l'influence des doctrines philosophiques de Descartes et de Bacon sur les Progrès de la Médecine* (1850), defended what he called "medical metaphysics." One of the most interesting elements of his line of argument consisted in his use of the Cartesian theory of innate ideas, which he considered metaphysically false and which lay at the center of the dispute between the followers of the Ideologists and Cousin. He used this concept to demonstrate how it could nevertheless serve in the understanding of true vital anthropology. As presented in the *Notae in Programma quoddam* (1647) in response to the Dutch doctor Henricus Regius, in particular, and by analogy with hereditary family illnesses, innatism provided the model of an active force "which has its attributes without which it would be impossible to conceive it." This enabled one to use Descartes to conceive what Descartes himself never thought and which some of his successors also forbade themselves from thinking: namely, a materialist vitalist dynamism:

> The doctrine of innate ideas, which is the basis of Cartesianism and has been given an irresistible proof, can, if applied with the discernment required by the different nature of the subject, be extremely conducive to the progress of the medical sciences.[17]

In the present case, false metaphysical dynamism was therefore recovered in order to conceive true physical dynamics. Undoubtedly this contained an extension of the Cartesian gesture which consisted in exhibiting, in false scholastic physics, that which could be reinvested in the adequate conception of the force which the soul possesses to set the body in motion.

The example of Jaumes is interesting because he went further than Cousin, by assuming the sometimes erroneous but always heuristic nature of Cartesian psychology. However, in this case, the principles of a renewal of physics were again to be found in metaphysics; in the return in force of that which Descartes himself had excluded from it in the work of Cousin and by a contamination of Baconism in the writings of Jaumes.

The analysis of the 1839 competition entries about Cartesianism will enable us to envisage other possible displacements that will be situated in the field of actual Cartesian physics. They bear witness to a clear awareness of the need for spiritualism to address the issues pertaining to natural philosophy, for it to confront Cartesianism with its Epicurean and Newtonian counterparts in order to dismiss it (for Rochoux) or, on the contrary, in order to reinforce it. So it is that metaphysical issues progressively shifted from the thinking substance toward the material substance, opening new perspectives and revealing new complexities within material substance.

I shall first examine how Damiron's report perceived these shifts and strove to invert their course; then how, while taking Damiron's remarks into consideration (while sometimes persisting in their initial choices), the works published diverged on these different points.

The 1839 Papers Seen Through the Prism of "Official" Philosophy: Damiron's Report

I will limit my remarks to the sections of the report dealing with the four competition entries mentioned at the start of this presentation (the other two submitted within the deadline were never published and gave rise to no significant remarks in the report).

The first paper, some ninety-five pages long, was written by Rochoux. Damiron criticized the author for being partisan in nature, conducting a Gassendi-like resurrection of Epicurean philosophy and his almost total silence concerning what should have constituted the core of his argument: metaphysical considerations. According to this representative of institutional spiritualism, Rochoux was wrong in saying nothing about the nature and origin of the various ideas on the soul or about free will considered as the faculty of judgment and action. In short, he couldn't win the prize because he said nothing about the fundamental issues of Descartes' metaphysics, without which nothing worthwhile could be said about physics: "This is a major omission, everything which should be there is not there."[18] Moreover, according to the author of the competition entry, Cartesianism was dead in the sense that in his *Censura philosophia cartesianae*, the skeptic, Pierre-Daniel Huet, distinguished two men in Descartes: on the one hand, the physician and the geometer, the experimenter unsurpassed by any modern, and, on the other hand, the philosopher [who was] much less fortunate in metaphysical speculations.

In the work submitted by Rochoux, the Académie, as represented by Damiron, rejected the specific form of "exclusivity" which consisted in identifying philosophy with the natural sciences and in excluding from it the principal issues pertaining to metaphysics. Since Cartesianism and its different receptions constituted the competition subject, this refusal manifested itself in a denunciation of the relevance of such a point of view in the writings of Descartes himself. The treatment of metaphysics was not only a prerequisite but also a prime requirement for whosoever wished to stem the deleterious effects of the Gassendi prism upon the interpretation of the philosophy of Descartes.

In one way, the procedure concerning Charles Renouvier is even more remarkable. For Damiron attributed to Descartes himself Cousin's preference for Descartes the metaphysician over Descartes the physician. Renouvier was therefore immediately corrected by a very clear formulation of the official institutional line:

> the author inadvertently lets slip judgements that are not always made with the strictest accuracy; for instance, he asserts several times that Descartes aspired to do natural philosophy. Yet clearly nothing is less Descartes' intention, as evinced by his *Meditations* and his *Discourse on the Method*; and there moreover are statements made by him on this subject that leave no doubt as to his real opinion. He therefore considered himself above all a metaphysicist; the physicist and geometrist hold a subordinate position in his eyes.[19]

Once that had been established, Renouvier enabled it to be understood that the "immoderate extension" of certain theses present in the writings of Descartes had engendered the philosophies of the adversaries. As Cousin had already highlighted in his *Philosophie de Locke* (1828), physiology constituted the most eloquent example of this:

> With respect to physiology, he remarks that if, as in physics, he has the vice of resorting too hastily to hypotheses, it should not be forgotten that beyond his merits for explaining and popularizing the discovery of the circulation of blood, his theory of man was so convincing that besides the Cartesian school, which adopted it as its own doctrine, it also became, with a few modifications, that of the opposing school, which adopting Descartes' organic mechanism merely changed the pineal gland into the cerebral centre and animal spirits into the senses and moreover extended these principles immoderately.

If, according to Damiron, this passage was "one of the most remarkable" of this paper, this was for two reasons. First, it enabled one to comply with this type of posterity of Cartesianism and, second, to exclude posterities that were receivable or in accordance with the original text, which remained well founded in dualist metaphysics. Renouvier's paper could thus receive an accolade because it could both indicate the dangers of certain physics-oriented readings of Descartes and circumvent them.

As for Francisque Bouillier, the starting point of his competition entry was a clear Cousin-inspired separation between Bacon, the natural philosopher, and Descartes, the metaphysician or promoter of grand ideas impregnating every system. However, according to Damiron, this entrant could also underestimate the importance of such decisive metaphysical issues as the proof of the existence of God, final causes, or innate ideas. The watchword remained unchanged and was theorized in the *Essai sur l'histoire de la philosophie au XVIIe siècle*, when dealing with the physician of the Cartesian school, Jacques Rohault: "In the school of Descartes, one is not a physician without being a metaphysician and reciprocally."[20] "Deserters" from Cartesianism, such as Regius, had to be hunted down, for they believed they could remain faithful to the master while reinvesting some of his physical propositions and thwarting prime philosophy.

Jean-Baptiste Bordas-Demoulin stands out from the others given the importance granted to physical issues and the insistence upon mathematics. In the name of the "tendencies" of the work of Descartes, he underlined the lack of comprehension of the activity of both bodily and spiritual substance and the uncertainty of Cartesian formulations concerning innate ideas, which gave rise to contrasted posterities of unequal legitimacy. However, in the eyes of Damiron, this entry was above all a chance to underline other nonnaturalist excesses: those of an "over-marked theological character," particularly concerning considerations of the effects of original sin: "these matters belong neither to the philosophy of Descartes, nor to philosophy itself."

Thus, physical considerations always corrected and founded metaphysics reconceived as psychology and distinct from theology. The reference to what was supposed to occur in the work of Descartes served as the standard against which to elaborate the entire history of philosophy.

Four possible avenues are implicit in Damiron's reading:

1. The direction taken by Rochoux, who dismissed both Cartesian metaphysics and Cartesian physics by promoting an Epicurean style of materialism.
2. The direction taken by Renouvier, who opened the way for a materialist-medical interpretation of Descartes himself, by underlining the importance of Spinoza and by promoting vitalism.
3. The direction taken by Bouillier, who defined a new Leibnitz-inspired animism, challenging both the duo-dynamism of the Montpellier School and the organic vitalism of the Paris School.

4. The direction taken by Bordas-Demoulin, who laid the foundations for a new general pathology studying the effects of the original Fall upon the body and soul, while claiming Malebranche to be his inspiration.

I shall rapidly examine these four points, considering them from the perspective of the published works. I shall indicate the later evolution of these points, whenever applicable.

The Content and Actual Displacements of the Papers

Rochoux's paper was published in 1843, entitled *Épicure opposé à Descartes*. It had a strong polemical presentation, "against the secret thinking of the commission"; this "secret thinking" was favorable not only to Descartes but above all to Descartes as institutionalized by Cousin. For whosoever intended to criticize this official Descartes and thus untie the knot linking psychology and the natural sciences in the whole of philosophy, the only alternative was to dismiss Descartes himself for internal contradiction or contradiction of experience. This was what Rochoux called "judging Cartesianism only by the words of the master" and not by what others say of them.

The first argumentative movement extracted what could serve Rochoux's project from Epicurean philosophy, as rehabilitated by Gassendi. The most important passage concerned physics and more precisely the sum of activity and spontaneous movement contained in the atom. This conception of the activity of matter enabled the establishment of a link with La Mettrie and Haller and the criticism of contemporary scientists, such as Claude Servais Mathias Pouillet in his *Éléments de physique expérimentale et de météorologie* (1829), who posed a problem: the claim for a force of inertia by virtue of which matter was supposed to be equally indifferent to rest and movement. Yet the machine was never at rest:

> every situation in which so many men have believed they saw inertia, matter at rest, is merely cases of equilibrium produced by the balance of opposed forces, so that this alleged inertia is actually a highly active struggle [. . .] a single principle, that of an atom endowed with motion or active matter, suffices to explain all natural phenomena. [. . .] every phenomenon of any kind constantly shows us force and matter indissolubly joined and always acting in unison.[21]

It could thus be expected that Rochoux would present Cartesianism as the reverse model of this philosophy of activity or material force. However, this did not happen exactly like that. Descartes did sow the seeds of truth in physics. But they were in contradiction with the metaphysical principles they were supposed to serve, particularly the theory of continuous creation.

Therefore metaphysics, as praised by Cousin and his followers, contaminated the very physics of activity which they condemned in the works of the adversaries of Descartes:

> Thus the supreme intelligence is obliged to pay attention incessantly to every individual particle of matter. There is no way to dispute it if one denies matter all force of its own. Descartes seems to have foreseen this objection: but he was able to anticipate it with one of the frequent contradictions for which Huet criticizes him so bitterly. Essentially, after stating that motion imparted by God is always in a straight line, he explains the changes of direction it is subject to with reference to the influence of matter. Does not matter therefore have some force of its own, if it can produce such a result? Thus the hypothesis of the inertia of bodies is overturned by the same man who until then had been forced to defend it.[22]

The particularity of Rochoux's work was thus the displacement of the debate from metaphysics to physics and the drawing of attention to the seeds of truth specific to the natural philosophy of Descartes, once it had been rid of the metaphysical principles which invalidated it. Rochoux was obliged to reject Descartes lock, stock, and barrel because the dominant interpretation of the master's work rejected any form of activity for matter and because that was where truth was to be found.

Another strategy consisted in taking seriously—within the writings of Descartes—the possibility of conceiving this activity within a philosophy which remained spiritualist. This was Renouvier's choice.

In the preface to his *Manuel de philosophie moderne*, Renouvier defined his approach as being an application of philosophy to those particular sciences called natural sciences, or as being a "natural philosophy"[23] conceived out of hitherto ignored Cartesian physics. To this, he added a thematization of the difference between spiritualism and materialism based upon the categories of activity and passivity with inertia on the side of matter. He also noted the danger of a naturalist approach which granted a form of autonomy to bodily substance. He responded to Damiron by inserting metaphysics into the Cartesian tree of knowledge. However, metaphysical issues, such as that of "pure notions," for instance, were never thematized for their own sakes. They were always approached from the relationship with physics.

Moreover, although he agreed with Cousin in recognizing that "Bacon can be considered neither as a philosopher nor as a metaphysician" and that Bacon wished to "call the physician a philosopher and the metaphysician a physician,"[24] he immediately went on to assert that Descartes himself deals with neither theology nor pneumatology, unless it is furtively, in some letters or in replying to objections and always to his great regret. By these means, Renouvier validated his thesis of a dissociation—for Descartes himself—between true science and the problematics which Bacon imputed to "metaphysics." By redefining the philosophical categories and by refusing the bridges between Bacon and Descartes, Renouvier upturned Damiron's reading, a reading which Damiron claimed to have based upon the Cartesian texts themselves. The text of the *Manuel de philosophie moderne* thus contained the recognition of another Descartes, different to the one found in the history of the Cousin school, alongside the claim for another philosophy arising out of the work of Descartes but not present as such within this work: namely, a form of vitalism.

Renouvier was visibly fascinated by what he called "the creative power with which Descartes himself endowed matter." According to Olivier R. Bloch, such an attribution to Descartes of the thesis of a material dynamism is, to say the least, "risky" if not paradoxical.[25] However, a restitution of the entire line of argument enables this situation to be understood.

The assertion of the *Principes de la philosophie* III, 47, according to which matter could take any form should at some stage be posed in clear metaphysical terms or otherwise forever be a potentially fruitful but "gratuitous" hypothesis. It allowed the recovering of Cartesian physics to serve the cause of vitalism. In answer to the question of the pre-eminence between metaphysics and physics, Renouvier replied with the question of the supremacy of real or concrete physics over general or theoretical physics. The external hierarchy between metaphysics and physics was displaced toward an internal hierarchy between two varieties of "physics," the "special" one allowing dynamics.

This raised the issue of what might become of this recovering of the activity of matter in a spiritualism in which theology was dominant. In his *Cartésianisme ou la véritable rénovation des sciences* (1843),[26] Bordas-Demoulin surprisingly proposed to turn to Malebranche rather than Descartes or even Leibniz. For the notion of intelligible extent was applicable to any force: God enabled it to be understood that no true activity could be conceived without quantity. And, contrary to what the dominant spiritualism thought, this quantity was in no way nullifying. On the contrary, it was at the base of the definition of any substance endowed with force. Therefore, crude bodies undoubtedly excluded any spontaneity. However, they did not exclude all manifestations of force. Bordas-Demoulin's solution thus consisted in breaking the excessive alternative between pure mechanism and pure dynamism by distinguishing, for extension, the two meanings of life which spiritualism had no difficulty in accepting. The single condition required to be able to found a "true spiritualism" was the uniting of quantity and force.

Just as Renouvier had displaced an external opposition (between metaphysics and physics) toward an internal dissociation (between real and abstract physics), so Bordas-Demoulin responded by rooting the activity of the soul itself in its "quantity" or "extent."

Finally, Bouillier's starting point was indeed the definition of Descartes as the "creator and father of true metaphysics."[27] But only to underline that his ignorance of the nature of created substances had directly led Descartes down the slope of a phenomenalization of the thoughts of the soul. Rochoux concluded from the inactivity of matter that, in order to account for experience, one had to have recourse to continued creation while deliberately reversing the order of the tree of knowledge and turning metaphysics into a default solution or a concealing make-do. Conversely, Bouillier asserted that, if one wished to be rational, the presupposition of continued creation implied the removal of all activity from creatures. Since metaphysics had primacy, the Malebranche reading of the second philosophy of Descartes was the only one possible. And so the consequences of Rochoux's reasoning could be extended to the human soul itself.

In a later work, entitled *Du principe vital ou de l'âme pensante*,[28] Bouillier proposed to correct Descartes on this point using Leibniz. In true dynamics, it is psychology, and not mechanics or physics, that had to designate the "prime science of force." The false notion of the soul accredited by Descartes had to leave the field of narrow rationality and open up to animism in order to save spiritualism. The vitalist or duo-dynamist pseudo-solutions were thus overcome by a reintegration of organic activity within psychology.

Conclusion

Whatever the differences between the spiritualists may have been,[29] the redefinition of philosophy, which was identified with metaphysics, which itself was reduced to psychology, transited via the proof of its clear distinction from the sciences. But this distinction was not the only thing to matter. There was also the recourse to lexicons, such as that of "particularity," in order to designate these sciences. For each time philosophers attempted to dominate the sciences and refuse their specific consistency, they employed this type of terminology. This terminology let it be understood that there existed a general, universal thought (a role which philosophy obviously assumed). It also let it be understood that the pre-eminence of this general, universal thought was sufficient to discredit "particular sciences" which could thus be assimilated to mere techniques.

One of the conclusions to be drawn from this state of affairs is that, in order to reintegrate the sciences into the tree of philosophy, one must either completely reject metaphysics or accept its redefinition through a confrontation with the concrete data of experience.

Twentieth-century Cartesian historiography would provide several examples along these lines. But the least that can be said is that Cartesian historiography left a much deeper impression upon minds than the historiography of its contradictors or detractors.[30]

Notes

1 Victor Cousin (1792–1867) entered the French Academy and the State Council in 1830. In 1832, he became a member of the council of Public Instruction where he was charge of philosophy. He was elected a member of the newly created *Académie des sciences morales et politiques* and appointed director of the *Ecole*

Normale Supérieure. In 1840, under the government of Thiers, he was minister of public instruction for some eight months. He presided over the *concours d'agrégation*—an essential exam for anyone wishing to enter academia—for 27 years. During the period we are concerned with, he possessed all the means necessary to fulfill his intellectual and political ambitions. But for this very reason, he was also strongly criticized.

2 François Joseph Victor Broussais (1772–1838) was a student of Bichat and Pinel, head of the hospital *Val de Grâce* in Paris, professor at the medical faculty, and inspector of public health. He was a member of the *Académie de médecine* founded in 1823 and of the *Académie des sciences morales et politiques* from 1832 onwards. He wrote extensively on the relation between the history of philosophy and the history of medicine. He had an influence comparable to Cousin's, whom he was constantly challenging in public. Auguste Comte (1798–1857) is generally presented as the founder of French positivism and sociology. He strongly criticized Cousin's spiritualism, by mobilizing the now famous argument that in psychology, the observer modifies what is observed. So psychology can by no way be an experimental science.

3 Joseph de Maistre (1753–1821) and Louis Gabriel Amboise (1754–1840) were both fierce opponents of the French Revolution. De Maistre was a feemason tilted towards esoterism; De Bonald was a defender of a return to the monarchy and to Catholic Church principles.

4 On the Cousinian interpretation of Descartes and the alternative propositions of the "Cousinians," see Delphine Antoine-Mahut, "The Leibniz-Stahl Controversy and the Renewal of French Spiritualism," in *Hegel and Schelling in Early Nineteenth-Century France*, eds. Kirill Chepurin, Adi Efal-Lautenschläger, Daniel Whistler and Ayse Yuva (Springer, International Archives for the History of Ideas, forthcoming); "Experimental Method and the Spiritualist Soul: The Case of Victor Cousin," in *The Philosophy of experience in 19th Century in France and Germany: Contextes, Methods, Problems*, eds. Delphine Antoine-Mahut and Silvia Manzo, special issue of Perspectives on Science, (forthcoming); "Maine de Biran's Places in French Spiritualism: Occultation, Reduction and Demarcation," in *Edition and English Translation of Maine de Biran, Rapport du physique et du moral chez l'homme, with various Studies of Maine de Biran's Philosophy*, eds. Darian Meacham and Joseph Spadola (London: Bloomsbury, 2016), 33–46; "Cartésianisme dominant et cartésianismes subversifs: Le cas de l'infirmier de Bicêtre jean-André Rochoux," in *La mer retentissante: Lectures de Descartes et Leibniz au XIXe siècle*, ed. Lucie Rey, *Corpus* 68 (2016): 25–56 (see also Lucie Rey's excellent Introduction to this volume); "La fabrique du cartésianisme néerlandais dans les histoires de la philosophie française au XIXe siècle," in *Les Pays-Bas aux XVIIe et XVIIIe siècles*, eds. Catherine Secretan and Delphine Antoine-Mahut (Paris: Champion, 2015), 107–24; "Reviving Spiritualism with Monads: Francisque Bouillier's impossible mission (1839–1864)," *British Journal for the History of Philosophy* 23, no. 8 (2015): 1106–27; and "Is the History of Philosophy a Family Affair? The examples of Locke and Malebranche in the Cousinian School," in *Philosophy and Its History: New Essays on the Methods and Aims of Research in the History of Philosophy*, eds. by Eric Schliesser, Justin Smith and Mogens Laerke (Oxford and New York: Oxford University Press, 2013), 159–77.

5 On the Cartesian framework of these various receptions, see Delphine Antoine-Mahut, *Descartes radical: Historicité d'un canon philosophique* (Paris: Vrin, forthcoming).

6 Jean-Philibert Damiron (1794–1862) became a member of the *Académie des sciences morales et politiques* in 1836 and obtained a chair in modern history at the Sorbonne in 1838. Victor Cousin put him in charge of writing up the reports on the dissertations presented at the 1839 competition. Damiron was the author

of an *Essai d'histoire de la philosophie en France au dix-neuvième siècle* (1828, last edition in 1838) and of an *Essai sur l'histoire de la philosophie en France au dix-septième siècle* (1846).

7 Jean-Baptiste Bordas-Desmoulin (1798–1859) is the author of a memoir valorizing Descartes' contribution to the scientific revolution, particularly in its mathematical dimension. Charles Renouvier (1815–1903) devoted his contribution attention to the physical dimension of the Cartesian contribution and to the various links between physics and metaphysics in Descartes and its ulterior receptions. Jean-André Rochoux was a nurse at the Bicêtre Hospital. His memoir is dedicated to Broussais and is openly intended to counter the dominant spiritualism. That is why Rochoux explains he does not apply to win a prize. That is also why he published his book himself in 1843.

8 With an exception for Bouillier's work on Cartesianism, which is still a reference today in Cartesian studies. And if we mention Renouvier today, it is not for his Cartesian contribution, but because as a neo-Kantian, he was a contributor to personalism or one of the promotors of French laicism.

9 Victor Cousin, *Fragments Philosophiques* (Paris: Charpentier, 1845), Avant-propos, VIII.

10 *Œuvres philosophiques de Maine de Biran*, 3 vols., ed. Victor Cousin (Paris: Ladrange, 1834–1841), 66.

11 Maine de Biran and Marie François Pierre Gontier, *Exposition de la doctrine philosophique de Leibniz* (Paris: L-G. Michaud, 1819), 19, left column.

12 *Philosophie de Locke*, 4th ed. (Paris: Didier, 1861), 46.

13 Victor Cousin, *Histoire générale de la philosophie depuis les temps les plus anciens jusqu'au XIXe siècle* (Paris: Didier, 1867), 390.

14 On this link between Spinoza and pantheism, see Pierre-François Moreau, "Spinoza et Victor Cousin," *Archivio di filosofia* 1 (1978): 327–31; "Spinozisme et matérialisme au XIXe siècle," *Raison présente* 51 (1979): 85–94; "Trois polémiques contre Victor Cousin," *Revue de métaphysique et de morale* 4 (1983): 542–8 and "Les enjeux de la publication en France des papiers de Leibniz sur Spinoza," *Revue de métaphysique et de morale* 2 (1988): 215–32; "'In naturalismo'. Leibniz, Spinoza et les spiritualistes français," in *Spinoza/Leibniz: Rencontres, controverses, réceptions*, eds. Pierre-François Moreau, Raphaële Andrault and Mogens Laerke (Paris: Presses de l'Université Paris-Sorbonne, 2014), 325–44.

15 Cousin, *Histoire générale de la philosophie*, 384, note 1. The passage is translated from the *Fragments de philosophie moderne.*

16 Alphonse Jaumes was Associate and Curator of collections at the Faculty of Medicine in Montpellier. *De l'influence* is a thesis presented to apply to the "Chaire de Pathologie et de Thérapeutique générales."

17 Alphonse Jaumes, *De l'influence des doctrines philosophiques de Descartes et de Bacon sur les progrès de la médecine* (Montpellier : J. Martel Aîné, Imprimerie de la Faculté de Médecine, 1850), 22.

18 Jean Philibert Damiron gives his Report in his Introduction to the *Essai sur l'histoire de la philosophie en France au XVIIe siècle* (Paris : Hachette, 1846). Here, 11. This publication, in head of the *Essai*, confirms that the Cousinian School considered it as an official position, statement.

19 *Id.*, 33–34.

20 *Id.*, t. II, L. IV, ch. V, 127.

21 Jean-André Rochoux, *Épicure opposé à Descartes* (Paris: Joubert, 1843), 11.

22 *Id.*, 53.

23 Charles Renouvier, *Manuel de philosophie moderne* (Paris: Paulin, 1842), 6.

24 *Id.*, 140–141.

25 Olivier René Bloch, "Marx, Renouvier et l'histoire du matérialisme," in *Matières à histoires* (Paris: Vrin, 1997), 342.
26 *Le cartésianisme, ou La véritable rénovation des sciences: ouvrage couronné par l'Institut, suivi de la théorie de la substance et de celle de l'infini, Volume 1.* Précédé d'un *Discours sur la réformation de la philosophie au dix-neuvième siècle, pour servir d'introduction générale, par François Huet* (Paris: J. Hetzel, Libraire-Éditeur, 1843).
27 Francisque Bouillier, *Histoire de la philosophie cartésienne* (Paris: Durand and Lyon, Brun et Cie, 1854), t. 1, 57.
28 Francisque Bouillier, *Du principe vital et de l'âme pensante, ou Examen des diverses doctrines médicales et psychologiques sur les rapports de l'âme et de la vie* (Paris: J-B. Baillère et Fils and Londres et Madrid, 1862).
29 On this point, see Delphine Antoine-Mahut, *L'esprit du spiritualisme français* (Paris: Vrin, forthcoming).
30 See this volume, *Introduction*.

3 Between Fake, Unfortunate, and Actual Dependence

The Tumultuous Relationships of Descartes' Physics and Metaphysics in the First Half of Twentieth-Century History and Philosophy of Science

Delphine Bellis

Daniel Garber's book, *Descartes' Metaphysical Physics*,[1] constituted the first systematic attempt to investigate, in Descartes' physics, subject matters which are *de facto* related to metaphysical assumptions, like the nature and existence of bodies, and the nature of motion and its laws. Garber's interpretative core principle was to take seriously the foundational relationships that Descartes claimed to exist between his metaphysics and physics. On that view, the link between metaphysics and physics, however problematic its exact nature may be, is to be found in the dependence of the laws of motion on God, which Garber did not consider a supplementary, somewhat unnecessary, foundational addition to their epistemological grounding.[2] Surprising as this may seem for contemporary Cartesian scholarship, this connection has not always been so much under the spotlight or taken at face value before Garber's study. Several factors account for the relatively meager or uncharitable attention those relationships have received. If a scholar like Joseph Millet, in his *Histoire de Descartes avant 1637* (1867), considered that Descartes' physics was "a flow" (*écoulement*) from his metaphysics,[3] most scholars denied that kind of straightforward relationship between Cartesian physics and metaphysics. Instead, they either questioned the very existence of any dependence relation or, in the case this relation was recognized, came to formulate it in more general foundational terms rather than as an example of deduction. According to that latter account, metaphysics might or might not affect physics' content: to the first interpretative group belong, for example, Hamelin, Gibson, and Koyré; to the second one, Adam and the first Gilson.[4] One question that then presses upon us and that might constitute a key to the diverse interpretations offered in this period is the following: what was the philosophical starting point of those authors and, in particular, what kind of conception of "metaphysics" did they endorse? If one can assume that, regarding their philosophical starting points, they were all somewhat influenced by the recent theoretical mutations undergone by physics at the beginning of the twentieth century, the same does not hold true for their understanding of metaphysics. Did they have a specific conception of metaphysics that could at least partly account for the way they interpreted the relations between Descartes' metaphysics and physics? And to what extent did that determine some recurring features in their reading of Descartes? We will see that the positivist conception of metaphysics played a crucial role in that respect. However, this common background conception did not lead to one single interpretation of the relationship between Descartes' metaphysics and physics. It will thus remain to explain how, from a widely shared conception of metaphysics, our commentators diverged in their general philosophical insights about metaphysics and physics, and especially in their historical analysis of Descartes' case.

Without providing any definitive answers to the aforementioned questions at this stage, we can identify at least two types of influential interpretative motives which led our commentators to downplay the constitutive role of Descartes' metaphysics for his physics, these motives being more often than not intertwined. (1) According to the first one, Descartes' physics appears prior to and autonomous *vis-à-vis* his metaphysics;[5] Descartes' metaphysics is then seen as an *a posteriori* product which had no impact on the content of his physics, but was centrally and instrumentally intended to increase physics' legitimacy in foundational terms. Louis Liard (1846–1917), in the wake of Auguste Comte's positivism and its focus on the clear watershed between science and metaphysics, can be considered as the initiator of that type of interpretation in the twentieth century. (2) This interpretation often goes together with the assumption—also held by Liard—that Descartes' physics derives mainly from his method, rather than from his metaphysics, or even that method constitutes the source of Descartes' whole thought, be it philosophical or scientific. Among the commentators who gave pride of place to that position, one finds Ernst Cassirer (1874–1945). On that account, method is related to Descartes' *mathesis universalis* which opens the way for Descartes' physics to be rooted not so much in his metaphysics as in his mathematics broadly understood.

However, even before Daniel Garber's book, a small number of philosophers and historians of science of the first half of the twentieth century became interested in the interplay between Descartes' physics and metaphysics. Now, the early twentieth century is characterized by deep transformations of physical science.[6] In that context highly favorable to a philosophical reflection on science, Descartes emerges as a recurring figure within philosophy and history of science. Despite all the historical and conceptual distance separating the French philosopher and scientist from early twentieth-century science, Descartes' status in the first half of the twentieth century is very characteristic of a reflection on science which develops, at the intellectual as well as institutional level, in conjunction with the history of science and the history of philosophy. It then became crucial to root a philosophical reflection on science in a survey of its specific forms throughout history; history of science emerged as a central subject of investigation for philosophy of science.

After I have attempted to review, in the first part of this chapter, the philosophical motives which explain why, in that period, the relationships between Descartes' physics and metaphysics have not always received the attention they deserve, I shall concentrate on a few cases which reveal how philosophical commitments related to science have fashioned various interpretations of the complex and diverse ways those relationships can be envisaged in the first half of the twentieth century: on the one hand, some philosophers and historians of science, like Gaston Milhaud (1858–1918) and Émile Meyerson (1859–1933), attempted to seriously assess the philosophical imprint of Descartes' metaphysics on his physics and to evaluate its significance from the perspective of the history of philosophy and science; on the other hand, philosophers of science like Pierre Duhem (1861–1916) and Gaston Bachelard (1884–1962) identified the dependence of Descartes' physics upon his metaphysics, but mainly in order to criticize it from the point of view of a normative philosophy of science.

The Interpretative Obstacles to the Investigation of the Relationships Between Descartes' Physics and Metaphysics

Liard's Mitigated Positivism

The survey I intend to draw will start a little bit ahead of 1900 because some major studies published by the end of the nineteenth century had a significant impact on the historiography of the relationship between Descartes' physics and metaphysics in the first half of the twentieth century. We will thus have to go back to 1882. That year, Louis Liard published his *Descartes*[7] and initiated an interpretative line which was to be followed by commentators like Charles Adam,[8] Étienne Gilson at the beginning of his career,[9] and more recently Pierre Costabel.[10] *Agrégé* in philosophy and *docteur ès lettres*, from 1884 to 1902 Liard held the post of director of higher education at the ministry of Education (*ministère de l'Instruction publique*) and was one of the most important reformers of the French universities by the end of the nineteenth century. In 1879 Liard had published *La Science positive et la métaphysique*,[11] which reveals the influence of Auguste Comte's positivism.[12] Liard's foreword to his study of Descartes clearly placed it in the light of contemporary philosophy and science.[13] Liard's conviction was that Descartes' general doctrines were still part of the core of contemporary sciences.[14] The unifying factor to be found within Descartes' whole philosophy (and thus between his metaphysics and his natural philosophy) lay in the seminal and pervasive role played by method, that is to say the method of clear and distinct ideas.[15] However, Liard considered that Descartes was the true ancestor of contemporary scientists precisely because he "sought and discovered the unity of phenomena within phenomena themselves"[16] and developed a physics that is "a positive explanation of the world," that is to say, a physics deprived of any metaphysical idea.[17] Hence, there is "a very clear-cut distinction between the metaphysical order and the scientific order."[18] Liard did not deny, however, that Descartes did connect his physics to metaphysical ideas such as God as the first principle of creation and conservation, and extension as the essence of matter.[19] But this connection is seen as somewhat extrinsic to Descartes' physics and, at the very least, is not constitutive but comes into play as an *a posteriori* justification.[20] Descartes indeed started elaborating his physics and drew the principles thereof before his first metaphysical investigations of 1629.[21] Extension and motion are clear and distinct ideas—that is mathematical ideas—which, as such, are the offspring of method, and not of the idea of God.[22] Even the Cartesian laws of motion are not intrinsically dependent on God's immutability according to Liard. Admittedly, Descartes adduced such an explicit justification in his *Monde* and *Principia philosophiae*, but this was not a scientifically required justification: those laws are most of all clear and distinct. Such a methodological justification would have sufficed to make them acceptable on scientific grounds. God's immutability came into play as an "overabundant proof" meant to satisfy "intellectual needs foreign to scientific problems and solutions."[23] Moreover, God's immutability does not entail, from an analytic point of view, a unique principle of conservation. Why should conserved motion be rectilinear rather than circular? There is no clue in God's idea that could lead the physicist toward the *Principles*' third law of motion by a simple metaphysical deduction.[24] God's immutability serves as—so to say—a regulative principle, but not as a determinative principle of the first laws of physics. In such a view, without any metaphysics Descartes' physics would have been what it is all the same.

For all that, Liard rejected the idea that Descartes' metaphysics would have served as a way to dissimulate the physics of an atheist.[25] Liard recognized that Descartes' first intention was metaphysical insofar as Descartes intended to unravel the causes of phenomena and not only their law-like connections. But he evaluated his physics against the background of what he considered to be the core of contemporary science, that is the search for the laws of phenomena rather than of the first metaphysical causes. From that normative very Comte-inspired point of view, Descartes' metaphysics is superfluous and can be reduced to an "unconscious" reaction against the logical progressive tendency of separation at work in Descartes' physics. Descartes can thus be seen at the same time as the promoter of a modern physics which has *de facto* conquered its autonomy from metaphysics and as the *arrière-garde* metaphysician who tried to react against the inevitable consequences of his scientific boldness.[26] Descartes' doctrines have remained the core of contemporary sciences because they lean toward a separation of physics from metaphysics. But in the light of contemporary science, Descartes' physics also appears as irremediably doomed to bear the weight of metaphysical considerations which have henceforth become superfluous. This interpretation could amount to reducing metaphysics to an ancillary and instrumental role *vis-à-vis* physics, albeit a superfluous one. But apart from what could appear to us as whiggish history of science, and more interestingly from a philosophical point of view, Liard was also forced to recognize that, if Descartes' metaphysics did not impact the content of his physics, it had another—philosophical—significance for science. Metaphysics constituted a shield against the risk of radical idealism. Indeed, physics might well be true, but does it refer to reality? If physics confines itself to the realm of the laws of phenomena, isn't there a risk of considering it as a mere creation of our mind? Liard ended up ingenuously reversing the claimed relationship between Descartes' physics and metaphysics: metaphysics was not the foundation from which the principles and first laws of physics could be derived, but it was called by physical science as a way to guarantee "the faith science has in itself."[27] By drawing attention to the uselessness of metaphysics for the development of physics in Descartes, Liard thus ended up highlighting the weaknesses of physical science itself, that is its "inability to demonstrate the reality of its objects," and therefore also the weaknesses of a strict form of Comtian positivism.[28] Through the historical examination of the relationships between Descartes' physics and metaphysics, Liard put himself in a position to distance himself from a strict Comtian positivism which might slip into a form of phenomenalism or radical idealism,[29] and reasserted a function and a proper object for metaphysics.[30]

Cassirer's Neo-Kantianism and the Central Role of the Mathesis Universalis

In the very year Liard published his *Descartes*, Paul Natorp issued his *Descartes' Erkenntnistheorie. Eine Studie zur Vorgeschichte des Kriticismus*.[31] Whereas Liard had been to a large extent inspired by Comtian positivism in his interpretation of Descartes, in the German-speaking world, neo-Kantianism was to be the major influence on the reading of Descartes' philosophy then. Natorp was soon to be followed by Ernst Cassirer. As members of the Marburg School, they focused on epistemology and logic at the expense of ontology, and this impacted their historical works on Descartes. They both considered the *Regulae* as the source text from which Descartes' mathematics-based method, and also his more mature philosophy as a whole, developed. Natorp considered the metaphysical grounding provided by the appeal to God's idea in Descartes' philosophy to be required only for assuring the existence of external reality.[32] Even if he acknowledged that the Cartesian laws of nature followed from God's nature,[33] his focus was on the concepts of the understanding conceived in their clarity and distinction as shaping the whole of knowledge (be it metaphysical or physical). From that point of view, extension was considered more as a concept than as a mode of being, hence more as an epistemological than as a metaphysical tool for physics.

Cassirer followed in Natorp's footsteps, as can already be seen in his "Descartes' Kritik der mathematischen und naturwissenschaftlichen Erkenntnis"

(1899). For him as well, the tight connection between philosophy and science was to be understood according to the common ground linking mathematics and method.[34] Just as for Liard, method was seen as the source of Descartes' physics. But even if Liard had considered that Descartes' method was centrally to be found in his *Regulae*, Cassirer introduced a new type of reading of Descartes, more sensitive to the evolution of his thought: Liard saw the *Regulae* as the methodological fountainhead that was to irrigate the whole of Descartes' works, including the more mature ones, and to which Descartes' thought as a whole—including the kind of reasoning adopted in his metaphysics—was to remain faithful, the metaphysical "superfluous" justifications added to physics notwithstanding;[35] on the contrary, Cassirer isolated the *Regulae* as an essential pre-metaphysical stage of Descartes' philosophy, against the background of which Descartes' later works could be critically evaluated because of their more metaphysical imprint. Cassirer identified in Descartes a "critical" core which was supposed to anticipate Kant's idealism. From that interpretative perspective, Descartes conceived the reduction of matter to extension as an epistemological reduction to space, that is to say to the object of geometry.[36] This epistemological rather than metaphysical reduction was supposed to take roots in the seminal text of the *Regulae* through the idea of a *mathesis universalis* and to remain the core of Descartes' approach to physical phenomena.[37] In so doing, Cassirer could elude the question of the problematic passage from metaphysics to physics and substitute to it the methodical or logical approach which was to be found before the *Discours de la méthode* was written. Matter, equated with three-dimensional extension or space, thus becomes "a pure concept of mathematics," while "pure geometry provides the general conditions of possibility of spatial organization."[38] Then, instead of formulating the problem of the constitution of physics in terms of the passage from the general principles of physics to specific material configurations in physics proper, Cassirer reformulated it in terms of the *logical* passage from pure mathematics (rather than metaphysics) to physics.[39] The "logical means" to operate that passage was supposed to be motion.[40] For Cassirer, motion was primarily and ideally to be understood as a logical concept, something thanks to which we can gain knowledge of nature, rather than as something created by God and implemented in the material world. This transformation of the principles of physics into logical entities reached as far as the concept of substance, which was reduced to a logical condition of extension.[41]

Nevertheless, this epistemological unification of mathematics, logic, and physics was not without inducing some tension within the residual metaphysical claim regarding the irreducible difference between the modes of being of geometrical and physical objects. Mathematics' reality remains purely ideal, whereas for Descartes bodies are attributed a reality external to the mind.[42] From the neo-Kantian point of view adopted by Cassirer, the residual impact of metaphysics upon physics could be assessed only in a negative way. The afterwards but nonconstitutive effect of metaphysics upon physics could be observed in the Cartesian tendency to reify space, to transform it into a thing, a substance.[43] This metaphysical imprint on physical concepts deprived them of their status of pure concepts of the understanding. Cassirer then identified a shift, in Descartes' thought, from an original, so to say proto-transcendental idealist conception which brought the concepts of the understanding to the fore, to a more metaphysical perspective. Whereas the "original" (*ursprünglich*) philosophical conception of the *Regulae* led Descartes to epistemologically idealize extension, in the later stage of his works the metaphysical problem of existence induced a materialization of the pure concepts of quantity, which was no less than a "reversal" (*Umkehrung*) of the dependence relations between truth and reality.[44] Descartes' metaphysics revealed a temptation to subordinate truth to reality instead of proceeding the other way around. According to Cassirer, through metaphysics, Descartes wanted to achieve the identification of being and concept, i.e., to reach being through our concepts, thus forgetting that our access to reality is allowed, but also mediated and therefore limited by our very conceptual resources. Thus, Cartesian metaphysics was here envisaged as the realist temptation whose imprint physics was likely to bear.[45]

The impact of this metaphysical temptation can be seen in the primacy granted to substance, which led to an exclusion of motion in the conceptual constitution of being—motion being reduced merely to a state or modification of an already given reality, namely extended substance.[46] Motion and rest, as states of a thing, were transformed into hypostases and this is what, according to Cassirer, led Descartes astray when he formulated his laws of matter. In the *Principia philosophiae*, Descartes did not show himself satisfied with "the ordinary sense" of motion, to wit "the action by which a body travels from one place to another."[47] Indeed, this amounted to a relative conception of motion, since it involved choosing a point of reference considered to be at rest and in relation to which a body could be said to be in motion or at rest. But on that account, the same body could be said to be either in motion or at rest, depending on the reference point chosen. To avoid this and reach "the truth of the matter," Descartes then proposed a definition of motion "in the strict sense of the term": "motion is the transfer of one piece of matter, or one body, from the vicinity of the other bodies which are in immediate contact with it, and which are regarded as being at rest, to the vicinity of other bodies."[48] Now, for Cassirer, Descartes should have held to the relative conception of motion, which was closer to a purely logical one. The metaphysical need to reach a definition of real motion in the *Principia philosophiae* obviously violated the principle of relativity of motion.[49] Despite the ideal conception of motion as a logical concept, Cassirer thus recognized that Descartes eventually drew motion back to a metaphysical principle, namely God as its cause. But this amounted to turning the logical concept of causality into the metaphysical notion of creation.[50] On that account, laws of nature were enacted by God and imposed as an external ordering on nature, whereas Descartes should have understood them as concepts which constitute nature as an object for thought.[51]

As a result, a double tendency or direction can be found in such texts as *Le Monde* whose physics stemmed from Descartes' method but in which metaphysical considerations were added to the epistemological deduction. Just as Liard had highlighted the twofold justification of the laws of nature through the metaphysical appeal to God and the clarity which they display to our mind, Cassirer was keen to show that, in *Le Monde*, not only was the validity of the laws of nature guaranteed by their divine creation, but the ground of that validity was also to be found in the natural, distinct way they are known to the mind.[52] In particular in *Das Erkenntnisproblem in der Philosophie und Wissenschaft der Neueren Zeit* (1906), Cassirer insisted on the idea that the new possible world taking shape in *Le Monde* from chapter 6 onwards derived from mathematical materials and innate ideas.[53] Cassirer even went as far as to contemplate the possibility of a deductive construction of physics from the principles, not of metaphysics, but of mathematics—a possibility Descartes would have failed to realize because he switched too quickly from the principles to various concrete cases (like the seven rules of collision in the *Principia philosophiae*) without taking the pain to elaborate a patient deduction and build a logical bridge between them.[54] Even within the post-*Regulae* stage of his philosophy, there was a kind of competition, in Descartes, between the metaphysical foundation of physical principles and their epistemological foundation, i.e., the way they are known by the intellect.

Therefore, there is an obvious tendency, in Cassirer, to downplay the foundational role of metaphysics for physics, in favor of an epistemological grounding. Descartes was mainly seen as one of the historical steps leading to the "Copernican Revolution" performed by Kant's system of transcendental philosophy. Cassirer's reading of Descartes is therefore a normative one: he distinguished between what was especially valuable in his philosophy (i.e., pre-Kantian idealist elements broadly speaking) and what impeded the development of a truly Cartesian critical philosophy. Metaphysics fell on the latter side and was seen either as a residue or interference coming from some metaphysical tendency (in "Descartes' Kritik der mathematischen und naturwissenschaftlichen Erkenntnis"), or more clearly as another "direction" (*Richtung*) followed by his philosophy only after 1628, as Cassirer made more explicit in *Das Erkenntnisproblem in der Philosophie und Wissenschaft der Neueren Zeit*.[55] On the whole, each time Cassirer considered the role played by metaphysics in the constitution of Descartes' physics, this was only to emphasize the limits of Descartes' critical approach. On that account, it is especially noteworthy that Descartes' metaphysics was regarded as an impediment to the development not so much to his physics in itself—which was constituted independently—as to his philosophy of science and epistemology as exposed in the *Regulae*. Cartesian metaphysics did not play any role in the constitution of physics. It did not have any impact either on its discovery or on its content and came into play only as an *a posteriori* justification.[56]

The Reintegration of Metaphysics Into Cartesian Physics From the Point of View of the History of Science

Despite the interpretative obstacles to a close examination of the role played by metaphysics in the development of Descartes' physics, some alternative readings arose which did not neglect the metaphysical import of this physics. Paradoxically, those readings, which tended to be more philosophically oriented, were formulated by thinkers who had a strong interest in the history of science. However this should come as no surprise since those thinkers considered that the history of science was crucial to the study of the history of philosophy and adopted philosophical positions opposed to positivism, which advocated a strict separation between philosophy and science and which, as a philosophy of science, was then in its heyday.[57]

Gaston Milhaud: A Contextual History of Science as Counterweight to Metaphysics in the Forging of Descartes' Physics

Author of a doctoral dissertation in philosophy entitled *Essai sur les conditions et les limites de la certitude logique* (1894), Gaston Milhaud was offered, in 1909, a chair in "*Histoire de la philosophie dans ses rapports avec les sciences*" at the Sorbonne.[58] Milhaud was especially concerned with the detrimental contemporary separation of philosophy and science for which he considered Victor Cousin and his school to be responsible.[59] This critical dimension went hand in hand with a thoroughly historical approach to science: philosophy was not supposed to analyze science from an external point of view, but it was from a historical analysis of scientific theories that one could understand how philosophy and science influenced one another. Milhaud's approach to the history of science was fundamentally continuist,[60] and this clearly impacted his interpretation of Descartes.

In 1921, his *Descartes savant* was published posthumously. The scope of the book was intended to grasp the whole of Descartes' scientific activities and not only their metaphysical dimensions, even if these were explicitly recognized. Milhaud adopted a normative stance on Descartes' science insofar as he sought to evaluate to what extent Descartes drove science into a fruitful path and what his contribution to the "actual progress of the positive sciences is."[61] That being said, Milhaud intended to conduct that evaluation through an impartial assessment of Descartes' scientific works, thus avoiding two pitfalls. The first one was a panegyric which would tend to eclipse his physics' weaknesses and view his whole work through the light of his groundbreaking philosophy; the second one, on the contrary, would consist in evaluating his physics with regard to his disparaged metaphysics.[62] Even if one could have feared Milhaud's evaluative perspective to give rise to a retrospective and very selective assessment of Descartes' physics, independent of its historical context, on the contrary Milhaud intended to resort to the history of science in order to place himself in a position to evaluate Descartes' originality. Therefore, history of science was meant to serve the purposes of philosophy of science.

Against the illusion of an individualist Cartesian physics based on the sole metaphysics of a thinker who wanted to isolate himself from the world and its prejudices, Milhaud sought to locate Descartes in a continuum with his predecessors. This amounted to counterbalancing the picture Descartes would have liked to offer of himself as a scientist who would have drawn all his knowledge from his mind, or in other words all his physics from his metaphysics and method: "[Descartes] seems to spontaneously arrive at his discoveries by a way proper to him, either by applying his Method, or by stating only the consequences of his Metaphysics."[63] Milhaud thus asked, "And then how much trust should we truly give to Descartes, who makes us believe that all his work appears, like a spontaneous reaction, from his Method and Metaphysics?"[64] Strictly subordinating Descartes' physics to his metaphysics is tantamount to subscribing to the picture of a philosopher whose thought would be a radical beginning, a "sort of spontaneous generation."[65] In order to counterbalance this biased view which Descartes himself had propagated, Milhaud intended to reconstitute the historical context that nourished the French philosopher's thought: "his work . . . is an immediate outgrowth (*dépendance*) of Method and Metaphysics, while at the same time it fits exactly with the suggestions of his predecessors and contemporaries."[66]

But how are we to understand physics' dependence on metaphysics? In no way should it be understood as a strong conceptual link, and even less so as a deduction *stricto sensu*. Rather, it comes under the diffusion of Descartes' personality or style, from one domain to the other, thus unifying all his intellectual products to a certain extent.[67] More than a link of dependence, there are some similarities between Descartes' metaphysics and physics. Those resemblances can be traced back to a common method which attempted to return to clear and simple truths and to follow a rigorous order. Admittedly, metaphysics allowed Descartes to prove the existence of the world from the consideration of God and to "take one step further" in order "to grasp only through divine immutability this world's laws of constancy, of conservation, of permanence."[68] The imprint of Descartes' metaphysics on his physics occurred through a kind of inertial motion which led him to keep going when he moved from metaphysics to physics. Milhaud considered this a sort of transgression of disciplinary boundaries, but only as a side effect of a tendency in the metaphysical realm. For Milhaud, this was not the core of the unique spirit that animated Cartesian physics. The latter was to be identified, not so much as something produced by an isolated individual, as for Descartes' metaphysics, but as something specific to the science of a given time in a continuum with that of previous times. History of science is what reestablishes a balance between the roles played by an individual metaphysics and by a historically situated intellectual context. From that perspective, the principle of inertia was not conceived by Milhaud as a consequence of Cartesian metaphysics and, especially, of the idea of conservation resting on divine immutability, but rather as something that was "in the air," something that appears in Copernican cosmology and Galileo's and Beeckman's works.[69] Descartes thus overestimated the dependence of his physics on his metaphysics.[70] Once set back against the background of the evolving sciences of his time, Descartes does not emerge as a "revolutionary" scientist anymore, even to the point that "it is truer to say that he was conservative,"[71] for his discoveries in optics or physics were largely in accordance with those of Snel or Galileo. On the whole, Milhaud deemed individual creativity to be fused into something broader that he called "a sort of trend which more or less dominates, at a given time, individual research and which, formed from the indefinite variety of the scientists' efforts, leads them to the same truths." This is what Milhaud also designated as "the natural movement of the sciences."[72] On that account, Descartes became one element in that movement, who deluded himself as to the original capacity of his metaphysics to produce his physics.

Meyerson and the Metaphysical Import of Descartes' Physics

More weight was to be given to metaphysics for the shaping of Descartes' physics by Émile Meyerson. A contemporary of Milhaud, he shared with the latter a continuist view of the history of science and a realist philosophy of science. Initially trained as a chemist, he then became a philosopher of science and as such was forcibly opposed to mainstream positivist views of his time which rejected ontological categories and the search for causes.[73] His strong commitment to a realist view of science, meant not only to describe (through laws) but also to *explain* (through causes), fed his vigorous interest in the relationships between physics and metaphysics. For Meyerson, explaining a phenomenon meant deducing it in a logical way from its antecedents.[74] Not only were ontological categories and principles essential to scientific practice, but more deeply they corresponded to some innate irreducible tendencies of the human mind. Even when metaphysics seemed to be purposely excluded from the realm of science, just as in the case of positivism, the tendency to consider that science gives us access to a mind-independent, objective reality was tantamount to an implicit metaphysical claim regarding the existence and reality of natural objects.[75] Metaphysics, conceived as the act of positing a reality that goes beyond mere phenomena and gives the latter their foundation, thus corresponded to a need of the human mind: "the scientific ontology itself can, in the last analysis, be considered as resulting from the need for explanation."[76] As he forcefully claimed, in the preface to *Explanation in the Sciences* (1921), "science is essentially ontological . . . it cannot dispense with a reality posited outside the self."[77] In other words, "[o]ne cannot do science or talk science without including as a substratum a body of presuppositions about being."[78] Science necessarily, albeit implicitly, claims "the ontological existence of objects."[79] Now, there was something somewhat paradoxical in that, while aiming—deliberately or not—to reach objective reality, science and philosophy pursued the same goal, namely they attempted "to make the whole of reality appear to depend upon thought itself."[80] Descartes' physics exemplified this approach for Meyerson.[81] Actually, this paradox was resolved by the metaphysical assumption that there is an "agreement between reason and reality."[82] According to Meyerson, science could not do without metaphysics, even if this did not necessarily dictate to which specific kind of metaphysics the physicist ought to subscribe.[83] Metaphysics was here to be understood according to its ontological meaning. This clearly impacted Meyerson's conception of the history of science, in which scientific change was determined by ontological change.[84] Science was supposed to start with the metaphysics of the common sense, and then to invent another ontology in order to destroy the previous one.[85]

This led Meyerson, in *Identity and Reality* (1908), to give pride of place to the principle of identity, characterized as the required application of the principle of causality to time, in the constitution of scientific theories.[86] The principle of identity is indeed "an integral part of our reason";[87] it is an "eternal and invincible tendency of the human mind."[88] But this is something the scientist is not necessarily even fully aware of.[89] Descartes was of specific interest to Meyerson because he exemplified several features characteristic of the intertwinement of physics with metaphysical issues. Indeed, in his *Metaphysical Meditations*, Descartes unified matter by identifying it with space. Now, according to Meyerson, "only spatial properties prove to fit the needs of our mind."[90] In Descartes, "thought and nature seem to merge."[91] This therefore corresponded to the metaphysical move of reduction of the world's diversity to identity that Meyerson deemed to be so central to physics. However, Descartes' reductionist metaphysics was also criticized by Meyerson insofar as it intended to deduce all physical phenomena from geometrical, homogeneous extension. Even if, according to Meyerson, "Descartes' program of reducing the physical to the spatial" was largely and successfully followed by Einstein, Descartes was laid astray when thinking that he could continuously deduce a whole cosmological system and the explanations of all physical phenomena from this metaphysical reduction of matter to space, without taking into account the diversity presented in experience.[92] Even if there was a fruitful tendency, in physical theories, to absorb all data into space,[93] this was a gradual procedure that the physicist accomplished unconsciously and which was somewhat at odds with her intended purpose to grasp reality. Whereas Descartes had operated this reduction in an explicit and philosophical, that is to say "immediate" way, physicists, in the course of history, have followed this path in a gradual manner, dealing with more and more abstract and mathematized objects, and without being fully aware of the slope along which they are sliding. Hence the interest of historical inquiry for Meyerson: despite all—positivist—claims to the contrary, history allowed him to disclose the implicit metaphysics at work in twentieth-century physics which, in comparison with Descartes', had receded from a fully and explicit philosophical status to an unconscious one.[94] Now, one of Meyerson's goals resided in accurately circumscribing the realm of legitimacy for metaphysics within science and thus throwing a *cordon sanitaire* around metaphysical impulses: "For only then, by understanding where metaphysics is inevitable, will the research scientist be truly protected against the temptation to abuse it."[95] The paradox remains, however, that Meyerson conceived of metaphysics as a spontaneous tendency to believe in the existence of the external world and of external objects, whereas Descartes' explicit metaphysics precisely dismantled this prejudice, only to retrieve the existence of the external world not as a spontaneous belief, but as the result of an intricate reflexive process.

In *Identity and Reality*, Meyerson was thus led to interpret mechanism in general, not only as a set of methodological principles in physics, but more fundamentally as satisfying a deeply rooted metaphysical need of the human mind[96] corresponding to the recurrent affirmation of the principle of identity. Since something had to remain, one had to posit the existence of extended particles in motion as the permanent building blocks of physical reality. Change is only the result of the modifications of their arrangement, that is to say of their relative displacement in space.[97] Atomism (i.e., the permanence of unbreakable parts of matter) combined with the unity of matter to give rise to mechanism.[98] Descartes was identified as the most representative proponent of mechanism,[99] the presence of which is however attested from antiquity to the early twentieth century.

Meyerson then explored several instances of a general principle of constancy or conservation, among which the principle of inertia. In that light, Descartes' physics became exemplary, especially because Descartes was one of the first proponents of this principle. For Meyerson, the principle of identity stood at the origin of the conservation of rectilinear uniform motion. But on Meyerson's own admission, inertia was "far from being an instinctive notion of our mind, which subsequent reasoning has only to disengage."[100] He means that it could by no means be produced by an *a priori* deduction.[101] And it could not be considered as an experiential truth either.[102] Faced with what appears at first sight as a conspicuous difficulty, Meyerson however considered Descartes' formulation of the principle to provide "the true foundation of inertia, that by which this principle is impressed upon our minds,"[103] namely God's immutability. Meyerson even became somewhat emphatic when he claimed:

> Here, it seems, is the true foundation of the principle. And although the possibility of an experimental proof exists, it is really this argument which produces our strong conviction, so different from that with which

we receive purely empirical formulae. . . . Consequently, Descartes has disclosed the essence of the principle in relating it to the "immutability of God," to the conviction that everything in nature persists. His, therefore, is the merit not only of having been the first loudly to proclaim the principle, but also to indicate its true foundation.[104]

In Meyerson's view, Descartes' principle of inertia, insofar as it was based on a metaphysical principle, namely God's immutability, exemplified the principle of identity which constitutes the bedrock principle of a tendency specific to human reason. Indeed, God is "a symbol of the general order of nature, and in this particular case, of the essential immutability of things—that is, of the causal principle,"[105] or, in other words, of "the identity of the universe in time."[106] But Meyerson's insight sheds some interesting light on the deductive status of Descartes' principle of conservation. Its explicit metaphysical dimension—at least as Descartes is supposed to have envisaged it—is somewhat called into question insofar as Meyerson showed that the Cartesian principle of inertia obeyed the principle of identity, was derived from it but could not be deduced from it strictly speaking: according to the principle of identity, something must remain identical, but exactly *what* should remain identical is not dictated by this principle.[107] As we have seen, this criticism was also formulated by Liard. Even if Descartes was supposed to have provided the true foundation for the principle of inertia, Meyerson maintained that this was not an *a priori* principle. It was not an *a posteriori* principle either because it was not discovered through experiential means. But it was supposedly obtained by a combination of both: it is indirectly derived from the principle of identity (it conforms to it), but the determination of what remains identical is provided by experience. For that very reason, Meyerson classified the principle of inertia among the plausible statements,[108] which "cannot be part of the 'metaphysical foundations of science.'"[109] Even if "Descartes always presents inertia as a pure deduction,"[110] Meyerson eventually admitted that this could not be any true deduction.[111] Consequently, providing a true *metaphysical foundation* for the principle of inertia—as Descartes was supposed to have done on Meyerson's own admission—did not amount to being able to *deduce* the latter from the former.[112] Thus, the problem of the exact type of logical relationship there is between Descartes' metaphysics and his physics is seen under new light. But what matters here is not so much whether Descartes' principle of inertia was rightly deduced from the metaphysical principle of God's immutability, as Descartes explicitly claimed, but rather that it expressed an *implicit* metaphysical principle, the principle of identity, which took here the specific form of the principle of substance (because Cartesian inertia converted motion into a state, which relates to an entity or a substance).[113] Thus, Meyerson's analysis of Descartes' principle of inertia showed, beyond Descartes' explicit attempt to ground it in the metaphysical principle of God's immutability, the deeper metaphysical foundation of this principle.

Inertia was not so much of an ontological as of an epistemological nature, that is to say it resulted from the intellectual constitution of our mind.[114] This also entailed a shift in how we might conceive of metaphysics, either as a grasp on what reality, in its more fundamental dimension, amounts to, or as the way the constitution of our mind leads us to reason.

The Detrimental Dependence of Descartes' Physics on His Metaphysics Seen From the Point of View of a Normative Philosophy of Science

As should have become clear by now, positivism played a central role in the way the relationships between Descartes' physics and metaphysics were interpreted in the first half of the twentieth century. Now, some philosophers of science developed an approach which was less opposed to positivism than were Milhaud or Meyerson. For all that, Descartes was not purely and simply neglected by them, but an examination of his doctrines remained crucial within a critical approach to the history of philosophy and science which was intended to be prescriptive regarding how contemporary science ought to proceed.

Duhem's Critique of Descartes' Reversal of Logical Order Between Physics and Metaphysics

In the preface to the second edition of *La théorie physique. Son objet, sa structure* (1st ed. 1906, 2nd ed. 1914), Pierre Duhem (1861–1916) claimed he wanted to put some order in the "chaos"[115] arisen from the radical changes brought by new physical theories and the subsequent philosophical debates. The object of physical theory ought to be specific to it, and therefore distinct from that of metaphysics. Hence the title of the first chapter of part I: "Physical Theory and Metaphysical Explanation." But as early as 1893, Duhem had tackled the issue in an article published in the *Revue des questions scientifiques* and titled "Physique et métaphysique." There he provided clear definitions of what he meant by "physics" and "metaphysics":

> To conform to contemporary usage, we give the name physics to the experimental study of inanimate things, considered in three phases: the observation of facts, the discovery of laws, and the construction of theories. We regard the investigation of the essence of material things, insofar as they are causes of physical phenomena, as a subdivision of metaphysics. This subdivision, together with the study of living matter, forms cosmology.[116]

And Duhem added:

> *Physics is the study of the phenomena arising from brute matter and of the laws that govern these phenomena.*

> *Cosmology seeks to understand the nature of brute matter, considered as the cause of phenomena and as the foundation* (raison d'être) *of physical laws* (emphasis in original).[117]

This distinction exists because, for Duhem, we have no "direct intuitive view of the essence of things."[118] In a Comtian way, metaphysics was thus understood as the search for the causes of phenomena beyond the reach of experience. Metaphysics is what transcends the observational methods of physics.[119] For that reason, it is highly speculative and doubtful and should not affect physical theory. At any rate, it could yield no consensus.[120] Hence, metaphysics and physics stand in a relation of reciprocal autonomy.[121] Duhem's sympathy for positivism is explicitly voiced in the second edition of *La théorie physique*: there he was keen to claim that his conception of physics met all requirements of the positivist method, and was positivist in its origins and conclusions.[122] For Duhem, metaphysical divergences concealed scientific continuity throughout history. His goal was thus to identify and isolate the metaphysical conceptions which had, so to say, contaminated scientific theories. Metaphysics indeed expressed an "invincible urge" common to all human beings.[123] Duhem is very well known to have rejected metaphysics from the realm of physics and to have attempted to safeguard physics' autonomy from any possible dependence on metaphysics. Metaphysics was not supposed to exert any constraints on the development of physics, which is based on autonomous self-evident principles and on experimental method.[124] The aim of physical theory was then only "to classify experimental laws."[125]

This normative stance led him to adopt a critical perspective on the history of science. History of science emerged as a way to reveal scientific continuity and progress, but also the dead ends to be avoided by twentieth-century physicists.[126] Given his views, Duhem obviously had to recognize that, throughout the history of science, metaphysics played a significant—and largely detrimental—role in the elaboration of physical theories. Descartes was for him a case in point and appeared as the emblematic figure of the thinker who placed physics under the dependence of metaphysics.[127] Descartes wanted to establish the nature of physical reality beyond experience and, for that reason, his physics became dependent on metaphysics conceived as a regional ontology which determined the nature of matter—as Descartes indeed did in the Fifth Meditation. Divergent conceptions of matter constituted the core metaphysical opposition between certain different attempts to provide explanations of the natural world. Metaphysical explanations intended to address the two following questions: "Does there exist a material reality distinct from sensible appearances? and What is the nature of this reality?"[128] Now, Duhem went on, "these two questions . . . do not have their source in experimental method. . . . The resolution of these questions transcends the methods used by physics; it is the object of metaphysics."[129] Therefore, Duhem focused on Descartes' conception of matter because this represented the core of the physics-oriented part of his metaphysics. Duhem restated Descartes' metaphysical reduction of matter to three-dimensional geometrical extension.[130] Now, the problem was twofold. One the one hand, this reductionist conception could not yield any physical law.[131] But on the other hand, Descartes gave the impression that his metaphysical principles could bring forth an entire cosmological representation.[132]

Consequently, Duhem aimed to show that, even though Cartesianism is representative of "any metaphysical doctrine which claims to terminate in a physical theory,"[133] this alleged dependence of physics upon metaphysics was a mere illusion. Duhem was intent on establishing that "[n]o Metaphysical System Suffices in Constructing a Physical Theory."[134] Metaphysics is too general to produce a whole physical theory.[135] On that issue, Duhem invoked the conservation of the quantity of motion supposedly deduced from God's immutability.[136] He voiced the following criticism which is not without echoing Liard's and Meyerson's:

> But an infinity of other expressions might just as well have satisfied these requirements: for the velocity we might notably have substituted the square of the velocity . . . instead of drawing from divine immutability the constancy of the Cartesian quantity of motion in the world, we should have deduced the constancy of the Leibnizian living force.[137]

The principles of physics that Descartes presented as "consequences" of his metaphysics were at best only compatible with it.[138] This analysis paved the way for a notion crucial to Duhem's philosophy of science, to wit that of hypothesis:

> What we have just said about Cartesianism can be repeated about any metaphysical doctrine which claims to terminate in a physical theory; in this theory there are always posited certain hypotheses which do *not* have as their grounds the principles of the metaphysical doctrine.[139]

Scientific theory, *qua* scientific, is elaborated and transmitted independently of explanations rooted in such or such metaphysical conception (like the Cartesian conception of matter). To take but an example, the Cartesian law of refraction remained valid, even though it was supposed to rely on a specific conception of the nature of light, namely a continuum of small balls of subtle matter conveying the pressure coming from luminous bodies.[140] Therefore, for Duhem, the diachronic continuity of scientific theories went hand in hand with their synchronic dissociation from metaphysical explanations.[141] On that view, Descartes was no less than the one who wrongly abolished the demarcation between metaphysics and physics initially entrenched in Scholasticism: "Nothing is more contrary than such a method to the Aristotelian conception, according to which a science, such as physics, rests on self-evident principles whose nature is investigated by metaphysics which cannot increase their certainty."[142] The core of Duhem's criticism of Descartes lay not so much in the fact that Descartes formulated a metaphysical proposition according to which matter was reduced to geometrical extension, or claimed that all phenomena could be accounted for by relying on that principle, but more that he regarded metaphysics as the very source of physics, and transformed metaphysics into some kind of proto-physics. It is very significant that Duhem wrote that the reduction of matter to extension by Descartes is a *physical*, rather than metaphysical, proposition:

> The first proposition in *physics* that Descartes established . . . grasps and expresses the very essence of matter. . . . The essence of matter thus being known, we shall be able, through the procedures of geometry, to deduce from it the explanation of all natural phenomena.[143]

Needless to say that, more than a blatant misinterpretation of Descartes' reshaping of the categories of metaphysics and physics, this statement criticized the reduction of matter to extension for trespassing on the physical realm. This is to be understood in the sense that physics was supposed to flow from metaphysics and actually did under Descartes' pen:

> Descartes pushed this proud principle to its extreme consequences. He was not content with asserting that the explanation of all natural phenomena may be derived completely from this single proposition: "The essence of matter is extension"; he tried to give this explanation in detail. He investigated the question of constructing the world with shape and motion by starting with this definition. And when he reached the end of his work, he stopped to contemplate it, and declared that nothing was missing in it.[144]

Duhem put forward a scathing attack on the reversal operated by Descartes between physics and metaphysics. Now, Duhem declared that

> if metaphysics precedes physics in order of excellence, it comes after physics in the order of logic. We cannot come to know the essence of things except insofar as that essence is the cause and foundation for phenomena and the laws that govern them.[145]

On the contrary, Descartes was paradigmatic of the reversal of logical order between physics and metaphysics that was initiated in the early modern period according to Duhem.[146] Duhem saw him as the very origin of this major intellectual shift which was to significantly impact seventeenth-century physicists in general.[147] Instead of simply "saving the phenomena," Cartesian cosmology claimed to produce them. The whole project pursued by Descartes in his *Principia philosophiae* fell under this full-blown criticism. Admittedly, Duhem was right to identify the realist significance of

Descartes' physics. However, Duhem certainly neglected that Descartes' "deduction" of the material world from some metaphysical principles was not to be understood in a purely logical way and involved some appeal to experience, as Descartes explicitly recognized in the *Discourse* and the *Principia*.[148]

One could therefore even go as far as to say that Duhem's notion of a natural classification of physical laws was meant to dismantle the reversal operated by Descartes between physics and metaphysics, while maintaining the autonomy of both fields. Indeed, even if Duhem's criticism of realist philosophies of science is rather well known, and even if he clearly excluded metaphysics from the realm of physics, he reintroduced some metaphysical import for science through the notion of "natural classification."[149] Admittedly, neither was physics derivable from metaphysics, nor the other way around for Duhem. But the coordination of physical laws was then supposed to reflect to some extent the order according to which reality was actually organized.[150] This notion reintroduced some kind of limited realism in Duhem's philosophy of science. Indeed, the history of science reveals a progressive unification of physical theories, a unity that points toward some reality uncovered by them: "In a word, the physicist is forced to recognize that it would be irrational to work towards the progress of physical theory if that theory were not the more and more clear, and more and more precise reflection of metaphysics."[151] The ideal physical theory functions as a kind of regulative ideal and

> would be the complete and adequate metaphysical explanation of material things. This theory, in fact, would classify physical laws in an order which would be the very expression of the metaphysical relations that the essences that cause the laws have among themselves.[152]

However, this "reflection" remained somewhat problematic, for "[t]he subordination that a theory establishes among various physical laws by classifying them does not oblige us to admit a similar subordination among the metaphysical laws of which the physical laws are the manifestation."[153] Moreover, the ideal physical theory, "like everything that is perfect, infinitely surpasses the scope of the human mind."[154] Contrary to Descartes, who introduced *a priori* intuition as a methodological procedure specific to metaphysics to grasp the essence of things, Duhem rejected this conception of metaphysics. Metaphysics as a specific intellectual task that would precede—in a Cartesian fashion—or *directly* follow from physics was excluded. But this is not to say that physics, in particular through the progress accomplished by its theories, did not convey some kind of metaphysical horizon or did not lead to some relevant metaphysical attitude. However, first this concerned a reality aimed at by laws and not bearing on the physical furniture of the world such as atoms—whose existence we could not establish according to Duhem. And second, this functioned as an *a posteriori* tool, and certainly not as a metaphysical guiding principle for the physicist. For Duhem, physics remained autonomous from metaphysics in its development, even if both happened to converge on a natural classification. On that account and in clear opposition to Descartes, metaphysics was reinstated as a *post-physical* discipline, as Duhem held it to be in Scholastic philosophy.

Bachelard or the Opposition Between the Scientific and the Metaphysical Mind

Duhem's criticism, addressing the pervasiveness of Cartesian metaphysics throughout Cartesian physics, was based on a continuist epistemology. No less harsh a criticism was leveled against Descartes' metaphysics by the discontinuist epistemologist Gaston Bachelard. As much as for Duhem, the history of science was a mainstay of Bachelard's philosophy of science. In his view, "epistemology" was intrinsically historical, but not because one could find—as Duhem attempted to do—forerunners of contemporary scientific positions in the past, but because science was a historical process by which it broke away from prescientific conceptions. History of science was intended to identify not so much continuity as gaps and ruptures between prescientific conceptions and truly scientific ones. Bachelard's philosophy of science advocated a backwards or recurrent reading of the history of science: only in light of recognized scientific conceptions was one enabled to trace their development and their moment of rupture with prescientific notions.[155]

This historical approach was conceived as a means to remedy the inadequacy of interpretations that philosophical theories hold on physical issues.[156] There was a tendency, in philosophy, to reify scientific concepts into rigid metaphysical categories, to adopt a realist interpretation of scientific notions.[157] As Dominique Lecourt writes: "Bachelard constituted his epistemology on the basis of the disqualification of all existing metaphysics."[158] Bachelard dismissed any claim from traditional metaphysics to understand what is at stake in physics.[159] On the contrary, "science in effect creates philosophy."[160] Science renders traditional metaphysical categories obsolete and opens the way to a transformation of traditional metaphysics into an authentic philosophy of science which is not deprived of any metaphysical dimension.[161] Scientific thinking "will show philosophers how to replace intuitive, immediate systems of metaphysics with systems whose principles are debatable and subject to experimental validations."[162] Science cannot be based on innate truths.[163] As a consequence, Bachelard fiercely opposed any reductionism such as Descartes' metaphysical reduction of matter to extension.[164] As Mary Tiles summarizes, he "den[ies] us the right to assume that the exact abstract structures of logic or mathematics are anywhere realized in reality."[165] The example of the piece of wax, in Descartes' *Meditations*, is a case in point. In a well-known passage of *The New Scientific Spirit*, Bachelard criticized the metaphysical reduction to intelligible extension operated by Descartes, and opposed it to the carefully produced drop of wax studied by the physicist in her lab and which opens up a multiplicity of complex phenomena to account for.[166] The search for simplicity might amount to some sort of sterilization of the physicist's objects. For Bachelard, twentieth-century science revealed that it was neither possible nor desirable to enclose physical notions within too rigid ontological categories. Some phenomena could be explained by conceiving light as particles; some others by conceiving it as waves.[167] If metaphysics prompted us to consider that these were two incompatible statements from an ontological point of view, we should rather reconsider our ontological categories than attempt to bend physics to unsuitable notions forged by outdated metaphysics. This applied to Descartes' metaphysical reduction of matter to the modes of figures and extension. Bachelard approvingly quoted Louis de Broglie's *Théorie de la quantification dans la nouvelle mécanique*:

> Descartes argued that it was important to explain natural phenomena in terms of figures and motions. The uncertainty relations express the fact that a rigorous description of this kind is impossible, since we can never know both the figure and its motion at the same time.[168]

As a consequence, "it has been a concern of contemporary science to overcome the contradictions of metaphysics."[169]

Bachelard's philosophy of science, as a "philosophy of no," characterized the new scientific spirit as a non-Cartesian epistemology.[170] By "Cartesian epistemology," Bachelard meant an epistemology "based as it is entirely on appeals to simple ideas."[171] Bachelard's non-Cartesian epistemology consisted in a rejection of some basic Cartesian notions such as the epistemology of simple, clear, and distinct ideas: science deals with complex objects, not with immediately simple ones, as the wax example revealed. But, for all that, this non-Cartesian epistemology was not merely a negation of Descartes' philosophy or some anti-Cartesianism, but extended to and included it to some extent.[172] Indeed Bachelard, while rejecting Descartes' metaphysical universal doubt,[173] was in favor of "a less pervasive skepticism than Descartes's, but a skepticism that for that very reason is clearer and more robust."[174] Bachelard's non-Cartesian epistemology was Cartesian to a large extent:

> The attitude that we must remain always in doubt of knowledge that had seemed certain in the past extends and indeed transcends Cartesian precaution and is truly worthy of being called non-Cartesian, as long as we remember that non-Cartesian philosophy complements Cartesian philosophy without contradicting it.[175]

The historical dimension of Bachelard's approach fits with his conception of science as approximate knowledge. Science proceeds by breaking with previous conceptions, but this can also involve rectifying them, disentangling them from the realist imagery from which they initially stemmed. Hence Bachelard's interest in scientific imagery, especially in atomistic imagery.[176] This led him to formulate a very harsh criticism of Descartes' metaphysics of space as the so-called "metaphysics of the sponge."[177] According to Bachelard, this image was an epistemological obstacle because it led Descartes to reject the hypothesis of the void and prevented him from elaborating a truly scientific account of condensation and rarefaction. Obviously, Bachelard was misled when thinking that the sponge was a truly metaphysical element of Descartes' thought; it could be no more than an illustration for Descartes.[178] Admittedly, the image did not belong to "metaphysics" according to Descartes' own criteria. But the classification of this imagery as "metaphysical" is telling of Bachelard's intent to distinguish, in Descartes, the philosopher's explicit metaphysics from the physicist's implicit metaphysics. The latter conveyed some implicit realism which imprisoned the physicist's mind within the boundaries of pre-established explanatory schemes. But above all, Bachelard might suggest that the metaphysics of the reduction of matter to extended geometrical parts in motion was eventually dictated by some imagination-driven representations. Needless to say that Bachelard's normative stance on the history of science could lead him astray in terms of interpretative accuracy. If Bachelard had undertaken to understand the formative role of Descartes' explicit metaphysics on this physics, he would have realized the somewhat arbitrary dimension of his classification of atomistic theories in *Les intuitions atomistiques*. Indeed, the typology he put forward in that book is at odds with Descartes' own corpuscular physics. Bachelard opposed two pictorial models: one of solid corpuscles, endowed with determined shapes, which collide and break without ever really merging; the other of a fluid universe in which all the bodies merge.[179] But Descartes' corpuscular physics by no means fits this typology because it precisely combined both models, and this mainly for metaphysical reasons that Bachelard was apparently intent on ignoring. Indeed, Descartes' metaphysical rejection of the void entailed the need for a very subtle matter, the matter of the first element, which offered no resistance and could fill all interstices in between corpuscles of the second and third elements, which did have a determined shape. What Bachelard characterized as "intuitions," in that case, might rather be seen as some rational solutions to accommodate Descartes' corpuscular physics to his metaphysical positions.

Conclusion

It will thus come as no surprise to conclude that the interpretations bearing on the relationships between Descartes' physics and his metaphysics from the end of the nineteenth century to the mid-twentieth century were highly influenced by the philosophical context related to the deep mutations undergone by physics at that time. These mutations extended so far and challenged so many received views that they aroused philosophical reflection on the method scientists were supposed to follow, the philosophical import of their theories, and the connection between those theories and other intellectual productions. But science also emerged as the intellectual production *par excellence* through which the extent and nature of human reason could be assessed. In that context, Descartes' thought became a crucial historical element in various attempts to understand human reason through its scientific achievements. It goes without saying that Descartes' philosophy and science were instrumentalized depending on the philosophy of science at play or the vision of history of science held by commentators.

Despite the variety of those interpretations encountered through the survey I have provided, one crucial interpretative key at that period emerges: this is the relation philosophers of science had to positivism. Maybe with the exception of Cassirer whose philosophy was nourished by neo-Kantianism—but whose interpretation of Descartes eventually came close to Liard's, the interpretations of a vast majority of philosophers of science, and even of historians of philosophy and science, were largely orientated by the conception of science and metaphysics conveyed by positivism. This was the case, be they supporters or opponents of positivism. Now, according to Auguste Comte's positivism, metaphysics was superseded by science and rendered obsolete by the latter. Therefore, any metaphysical reflection intended to impact scientific theories could only be seen as an anachronism and a kind of logical aberration. For Comte, positivist science had to abandon any search for hidden causes beyond observation and reasoning. Etiology was dismissed in favor of the search for laws expressing constant relations in nature. Ernst Mach's neo-positivism and its focus on scientific laws as abridged descriptions of phenomena was another major influence (especially for Duhem). Mach also firmly rejected the introduction of anything metaphysical into scientific theories.

All the authors I have studied therefore share a conception of metaphysics—at least as far as it relates to physics—characterized by three subject matters: metaphysics bears on (1) the existence of the material world as accounted for by physics—metaphysics is thus to be understood as a kind of realist philosophy of science; (2) the essence or nature of the entities composing the material world—from that perspective, metaphysics is conceived as a regional ontology; (3) the dependence of the laws of nature on God—although this aspect appears less prominently.[180] On that account, they are not unfaithful to Descartes' conception of metaphysics, for he established the nature and existence of the material world in the fifth and sixth *Metaphysical Meditations*. Also, in *Le Monde*, the considerations of God's immutability and of his constant manner of acting upon the physical world were labeled "metaphysical."[181]

However, two features distinguish Descartes' metaphysics from the conception held by those historians and philosophers of the first half of the twentieth century. First, the characterization of metaphysics assumed by all those authors has a clear focus on its ontological dimension, which they placed in opposition with or distinguished from epistemology. However, this amounted to creating a kind of artificial divide between ontology and epistemology, which could not be more remote from Descartes' own profound transformation of metaphysics.[182] This led Liard and Cassirer to partially ignore the metaphysical dimension of Descartes' physics that could be drawn from his epistemology. For positivists, as well as mitigated positivists or anti-positivists, if science happened to deal, for better or for worse, with metaphysics, this first and foremost meant with ontology. On that view, the scientist could be tempted to claim the existence of some kind of reality beyond phenomena and their laws. The introduction of metaphysics into science amounted to an attempt to relate scientific theories to this posited reality, to deduce from scientific laws the nature and essence of physical bodies. This attempt could (1) remain at a general level, stating what the nature of bodies in general was, or (2) bear on the specific type of bodies (atoms for example) which were supposed to exist beyond the observational level. The first more general attempt (1) could certainly be deemed metaphysical in Descartes' sense and shows that Descartes' metaphysics, contrary to the Kantian or Comtian notion of metaphysics, is not confined to what goes beyond experience but is related to the empirical realm insofar as it bears on the external world.[183] However, the latter kind of so-called "metaphysical" endeavor (2) certainly did not belong to metaphysics for the author of the *Meditations*: Descartes would have never characterized as such the determination of the three types of corpuscles on which many of his physical explanations are based. This is but one example of the problematic application of the notion of metaphysics inherited from a positivist or neo-positivist context to Descartes' thought.

Second, the other divergence between Descartes and his commentators concerning their conception of metaphysics is related to the status of those metaphysical statements identified by some of these philosophers as being at work in Descartes' physics. Whereas Descartes conceived metaphysics as an explicitly reflexive intellectual approach, his commentators of the first half of the twentieth century all saw it rather as a psychological, partially uncontrolled and spontaneous tendency which was characteristic of either Descartes' character or human mental structures. This conception which very much amounted to psychologizing what they identified as metaphysics could be expressed in slightly different ways. For historians of philosophy like Milhaud, Liard, and Cassirer, Descartes' metaphysics was related to Descartes' specific psychology. For Milhaud, metaphysics was the expression of Descartes' singular character, which attempted to isolate itself from the world and its prejudices. Liard saw metaphysics as Descartes' "unconscious" reaction against the autonomization of physics from metaphysics, a development initiated by Descartes himself. Cassirer viewed metaphysics as a late tendency that Descartes followed, moving away from the *Regulae*'s epistemology. For more systematic philosophers of science like Meyerson, Duhem, and Bachelard, metaphysics was conceived more generally as the product of psychological characteristics shared by most human beings. For Meyerson, metaphysics expressed certain spontaneous, implicit tendencies of the human mind. Even Duhem identified in human beings an "invincible urge" to seek the nature of material things. Bachelard also interpreted scientific pitfalls as metaphysically driven, tacit, covert tendencies of the mind. This is, of course, largely at odds with the arguments explicitly presented in Descartes' *Meditations* or *Principia philosophiae* and with his very conception of a self-reflexive metaphysical approach.[184] But the strength of those philosophical interpretations lay also in their ability to identify, albeit in a critical and normative way, specificities of the articulation between Descartes' metaphysics and physics, especially its realist significance and the reversal operated, in comparison with Scholastic philosophy, between metaphysics and physics. Rather than advocating, as Bachelard and Duhem did, a return to an order in which metaphysics always comes after physics, the task remained to account for the philosophical and historical specificity of the relation between metaphysics and physics for Descartes, as well as the difficulties of such a relation from the point of view of seventeenth-century philosophy.[185]

Notes

1 Daniel Garber, *Descartes' Metaphysical Physics* (Chicago: University of Chicago Press, 1992).

2 *Ibid.*, 263–305. Garber thus considerably downplayed the role supposedly played by method in the elaboration of Cartesian physics or metaphysics, relating it to the early phase of Descartes' works and reducing it to a mainly rhetorical function in the *Discours*: see *ibid.*, 44–9. Even if Garber was ready to envisage that the exact modalities of dependence of physics on metaphysics were far from being straightforward, he never really questioned the idea that there was some kind of deductive continuity between Cartesian metaphysics and physics. For example, Garber described the content of Descartes' *Principia* as follows: "after a few brief sections, exhorting the reader to reject his former beliefs, Descartes begins at the beginning, with the *Cogito* in this case, and deduces on down, showing step by step how the Cartesian world, vortices, planets, and all, can be deduced from this initial intuition" (*ibid.*, 47).

3 "Physics is, in effect, a sort of flowing of metaphysics. . . . Our reason contains in potentiality the whole of physics, and we can draw it out only by the work of our mental activity." (Joseph Millet, *Histoire de Descartes avant 1637* (Paris: Librairie académique, Didier & Cie, 1867), 458–59, my trans.) However, as Keeling pointed out, these are two different things to say that "every natural law is deducible from the primary propositions of Metaphysics" and that "this or that natural law *was so deduced*" (emphasis in original, Stanley Victor Keeling, *Descartes* (Oxford: Oxford University Press, 1968), 136).

4 Within those who think that the foundational legitimacy brought to physics by metaphysics is not fake, but does not affect its content, are Charles Adam (Charles Adam, *Vie et Œuvre de Descartes*, supplément à l'édition de Descartes, vol. XII (Paris: Léopold Cerf, 1910), 143–44, 370) or Étienne Gilson (Étienne Gilson, *La Liberté chez Descartes et la théologie* (Paris: F. Alcan, 1913)). On that account, metaphysics is seen as having the purely instrumental function of providing a more solid ground for physics. Among those who, on the contrary, consider that physics' content is affected by the foundation provided by metaphysics, one can include, for example, Octave Hamelin, A. Boyce Gibson, and Alexandre Koyré.

Hamelin considered physics to be linked to metaphysics in a twofold way which affected its foundations, as well as its content: (1) indirectly through method because the method of certainty eventually relies on the metaphysics of the *cogito*; (2) directly because the fundamental laws of physics require the metaphysical foundation that is to be found in God, especially for the conservation of the quantity of motion: see Octave Hamelin, *Le Système de Descartes* (Paris: Alcan, 1921), 95, 98, 310, 313–14. Contrary to Cassirer, he refused to split Descartes' philosophy into a pre-metaphysical stage corresponding to Descartes' *Regulae* and scientific investigations and a later metaphysical stage. For him, the *Regulae* already contain some metaphysics in the reduction of body to extension (he did not seem to contemplate the possibility that this reduction be only epistemological rather than metaphysical). He envisaged Descartes' philosophy as a system in which "everything holds together." Gibson relied a lot on Descartes' letter to Mersenne dated 15 April 1630 (AT I, 144) to reject the idea that his metaphysics and physics would be detachable from one another or that the former would be subservient to the latter. According to him, metaphysics and physics mutually support each other: see A. Boyce Gibson, *The Philosophy of Descartes* (London: Methuen & Co. Ltd, 1932), 47–55. Alexandre Koyré also leant toward a strong link between Descartes' physics and his metaphysics in foundational terms. Contrary to Liard, he did not conceive science as the mere establishment of relations between phenomena known through experience. As he made it clear in the foreword to *From the Closed World to the Infinite Universe* (Alexandre Koyré, *From the Closed World to the Infinite Universe* (Baltimore: The Johns Hopkins University Press, 1957), viii), the scientific revolution is first of all an intellectual revolution based on the geometrization of space and on the dissolution of the ancient cosmos (see also Alexandre Koyré, *Galileo Studies*, trans. John Mepham (Hassocks, Sussex: The Harvester Press, 1978), 3). Koyré subscribed to a kind of mathematical Platonism. Hence the importance of metaphysics for the forging of Descartes' physics, since the reduction of matter to geometrical extension is precisely operated by metaphysics. According to Koyré, Descartes had the "need to give metaphysical foundations to the new science" (Alexandre Koyré, *Entretiens sur Descartes* (New York and Paris: Brentano's, 1944), 93, my trans.). Metaphysics is required to provide a justification of our clear and distinct ideas through God's existence (Alexandre Koyré, *Entretiens sur Descartes*, 109). This is what guarantees that our mathematical ideas can apply to nature: "A Cartesian science, which postulates the true value of 'mathematism,' which builds a geometrical physics, cannot do without metaphysics" (Alexandre Koyré, *Entretiens sur Descartes*, 99, my trans.). However, this does not mean that such a geometrization ought to result from a metaphysical endeavor, as is obvious in Galileo's case. Therefore, Koyré did not really feel the need to explore in depth the relationships between Descartes' physics and metaphysics. In general, none of the aforementioned authors deeply delved into the issue of those relationships by providing a detailed analysis of them. My paper will therefore not focus on them.

5 This is obvious concerning Descartes' scientific endeavors prior to 1629. On that year, Descartes is supposed to have written a first short metaphysical treatise: see CSMK, 53; AT I, 350. In a letter to Mersenne dated 15 April 1630, Descartes declared: "I can say that I would not have been able to discover the foundations of physics if I had not looked for them along that road [i.e., in metaphysics]" (CSMK, 22; AT I, 144). This suggests that from 1629 to 1630 onwards, there existed some kind of dependence relation between Descartes' physics and his metaphysics. However, this does not rule out that Descartes might have changed his mind as to the nature of this relation over time, between *Le Monde*

(1630–1633) and the *Principia philosophiae* (1644) or, at least, that in the *Principia* he might have made this relation explicit in a way he had not in *Le Monde*. In the latter text, Descartes did not set out his metaphysics as he later would in the *Discours de la méthode* and the *Meditationes*. However, he suggested that considering bodies as extended parts of matter in motion was consistent with experience, easy to imagine and conceive, and sufficient to account for phenomena: see CSM I, 83, 89, 91–2; AT XI, 7, 25–6, 33–6. He also intimated that the laws of motion depended on the nature of God: see CSM I, 92–3; AT XI, 37–8. In a nutshell, in *Le Monde*, Descartes did not make explicit the metaphysical commitments underlying his physics, but this does not imply that such commitments did not exist and had not already shaped his physics.

6 We can recall at the very least Planck's quanta theory, Einstein's works on relativity, Louis de Broglie's wave mechanics, and the foundation of quanta mechanics by Heisenberg.

7 Louis Liard, *Descartes*, 3rd ed. (Paris: Félix Alcan, 1911).

8 See Adam, *Vie et Œuvre de Descartes*.

9 See Gilson, *La Liberté chez Descartes*; Jean-Luc Marion, "L'instauration de la rupture: Gilson à la lecture de Descartes," in *Étienne Gilson et nous*, ed. Monique Couratier (Paris: Vrin, 1980), 13–34; Delphine Bellis, "Météores cartésiens et météores scolastiques: la lecture philosophique d'Étienne Gilson," in *Gilson et Descartes à l'occasion du centenaire de 'La liberté chez Descartes et la théologie'*, eds. Dan Arbib and Francesco Marrone (Lecce: ClioEdu Edizioni ("Examina Philosophica—I quaderni di Alvearium," 2), 2015), 41–56.

10 See Pierre Costabel, "Physique et métaphysique chez Descartes," in Pierre Costabel, *Démarches originales de Descartes savant* (Paris: Vrin, 1982), 181–90.

11 Louis Liard, *La Science positive et la métaphysique*, 3rd ed. (first published in 1879) (Paris: Germer Baillière et Cie, 1893).

12 See Alain Renaut, "Une philosophie française de l'université allemande: Le cas de Louis Liard," *Romantisme* 88 (1995): 85–100.

13 Liard, *Descartes*, "Avant-propos," s.p., my trans.: "But we thought that it would not be without interest for contemporary science and philosophy to contemplate Descartes' doctrine in the light they provide."

14 *Ibid.*, 3, my trans.: "Descartes' general doctrines . . . have remained in part the spirit of contemporary sciences."

15 *Ibid.*, 5.

16 *Ibid.*, 70–1, my trans.

17 *Ibid.*, 69, my trans.: "What characterizes his physics and makes it something entirely new, unprecedented, is the absence of any metaphysical idea."

18 *Ibid.*, 66–7, my trans.

19 *Ibid.*, 90–2.

20 Interestingly enough, siding with Lucien Lévy-Bruhl, Charles Adam, and Étienne Gilson on that issue, the same idea is present, though more confusedly expressed and certainly with a critical turn, in Catholic thinkers like Lucien Laberthonnière or Jacques Maritain who criticized Descartes for having reduced metaphysics to this auxiliary status: see Jacques Maritain, *Trois Réformateurs: Luther—Descartes—Rousseau* (Paris: Plon, 1925), 117; Jacques Maritain, *Le Songe de Descartes* (Paris: R-A. Corrêa, 1932), 133–34; Lucien Laberthonnière, "La théorie de la foi chez Descartes," *Annales de philosophie chrétienne* 12 (1911): 401, quoted in Henri Gouhier, *La Pensée religieuse de Descartes* (Paris: Vrin, 1924), 18; Lucien Laberthonnière, *Études sur Descartes I* (Paris: Vrin, 1935), 131, 248.

21 Liard, *Descartes*, 93, 95–6.

22 *Ibid.*, 100–1, 105.

23 *Ibid.*, 102, my trans.

24 *Ibid.*, 103–4.
25 *Ibid.*, 106–8.
26 *Ibid.*, 108–9.
27 *Ibid.*, 275, my trans.
28 *Ibid.*, 280, my trans. The same point is made by Stanley Victor Keeling for whom metaphysics "confer[s] upon [physics] the existential reference that it lacks": see Keeling, *Descartes*, 81–4, 133.
29 See Liard, *La Science positive*, 49, my trans.: "the only things we experience are our states of consciousness. Strictly speaking, we do not perceive a world external to us, but we are aware of sensations of resistance, color, taste, sound, etc."
30 In *La Science positive et la métaphysique*, Liard had already set the limits of a too radical form of positivism by reasserting the necessity of metaphysics as corresponding to the movement of the mind going from what is conditioned to what is unconditioned. There he already resorted to Descartes' metaphysics as escaping Comte's criticism of the metaphysical age as a mere reification of abstract entities: see Liard, *La Science positive*, 45, 54, 348–52, 355–56.
31 Paul Natorp, *Descartes' Erkenntnistheorie: Eine Studie zur Vorgeschichte des Kriticismus* (Marburg: N. G. Elwert'sche Verlagsbuchhandlung, 1882).
32 *Ibid.*, 75.
33 *Ibid.*, 113.
34 Ernst Cassirer, *Descartes' Kritik der mathematischen und naturwissenschaftliche Erkenntnis*, in *Gesammelte Werke*, Bd 1 (Hamburg: Felix Meiner Verlag, 1998), 3.
35 Liard, *Descartes*, 108.
36 Cassirer, *Descartes' Kritik*, 19. One finds a similar approach in Léon Brunschvicg's *Les Étapes de la philosophie mathématique* (Paris: Alcan, 1912; new ed. Paris: Blanchard, 1993). Brunschvicg did not consider as central the connection between Descartes' physics and metaphysics, but rather the "reform of physics brought about by mathematics," a reform that gave birth to "a geometer's physics" (*Les Étapes de la philosophie mathématique*, 107, my trans.). If physics and metaphysics seemed to bear a relationship to one another, this was, for Brunschvicg, through the notion of space. Nevertheless, extension and motion which depend on the notion of space were considered by him more as geometrical than as metaphysical notions (*Ibid.*, 112–3). In the end, the overestimation of mathematics for Descartes' thought as well as Brunschvicg's own critical idealism led him to claim the discontinuity of Descartes' physics and metaphysics because of the gap between mind and extension. On the contrary, Edwin Arthur Burtt considered that "the new scientific metaphysics is to be found in the ascription of ultimate reality and causal efficacy to the world of mathematics, which world is identified with the realm of material bodies moving in space and time" (Edwin Arthur Burtt, *The Metaphysical Foundations of Modern Science*. 1924. (rev. ed. London: Routledge & Kegan Paul, 1980), 303). Contrary to Burtt, Cassirer and Brunschvicg did not consider that the central role ascribed to mathematics in Descartes' physics necessarily amounted to a metaphysical approach but could—at least in the *Regulae*—be described in epistemological rather than ontological terms. That Descartes did not stick to that epistemological stance was, however, recognized by Cassirer.
37 Cassirer, *Descartes: Lehre-Persönlichkeit-Wirkung*, in *Gesammelte Werke*, Bd. 20 (Hamburg: Felix Meiner Verlag, 2005), 10–11.
38 Cassirer, *Descartes' Kritik*, 21, my trans.
39 This passage from logic to physics is supposed to occur "by itself and without any friction": Cassirer, *Descartes: Lehre-Persönlichkeit-Wirkung*, 24, my trans.
40 Cassirer, *Descartes' Kritik*, 21–2.

41 *Ibid.*, 33. In *Das Erkenntnisproblem in der Philosophie und Wissenschaft der Neueren Zeit*, Cassirer even went as far as to claim that, in the *Regulae*, extension was conceived independently of the metaphysical notion of substance, and as a "mental substratum" put under natural phenomena in order to make them measurable and understandable: see Ernst Cassirer, *Das Erkenntnisproblem in der Philosophie und Wissenschaft der Neueren Zeit*, in *Gesammelte Werke*, Bd. 2 (Hamburg: Felix Meiner Verlag, 1999), 340.
42 Cassirer, *Descartes' Kritik*, 38–9.
43 *Ibid.*, 47.
44 *Ibid.*, 62–3.
45 Cassirer, *Das Erkenntnisproblem*, 401–2.
46 Cassirer, *Descartes' Kritik*, 41–2, 44–5.
47 CSM I, 233; AT VIII-1, 53.
48 CSM I, 233; AT VIII-1, 53.
49 AT VIII-1, 53–6; Cassirer, *Descartes' Kritik*, 48.
50 Cassirer, *Descartes' Kritik*, 59.
51 *Ibid.*, 60.
52 *Ibid.*, 26.
53 Cassirer, *Das Erkenntnisproblem*, 396–7.
54 *Ibid.*, 399–401.
55 *Ibid.*, 369.
56 *Ibid.*, 370.
57 On Milhaud and Meyerson, see René Poirier, "Meyerson, Milhaud et le problème de l'épistémologie," *Conférence de la Société Française de Philosophie* (26 novembre 1960): 127–55. http://s3.archive-host.com/membres/up/784571560/GrandesConfPhiloSciences/philosc05_poirier_1960.pdf. One could add Edwin Arthur Burtt to this group of anti-positivist historians of philosophy and science. See *Metaphysical Foundations*, 277: "there is no escape from metaphysics."
58 See Anastasios Brenner, "Le parcours épistémologique de Milhaud," in *Science, Histoire et Philosophie selon Gaston Milhaud: La constitution d'un champ disciplinaire sous la troisième République*, eds. Anastasios Brenner and Annie Petit (Paris: Vuibert, 2009), 19–30.
59 See his lecture in honor of Paul Tannery pronounced in 1905 and published in 1906: Gaston Milhaud, "Paul Tannery," *La Revue des idées, Études de critique générale*, no. 25 (15 janvier 1906): 3.
60 On the permanence of conceptual frameworks, categories, speculative tendencies, and principles throughout the history of philosophy and science, from antiquity to the beginning of the twentieth century, see *ibid.*, 5–6. In particular, Milhaud considers that there is a recurrence of a principle of conservation which takes the form of a unique element in pre-Socratic thought, of the constancy of the sum of things in Lucretius, of the principle of the conservation of the quantity of motion in Descartes, of that of *vis viva* in Leibniz, of that of mass in Lavoisier, and of that of energy by the end of the nineteenth century.
61 Milhaud, *Descartes savant* (Paris: Alcan, 1921), 11, my trans.
62 *Ibid.*, 9–10.
63 *Ibid.*, 228, my trans.
64 *Ibid.*, 246, my trans.
65 *Ibid.*, 228.
66 *Ibid.*, 248, my trans.
67 *Ibid.*, 247.
68 *Ibid.*, 248, my trans.
69 *Ibid.*, 243.
70 Gaston Milhaud, "Descartes et Newton," in Gaston Milhaud, *Nouvelles Études sur l'histoire de la pensée scientifique* (Paris: Alcan, 1911), 225–6, my trans.:

"Then if Descartes is obsessively committed to linking each of his discoveries to his Method or his Metaphysics, should this really lead us to overlook, as much as he did, at what point they are independent [of his Method and his Metaphysics], and at what point they are directly linked, by this sort of intuition that characterizes genius, to the natural trend of scientific thought?" Milhaud also emphasized that, in practice, Descartes had to resort to experience, even if he had in mind an ideal of deduction of his physics from his metaphysics: see *Descartes savant*, 211–2.

71 Milhaud, *Descartes savant*, 245, my trans.

72 *Ibid.*, 15, my trans. See Simone Mazauric, "Milhaud, Descartes et l'histoire des sciences," in *Science, Histoire et Philosophie selon Gaston Milhaud*, 117–31. See also Milhaud, *Descartes savant*, 249.

73 See David Lévy, "Émile Meyerson," in *L'épistémologie française, 1830–1970*, eds. Michel Bitbol and Jean Gayon (Paris: Presses Universitaires de France, 2006), 357–74.

74 See Émile Meyerson, *Explanation in the Sciences*, trans. Mary-Alice and David A. Sipfle (Dordrecht: Kluwer, 1991), 1.

75 *Ibid.*, 27: "The whole of science rests on the bed-rock—not exposed of course (since attempts have been made to deny the existence of such a foundation), but nevertheless solid and deep—of belief in a being independent of consciousness."

76 *Ibid.*, 1.

77 *Ibid.*

78 *Ibid.*, 399.

79 *Ibid.*, 422.

80 Émile Meyerson, *The Relativistic Deduction: Epistemological Implications of the Theory of Relativity*, trans. David A. and Mary-Alice Sipfle (Dordrecht: Reidel, 1985), 180; Meyerson, *Explanation in the Sciences*, 428: "the conflict, the irreconcilable opposition between the *inevitable* belief in external reality and the *irresistible* belief in the rationality of this same reality."

81 Émile Meyerson, *Identity and Reality*, trans. Kate Loewenberg (New York: Gordon and Breach (Classics in the History and philosophy of Science), 1989), 388: "And as to Descartes, it is certain that he sought for the essence of things as eagerly as the scholastics, but, unlike these latter, he found it in space. . . . "

82 Meyerson, *Explanation in the Sciences*, 84.

83 *Ibid.*, 429.

84 Sandra Laugier, "Science and Realism: The Legacy of Duhem and Meyerson in Contemporary American Philosophy of Science," in *French Studies in the Philosophy of Science*, eds. Anastasios Brenner and Jean Gayon (Dordrecht: Springer, 2009), 97: "Meyerson was the first to propose a model of the evolution of science in which ontological change defines scientific change. . . . Moreover, it is these changes of ontology that allow him to describe, in *Identity and Reality* and *Explication in the Sciences*, the conceptual changes made in the history of the sciences."

85 See Meyerson, *Explanation in the Sciences*, 68.

86 Meyerson's principle of identity falls within some general psychological tendencies of the human mind which tend to reduce diversity to identity. Its meaning is thus broader than what is usually understood by the logical principle of identity.

87 Meyerson, *Identity and Reality*, 95. In *Explanation in the Sciences*, Meyerson, on that point, will explicitly oppose Milhaud who considered the principle of conservation as a convention or as a "guiding principle" for the physicist: see Meyerson, *Explanation in the Sciences*, 508–9.

88 Meyerson, *Identity and Reality*, 102. See Laugier, "Science and Realism," 100.

89 Meyerson, *Explanation in the Sciences*, 11: "Man does metaphysics as he breathes, involuntarily and, above all, usually without realizing it"; *ibid.*, 395: "all scientific explanations are naturally, unconsciously, necessarily ontological."

90 *Ibid.*, 135.
91 *Ibid.*, 426.
92 *Ibid.*, 271, 382. *Ibid.*, 439: "Cartesian deduction differs from that practiced by the science of today in that it claims to be continuous"; *ibid.*, 459: "But we also know that we cannot, we shall never be able to deduce everything . . . that it is impossible to reduce the science of nature to a coherent, rational conception: this is why we are forced to recognize that Descartes was mistaken"; see Meyerson, *Relativistic Deduction*, 129, 194.
93 This positive dimension is more clearly emphasized in *The Relativistic Deduction* than in *Identity and Reality*.
94 See Meyerson, *Relativistic Deduction*, 180–1.
95 Meyerson, *Explanation in the Sciences*, 526.
96 Meyerson, *Identity and Reality*, 91.
97 *Ibid.*, 93.
98 *Ibid.*, 234–58.
99 *Ibid.*, 87.
100 *Ibid.*, 137.
101 *Ibid.*, 140.
102 *Ibid.*, 141.
103 *Ibid.*, 145.
104 *Ibid.*, 146.
105 *Ibid.*, 201.
106 *Ibid.*, 199.
107 *Ibid.*, 147.
108 *Ibid.*, 147–8; Meyerson, *Explanation in the Sciences*, 440: "the principle of inertia is merely *plausible*, that is, it contains an affirmation in which, by virtue of the particular structure of our understanding, we are disposed to believe on the basis of proofs we would under other circumstances judge insufficient."
109 Meyerson, *Explanation in the Sciences*, 442.
110 Meyerson, *Identity and Reality*, 144.
111 *Ibid.*, 147: "This latter principle [i.e., the causal principle] certainly is *a priori*. But we have seen that this is not a statement from which we can directly deduce precise propositions; it is that which renders nugatory all *a priori* demonstrations from Descartes to d'Alembert and Spir."
112 From that point of view, Meyerson anticipated Desmond Clarke's interpretation which distinguished between metaphysics as providing foundation for physics which he accepted, and metaphysics as the source from which to deduce physics logically, which he rejected: see Desmond Clarke, *Descartes' Philosophy of Science* (Manchester: Manchester University Press, 1982), 13, 77–107.
113 Meyerson, *Identity and Reality*, 154–5: "certain scientific statements constitute only a particular expression of the metaphysical principle of substance"; *ibid.*, 161: "It is thus that motion having become a state, transforms itself into an entity, a substance." In the *Relativistic Deduction*, Meyerson slightly modified his interpretation, stating that, while motion is considered as a state, velocity is seen as a substance in the formulation of the Cartesian principle of inertia: see Meyerson, *Relativistic Deduction*, 34–5.
114 Meyerson, *Identity and Reality*, 147: "every proposition stipulating identity in times . . . finds our minds prepared, it seduces them, and is immediately adopted." As Sophie Roux has convincingly shown, there is however a shift, between *Identity and Reality* and the later *Explanation in the Sciences*, from a (at least partial) ontological to a more plain epistemological conception of identity, that is to say from identity conceived as the real permanence of some things in the world to identity as corresponding to a formal equivalence the

mind posits between propositions: see Sophie Roux, "Histoire de la physique classique et historicité des sciences chez Meyerson," in *L'histoire et la philosophie des sciences françaises à la lumière de l'œuvre d'Émile Meyerson (1859–1933)*, eds. Eva Telkes-Klein and Elhanan Yakira (Paris: Honoré Champion, 2010), 106. But the epistemological dimension of identity is already forcefully stated as early as *Identity and Reality*.

115 Pierre Duhem, *The Aim and Structure of Physical Theory*, trans. Philip P. Wiener (New York: Atheneum, 1962), xvii.

116 Duhem, "Physics and Metaphysics," trans. Roger Ariew and Peter Barker in Pierre Duhem, *Essays in the History and Philosophy of Science* (Indianapolis: Hackett, 1996), 30. Cosmology is therefore fundamentally distinct from astronomy, which, for Duhem, sought mainly to describe celestial phenomena according to mathematical models, and not to inquire into the reality and nature of celestial bodies.

117 *Ibid.*, 31.

118 *Ibid.*

119 See Duhem, *Aim and Structure of Physical Theory*, 10.

120 Pierre Duhem, *La Théorie physique: Son objet, sa structure* (Paris: Vrin, 1997), 17, my trans.: "It is clear that by making physical theory depend on metaphysics, one does not at all contribute to assure it the advantage of universal consensus." This passage was omitted in the English translation.

121 See Duhem, *Aim and Structure of Physical Theory*, 274.

122 *Ibid.*, 275–82.

123 Pierre Duhem, "Quelques réflexions au sujet des théories physiques," *Revue des questions scientifiques* 31 (1892), trans. Roger Ariew and Peter Barker, "Some Reflections on the Subject of Physical Theories," in *Essays in the History and Philosophy of Science*, 15.

124 Duhem, "Physics and Metaphysics," 34–5.

125 *Ibid.*, 36.

126 See Duhem, *Aim and Structure of Physical Theory*, 268, 270. Duhem agreed with Ernst Mach on the value of the history of science for physics' students as well as for physicists themselves: see Pierre Duhem, "Analyse de l'ouvrage de Ernst Mach, *La mécanique: Étude historique et critique de son développement*," *Bulletin des sciences mathématiques* 27 (1903), trans. Roger Ariew and Peter Barker in Pierre Duhem, *Essays in the History and Philosophy of Science*, 118.

127 As Anastasios Brenner insightfully notices, "Cartesian physics enjoys a privileged status: it marks the extreme point of reaction against Scholastic philosophy; it also reveals certain metaphysical presuppositions of modern science" (*Duhem. Science, réalité et apparence* (Paris: Vrin, 1990), 97, my trans.).

128 Duhem, *Aim and Structure of Physical Theory*, 10.

129 *Ibid.*, 10.

130 *Ibid.*, 13.

131 *Ibid.*, 17.

132 The problem bears on the possibility to deduce such a representation, but also on the coherence of the representation, for the homogeneity of matter seems to exclude its division into differentiated corpuscles: see *ibid.*, 13.

133 *Ibid.*, 18.

134 *Ibid.*, 16.

135 *Ibid.*

136 *Ibid.*, 17–20.

137 *Ibid.*, 18.

138 *Ibid.*

139 *Ibid*. The choice of some hypotheses is the second of the four fundamental operations performed by the physicist: the first one is the definition and measure of magnitudes; the third one the mathematical development; the fourth one the confrontation of theory with experience.
140 On Descartes' law of refraction, see *ibid.*, 33–4.
141 *Ibid.*, 34.
142 *Ibid.*, 43.
143 *Ibid.*, 43–4 (my emphasis).
144 *Ibid.*, 44.
145 Duhem, "Physics and Metaphysics," 32.
146 *Ibid.*, 43.
147 *Ibid.*, 46.
148 See AT VI, 63–5; AT VIII-1, 100–1.
149 On this notion, see Andrew Lugg, "Pierre Duhem's Conception of Natural Classification," *Synthese* 83, no. 3 (1990): 409–20.
150 See Duhem, *Aim and Structure of Physical Theory*, 31.
151 Duhem, "La valeur de la théorie physique, à propos d'un livre récent," *Revue générale des sciences pures et appliquées* (1908), trans. Roger Ariew and Peter Barker, in Pierre Duhem, "Logical Examination of Physical Theory," *Synthese* 83, no. 2 (1990): 187. The history of science is indicative of the path that science has been following up to the present. From that historical insight, it becomes somewhat possible to prolong that path toward the final, ideal state of the physical theory: see Duhem, *Aim and Structure of Physical Theory*, 303.
152 Pierre Duhem, "L'École anglaise et les théories physiques," *Revue des questions scientifiques* 34 (1893), trans. Roger Ariew and Peter Barker, "The English School and Physical Theories: On a Recent Book by W. Thomson," in *Essays in the History and Philosophy of Science*, 68.
153 Duhem, "Physics and Metaphysics," 37.
154 Duhem, "The English School and Physical Theories," 68.
155 See Gaston Bachelard, "L'actualité de l'histoire des sciences," in Gaston Bachelard *L'engagement rationaliste* (Paris: Presses Universitaires de France, 1972), 138, my trans.: "history of science necessarily amounts to determining the successive values of the progress of scientific thought." See Stephen Gaukroger, "Bachelard and the Problem of Epistemological Analysis," *Studies in History and Philosophy of Science* 7, no. 3 (1976): 233–4.
156 On this point, see Dominique Lecourt, *L'épistémologie historique de Gaston Bachelard* (Paris: Vrin, 1978).
157 Gaston Bachelard, *Le matérialisme rationnel*, 3rd ed. (Paris: Presses Universitaires de France, 1972), 141, my trans.: "And every philosophy, explicitly or tacitly, steadily or covertly, uses the *realist function*. Every philosophy deposits, projects or assumes one reality"; Lecourt, *L'épistémologie historique*, 23, my trans.: "when a philosopher reads a word, he has a tendency to see in it a being; the scientist sees in it a concept whose whole being resolves itself into the system of relations in which it is inscribed."
158 Lecourt, *L'épistémologie historique*, 84, my trans.
159 Gaston Bachelard, *La philosophie du non*, 5th ed. (Paris: Presses Universitaires de France, 1970), 1–2, my trans.: "In particular, if one attempts to shed light on science's problems through metaphysical reflection, if one intends to mix theorems with philosophems, one is faced with the necessity of applying a necessarily finalist and closed philosophy on an open scientific thought." That Bachelard was influenced by Comte can be seen in his law of the three stages of the scientific mind which was only intended to be more accurate than Comte's: see Gaston Bachelard, *The Formation of the Scientific Mind*, trans. Mary McAllester Jones (Manchester: Clinamen Press, 2002), 20.

160 Gaston Bachelard, *The New Scientific Spirit*, trans. Arthur Goldhammer (Boston: Beacon Press, 1984), 3.
161 Bachelard, *La philosophie du non*, 1–2, my trans.: "The mind can change its metaphysics; it cannot do without any metaphysics."
162 Bachelard, *New Scientific Spirit*, 2–3.
163 *Ibid.*, 171: "Innate truths naturally have no place in science. Reason has to be shaped in the same way as experience."
164 *Ibid.*, 62: "Materialism also tends to limit the notion of matter in still another sense, by precluding the so-called action at a distance, or in other words by refusing to allow a material object to produce effects at points in space other than where it is located. Thus materialism shades by degree into realist atomism. Descartes' protestations to the contrary are scarcely convincing if matter alone has the attributes of being extended, composed of solids, associated with strictly local properties, and defined by a form, indeed inextricably bound up with a form"; *Ibid.*, 138: "The Cartesian method is reductive rather than inductive. But reduction distorts analysis and hinders the extension of objective thought."
165 Mary Tiles, "Technology, Science, and Inexact Knowledge: Bachelard's Non-Cartesian Epistemology," in *Continental Philosophy of Science*, ed. Gary Gutting (Malden: Blackwell Publishers, 2005), 172.
166 Bachelard, *New Scientific Spirit*, chap. VI, 165–9.
167 This example is analyzed at length in chapter IV of Bachelard, *New Scientific Spirit*, 85–98.
168 *Ibid.*, 138.
169 *Ibid.*, 4.
170 *Ibid.*, 135–77.
171 *Ibid.*, 17.
172 This can be deduced from the way Bachelard defines non-realism and non-materialism as "an opening of realism, of materialism" and adds that non-realism "is a realism" (Bachelard, *La philosophie du non*, 15, my trans.). By the end of the work, he makes explicit what he means by a negation: "Negation must stay in touch with the first formation. It must allow a *dialectic generalization*. Generalization by the 'no' must include what it negates. Actually, the whole development of scientific thought for a century arises from such dialectic generalizations which envelop what they negate" (emphasis in original, *ibid.*, 137, my trans.). On this, see Vincent Bontems, "Le 'non-cartésianisme.' La méthode non-cartésienne selon Gaston Bachelard et Ferdinand Gonseth," in *Qu'est-ce qu'être cartésien?* ed. Delphine Kolesnik-Antoine (Lyon: ENS Éditions, 2013), 567–80.
173 See Gaston Bachelard, *Le rationalisme appliqué* (Paris: Presses Universitaires de France, 1949), 50–1. The risk pointed out there is one of destroying the world without being in capacity to recover or rebuild it.
174 Bachelard, *New Scientific Spirit*, 147.
175 *Ibid.*, 163.
176 Bachelard, *La philosophie du non*, 139, my trans.: "Indeed, it seems to us impossible to understand the atom of modern physics without evoking the history of its imagery, without recollecting the realist and the rational forms. . . . In some way, what is removed from the image must be found in the amended concept. Therefore I would be willing to say that the atom is exactly the *sum of the criticisms* to which its primary image was subjected."
177 Bachelard, *Formation of the Scientific Mind*, 86: "In addition, examples can be found where very great minds are stuck, so to speak, in primary images. For Descartes, doubting the clarity and distinction of the image offered to us by the sponge means making explanations unjustifiably *over subtle*. . . . Descartes' confidence in the clarity of the image of the sponge is very symptomatic of this

inability to bring doubt to bear on the detail of objective knowledge, to develop a discursive doubt that would wrench asunder all reality's bonds and images' every angle . . . a sponge shows us sponginess. It shows us how one particular kind of matter 'is filled' with another. This lesson in heterogeneous fullness suffices to explain everything. The metaphysics of space in Descartes is the metaphysics of the sponge."

178 See Frédéric de Buzon and Vincent Carraud, *Descartes et les Principia II: Corps et mouvement* (Paris: Presses Universitaires de France, 1994), 52–4.

179 Gaston Bachelard, *Les intuitions atomistiques (Essai de classification)* (Paris: Vrin, 1975), 19–21.

180 Bachelard is the only one, among the authors I have studied, not to deal with this issue.

181 CSM I, 93; AT XI, 38.

182 On the profound transformation of the very definition of metaphysics by Descartes and the inclusion of epistemology in that field—inclusion which was completely foreign to Scholastic metaphysics, see Gary Hatfield, "Metaphysics and the New Science," in *Reappraisals of the Scientific Revolution*, eds. David C. Lindberg and Robert S. Westman (Cambridge: Cambridge University Press, 1990), 111–2.

183 See *ibid.*, 110–6.

184 This was already noted by Hatfield concerning E. A. Burtt: see *ibid.*, 93–4.

185 I would like to thank Raphaël Chappé for his helpful suggestions on an earlier version of this chapter, and Jonathan Regier and Charles Wolfe for their emendations to my English text.

Part II

Physics and Metaphysics in Descartes' Time

4 The Metaphysical Roots of Physics, and the Alleged Link Between Taurellus, Gorlaeus, Regius, and Descartes

Christoph Lüthy

Introduction

In identifying the philosophical sources that influenced Descartes, scholars have pointed to various obvious sources: scholastic authors, representatives of the pure and mixed mathematical sciences, plus a series of individuals with whom Descartes corresponded or who are mentioned explicitly or implicitly in his correspondence. But there exists a further category of authors, whose influence on Descartes is not usually examined, even though at the beginning of the *Querelle d'Utrecht*, the theologian Gisbertus (or Gijsbert) Voetius pointed his angry finger at them, and despite the fact that a number of early historiographers of philosophy believed to recognize their trace. In is chapter, I would like to draw attention to a tradition of Protestant metaphysics, and specifically to authors who believed, often for theological reasons, that Aristotelian metaphysics was either bogus or at least ill-conceived, and that a new *philosophia prima de ente*, a first, ontological philosophy, ought to be developed either to replace Aristotle's metaphysics entirely, or at least to reformulate it thoroughly. As will be argued in the following, in sixteenth-century Protestant German lands, and—importantly—even before Francisco Suárez' *Disputationes metaphysicae* (1597) had been published, which is usually seen as the beginning of modern metaphysics, attempts were underway in German-speaking lands to develop an ontological first philosophy. This ontology subsequently blended with Suárez-style metaphysics and spread into the Dutch Republic, where it played an important role in the years that Descartes was living there.

Why these authors might represent an interesting additional ingredient to our understanding of the origins of Descartes' metaphysics can be gauged, *ex negativo*, from a passage found in Daniel Garber's admirable *Metaphysical Physics*:

> [i]t is fair to say that [Descartes'] view of the order of knowledge may well have presented a significant departure from the mainstream of the scholastic tradition. Though there were many differences between different scholastic writers, there was wide agreement that knowledge of

> physics is largely independent of knowledge of metaphysics, however precisely either discipline is defined. And so, they claimed, one can (and, in fact, ought to) study physics before undertaking the more elevated studies of God and being as such that pertain to first philosophy. In demanding that physics must be grounded in some sense in metaphysics, in knowledge of God and the soul, Descartes is stepping clearly outside that tradition. And so when by the 1630s Descartes came to hold to the priority of metaphysics in the strong sense he held it, his view would likely have been recognized as a clear departure from the received view.[1]

As for the traditional place of metaphysics, it is generally presumed that Aristotle had not given a name to those books of his that dealt with "the science that we are seeking."[2] According to the standard account, the name "metaphysics" was attached to them later, to define the place where they had to be placed in the Aristotelian corpus, namely "*meta ta biblia physiká*," "after the books on physics."[3] In keeping with that classificatory name, the scholastic tradition placed, and also taught, metaphysics after physics. To be sure, the superior status and nobility of metaphysics was not contested, nor its primacy in the "order of nature," as it was called; still, "in the order of learning," the general view was that metaphysics could only be tackled and understood once one had mastered natural philosophy. Moreover, precisely because metaphysics was the most abstract and general science, Aristotle himself seems to have doubted that natural philosophy could "stand to derive much of its contents from first philosophy," to use Gary Hatfield's expression.[4]

This characteristic placement of metaphysics as the last, albeit highest, of the disciplines is also common to the textbooks of seventeenth-century scholastics, although there are a few exceptions to this rule. As Roger Ariew has shown, Théophraste Bouju, for example, distinguished in his philosophy textbook of 1614 between a general metaphysics and a particular metaphysics, whereby he placed general metaphysics after logic, but before physics; and particular metaphysics after physics. Bouju's reasoning was as follows:

> Universal Metaphysics comes after Logic, because by treating what is common to all the other sciences, it is useful to enter into the matter in this way, so that, beginning with universal things, one does not have to repeat anything several times. . . . The book of particular Metaphysics follows those of the Physics immediately after the book about the rational soul, because it treats the other immaterial substances, which are more excellent than the rational soul.[5]

When Descartes places metaphysics before physics, however, this happens for very different reasons. Neither does he link metaphysics to logic, nor is his primary motivation that of deriving the particular from the general. Even though in book 2 of the *Principia*, he does seem to provide the general principles of physics, for epistemological reasons he usually combats the idea that one knows the particular from the general.[6] Moreover, Descartes treats all substances—God, the *res extensa* and the *res cogitans*—and thus (unlike Bouju) clearly also the "immaterial substances," before he turns to his physics. Finally, and very much related to this last point, his metaphysics integrates something that had not previously belonged to this discipline, namely a general account of *res cogitans* as the knowing mind, of knowledge in general, and of certain knowledge in particular. In the celebrated tree metaphor from the introduction to the French edition of the *Principia* (1647), this epistemologically founded metaphysics is described as the root from which physics and all other disciplines grow. In keeping with that view, the *Principia* itself (1644) starts with that novel mixture of epistemology and metaphysics that was first adumbrated in the *Discourse on Method* (1637) and more fully explicated in the *Meditations* (1641).

So, with respect to the canonical Aristotelian tradition, then, Garber's observation about Descartes' reversal of the relation of physics and metaphysics is therefore certainly correct. With respect to the metaphysical principles themselves, it is known that Descartes was acquainted with Suárez' *Disputationes metaphysicae*, as he cites passages from this work in a letter of 1630.[7] Unsurprisingly, scholars have sought to find traces of Suárez' influence on Descartes' metaphysics, finding such traces specifically in his terminology.[8]

While this influence will certainly not be denied here, this chapter wishes to draw the reader's attention to another possible line of influence. Not only did what in this chapter I call by the name of 'Protestant metaphysics' originate more than twenty-five years before Suárez' *Disputationes metaphysicae*, but, more importantly for our argument, it exerted a traceable influence on the Dutch environment in which Descartes worked, and through them probably also on Descartes himself, albeit maybe only in an indirect manner.

It would be worthwhile examining a whole series of authors here, including Bartholomeus Keckermann, Rudolph Goclenius, Clemens Timpler and Otto Casmann, who either directly taught the late sixteenth- and early seventeenth-century northern Dutch intellectual elite at the Academy of Steinfurt, which lay close to the Dutch border, or else exerted a traceable influence on the teaching of philosophy at the first Dutch universities. Alas, the scholarly literature has said to date very little about that fascinating connection.[9] In this chapter, we will however limit ourselves to one specific author, Nicolaus Taurellus, and to an equally specific intellectual pedigree that goes from Taurellus to David Gorlaeus and from there to Descartes' friend Henricus Regius. We do so for the simple reason that Gisbertus Voetius mentioned that line explicitly at the start of the so-called *Querelle d'Utrecht*.

What was the nature of Voetius' claim? The *Querelle*, as we know from Theo Verbeek's work, gradually built up during a series of disputations in late 1641, as Henricus Regius, professor of medicine at the University of

Utrecht, had students defend what to us appear as distinctly Cartesian doctrines in public disputations (although historically, the interesting point is that many of these doctrines had not yet been published by Descartes himself, whose *Principia* would only appear in 1644, so that the question of Regius' original contribution to their development remains unclear).[10] Moving from physiology as it was understood in the field of medicine, which was Regius' official teaching duty, to physiology as it was understood by natural philosophers (namely as an equivalent of 'natural philosophy'), the series of disputations culminated in a veritable introduction to a corpuscular philosophy, which explicitly rejected substantial forms. The striking resemblance between that philosophy and what Robert Boyle would a couple of decades later describe as the essence of the 'mechanical philosophy' can be seen from the following ditty, which Regius had one his students recite during one of the disputations, for the sake of easy memorization:

> Mind, measure, quiet, motion, position and figure
> Are together with matter the origin of all things.[11]

As the disputations proceeded, getting ever bolder in their departure from the received principles of natural philosophy, they attracted an increasing number of curious onlookers. Public scandal eventually erupted when Regius, on 8 December 1641, had a student—in fact, a student of theology!—defend the thesis according to which man was not to be understood as a union of body and soul, but as an accidental composite being, an *ens per accidens*.[12] At that point, the rector of the University, the theologian Gisbertus Voetius, stepped in, had this specific thesis officially condemned, and then organized his own series of theological disputations to rebut Regius' Cartesian physics. According to Voetius, the perilous notion of man as a composite being, which ran counter to the idea of a formal union of body and soul, annihilated the anthropological basis of Christian theology. The *Querelle* ended, as is well known, with a wholesale ban on Descartes' philosophy in Utrecht.[13]

What is relevant for our current purposes is that when Voetius voiced outrage at the description of man as an *ens per accidens*, Regius initially defended himself, somewhat sheepishly, by saying that the contested thesis was neither his own idea nor Descartes', but that he had taken it from a work by David Gorlaeus. Nowadays fairly forgotten, Gorlaeus was a Dutch author whose *Exercitationes philosophicae*, published posthumously in 1620, was quite popular at the time. For Voetius, however, Regius' invocation of that author made things only worse.[14] As it turned out, Voetius remembered Gorlaeus (who had died in 1612) from the time when they had both studied philosophy at Leiden, and he claimed to know precisely from which poisoned well Gorlaeus had taken his philosophical heresies. In the published version of the corollary to one of his own theological disputations, which were held later in the same month of December 1641, Voetius exposed the evil pedigree of the notion of man as an *ens per accidens* as follows. It went, he explained, all the way back to a 1573 book by a certain Nicolaus Taurellus. The convoluted sentence, which will be unpacked in the course of this chapter, reads as follows:

> The paradoxical claim [about man being an *ens per accidens*] made in his *Triumph of Philosophy*, in the prefaced axioms D.4 and D.5 by Nicolaus Taurellus (who was called an atheist physician by the Heidelberg theologians in their judgement on [Conrad] Vorstius' *De Deo*, which they sent to the delegates of the Synod of Holland in 1610), and which, due to the imprudence of youth, our compatriot David Gorlaeus took up in his *Exercitationes philosophicae* . . . is contrary to not just to physical, but also to metaphysical, pneumatological and theological truth.[15]

Voetius' complex historical reconstruction happens to be correct. That the last part of the pedigree is legitimate is obvious: Regius himself had denounced Gorlaeus as his source, and the latter had indeed described man as an *ens per accidens*. In fact, Regius' unsuccessful attempt at shifting the blame away from himself and Descartes is all the more credible as his published work contains various explicit references to Gorlaeus' *Exercitationes*. As for the earlier part of Voetius' reconstruction, it is equally true. That Gorlaeus had taken the dualistic notion of man as a compound of soul and a material body from Taurellus will be shown below.[16]

Descartes himself, who had not only attempted to monitor from a distance the disputations themselves, but also its unpleasant aftermath, continued to send detailed instructions to Regius about how he should behave. His advice, once the dispute had erupted, was above all political, and had to do with good behavior, which included above all the attempt to abstain from provoking the establishment through the use of a novel vocabulary. Curiously, however, his longest letter, of January 1642, contains a draft for an apology-cum-rebuttal, which sends a startling double message. On the one hand, Descartes recommends that Regius should publicly humble himself before Voetius, abstain for a certain period from further disputations, stop pressing for a novel terminology, and instead cunningly offer his new wine in seemingly old wineskins. On the other, he drafts a rebuttal of Voetius' claims, and in particular regarding the latter's insistence on the necessary connection between a theory of substantial forms and orthodox theology. According to Descartes' draft, however, the real theological heterodoxy lies hidden in the notion of substantial forms, and whoever does not recognize that, is a moron. Moreover, he proposed to Regius to declare that even though neither he nor Regius embraced Taurellus' or Gorlaeus' philosophy, and did believe in the full interaction of soul and body, they did think that it could be helpful to speak of man as an *ens per accidens* in so far as it avoided the misunderstanding that body and soul were not distinct.[17]

Incidentally, had Regius followed the first part of Descartes' advise, the *Querelle* would have remained just one of those short-lived local quarrels that characterized life at early-modern universities. The decision to rebut Voetius' views in writ instead of merely apologizing and otherwise keeping a low profile, produced of course exactly what the two *novatores* had wished to avoid: a further escalation, which culminated in the official condemnation of Descartes' philosophy.

But let us return to the questions with which this chapter is concerned. First, what might the Taurellus-Gorlaeus-Regius link imply about the relation between metaphysics and physics in the Netherlands in the period in which Descartes worked there? Second, does the answer to the first question shed any additional light on Descartes' own view of metaphysics and its place in the order of the philosophical disciplines?

Nicolaus Taurellus

The answer to the first question requires a chronological approach, which is why we must start with the *Philosophiae triumphus*. Its author, Nicolaus Taurellus (or Niklaus Öchslein in the original German), needs a brief introduction, because although he was the "first Lutheran metaphysician," as Ulrich Leinsle has called him in his beautiful study, *Das Ding und die Methode*, he is no longer a philosophical household name.[18] Taurellus was born in 1547 in Montbéliard, a city that in those days belonged to the Grand Duchy of Württemberg. He studied in Tübingen under the famous Jacob Schegk, whom he would praise for the rest of his life for having taught him an independent way of philosophizing.[19] In fact, according to Taurellus' own testimony, the *Philosophiae triumphus* is a much-expanded version of theses he had defended under Schegk for his *magister artium* in 1565 and published in 1567.[20] Taurellus thereafter first turned to theology, but became quickly disgusted by the endless quarrels and feuds between the various theological currents and sects. He therefore switched to medicine, which he studied in Basel, where he obtained his doctorate in 1570. His first teaching position was however in ethics, a chair he inherited from Theodor Zwinger, who had himself moved on to a medical chair. In 1580, Taurellus received a call to Nuremberg's newly established University of Altdorf, where he served as professor of medicine and natural philosophy until his death in 1606.[21]

Although Taurellus' oeuvre is rich and varied, ranging from medical topics to emblematics, physics, and metaphysics, it is above all his first book to which we must direct our attention, namely the *Triumph of Philosophy*, which Voetius cited as the original source of the thesis that man is an accidental being. This book, published in 1573, when Taurellus was a medical doctor and professor of ethics aged 26, carries a title that is as long and pompous as it is programmatic:

> The Triumph of Philosophy, that is, a metaphysical method of philosophizing, in which human reasons are thus deduced from divinely instilled ideas that through most solidly constructed demonstrations the truth of the matter will openly shine forth, and Philosophy will burst forth victorious, after having been buried for a long time through the authority of the philosophers. For, in six hundred questions [on issues] in which Philosophy used to appear to be battling with revealed truth, she will be so truly reconciled with it, that she must not only be said to serve faith, but provide its very foundation.[22]

As the title indicates, Taurellus intends in his book to develop a new "metaphysical method of philosophizing," which "through . . . demonstrations" "deduced" "human reasons" "from divinely instilled ideas." At least at the level of the title, Taurellus appears to propose a methodology that shares certain traits with Descartes', namely one in which metaphysical ideas—innate or otherwise provided by a divinely sanctioned 'natural light'—form the basis of a philosophy that arrives at its conclusions deductively and by demonstration. This *prima facie* similarity does not of course amount to the claim that Taurellus anticipated Descartes; and the latter's claim that he had never heard of Taurellus may in fact be correct.[23] Indeed, as far as style of presentation, theological preoccupation and terminology are concerned, the two authors have little in common. At the same time, the programmatic title cannot but strike us for certain parallels with a number of Descartes' methodological intuitions. Moreover, Taurellus' *Triumph of Philosophy* offers us a metaphysics that combines epistemology, proofs of God and principles of matter theory that are not only unique, but cannot but remind us of Descartes' combination. At the same time, Taurellus' very title also indicates an important difference from Descartes, who was not primarily motivated by the desire to resolve the age-old tensions between philosophy and faith, even though the full title of his metaphysical chef-d'oeuvre was of course *Meditation on the First Philosophy, in which the Existence of God and the Immortality of the Souls Are Demonstrated*. Moreover, several readers, maybe most famously Henry More, initially thought that his philosophy did in fact provide the basis for a new theology.[24]

But while the theological tone of Descartes' *Meditations* may partially have been a ruse, the search for a solid foundation for Christian doctrine was central to Taurellus' philosophical engagement.[25] In keeping with his official appointment at Altdorf (1580–1606), he officially described himself as "professor of Aristotelian philosophy and of Galenic medicine," he really understood himself first and foremost as a "Christian philosopher."[26] As such, his greatest ambition remained, once the initial controversy over his daring *Philosophiae triumphus* had abated, to reform philosophy in such a way that it could provide a solid basis for theology. The full title of his 1596 *Synopsis metaphysices*, which translates as "A Survey of Aristotelian

Metaphysics Emended and Completed According to the Norms of the Christian Religion," captures the essence of this ambition quite succinctly.[27] This last-named work is, by the way, yet another ontological treatise that preceded Suárez' *Disputationes metaphysicae*, but whose fortune has remained unexplored—even though in 1734, one Jacob Wilhelm Feuerlin decided to attempt a rehabilitation of both the treatise and its author.[28]

Of all of Taurellus' works, it was the bold *Philosophiae triumphus* that was to have the most direct and traceable impact on the Dutch milieu in which Descartes was to live and work. It is significant that in 1617, only a few months before Descartes first visited the Netherlands, a second edition of the *Philosophiae triumphus* was produced for the Dutch market by an unlikely publisher and in an unequally unlikely place, Arnhem.[29] Why it was specifically the Dutch who should have been interested in additional copies of this book will become clear below. For the moment, it will suffice to know that this reprint was related to exactly the historical lineage that Voetius mentioned in his historical reconstruction of the pedigree of man as *ens per accidens*, that is to say, with the theologian Conrad Vorstius and his alleged debt to Taurellus.

But what was Taurellus' main argument in his voluminous treatise? Like Peter Ramus (Pierre de la Ramée), Taurellus disliked Aristotle's metaphysics, which he thought was a hodgepodge of theology, logic, ontology and wisdom.[30] Unlike Petrus Ramus, however, he did not want to replace metaphysics by a new science of dialectics, but instead to reform it—a wish that we have already encountered in a concise form in the title of his later *Synopsis metaphysices*. To this end, Taurellus devised a *metaphysica universalis*, which had as its goal the discovery, definition and demonstration of the qualities of being inasmuch as they were shared by all *entia*. Given that it dealt with "being as being" (*ens qua ens*), this universal metaphysics had to precede all other sciences, including theology.

The starting point of Taurellus' metaphysical reform is that faith—the right kind of faith, of course—requires knowledge. "Who does not know God, will not believe in Christ."[31] By knowledge, Taurellus did not mean the believer's acquaintance with the contents of Sacred Scripture, but a philosophical understanding of the Deity.[32] The aim of Taurellus' ambitious *Philosophiae triumphus* is therefore the establishment of the right kind of knowledge. Given that philosophical knowledge can never be at odds with theological truths, and that there cannot be any double truth, it follows that where philosophy has hitherto been found to clash with theology, it must be revised.[33] In this sense, Taurellus performs exactly the same "veritable hermeneutic circle from the *ratio metaphysica* to the *depositum fidei*" that Costantino Esposito discerned in Suárez' *Disputationes metaphysicae*, published twenty-five years later.[34] Therefore, for Taurellus, a correct metaphysical first philosophy helps us understand God, while a combination of reason and revelation at the same time helps us correct those errors that

Greek philosophers had committed, for example by proclaiming the eternity of the world or the mortality of the soul.

Importantly, on the premise that "it is with the same mind that we believe and understand," Taurellus starts his metaphysical treatise with a study of the mind![35] The first treatise of his book is in fact called "On the Powers of the Human Mind" (*De viribus humanae mentis*). Although his examination of the mind is a far cry from Descartes' psychological self-examination, it is certainly striking so see how Taurellus' *Triumphus*, like Descartes' *Meditations*, moves from a description of our mental powers and their ability to cognize the truth to the metaphysical principles of physics, and that he proceeds by way of an examination of the type of knowledge that God has granted us in this life.

The conceptual movements of the two authors and their revolutionary works are, however, quite different, as are their intellectual motivations. For Taurellus, the ultimate aim is to arrive at a correct knowledge of God. A sound theology requires the right kind of first philosophy, as "one may say of philosophy that it constitutes the foundation of theology."[36] It is for this reason that a fully developed science of the *ens qua ens*, which Taurellus variously calls *prima philosophia* or *metaphysica universalis*, can and must precede all philosophizing and theologizing. But can our mind arrive at an understanding of the *ens qua ens*? Yes, it can. Thanks to the fact that our reason (*ratio*) is still intact, despite Adam, Eve, the apple and its fatal consequences, our *facultas intelligendi* still does allow us to reach a correct understanding of things. In fact, whereas before the Fall, the contemplation of God was the foundation of felicity, felicity is now obtained through philosophical knowledge. The cognition that we can obtain of God's being is due to the adequacy of both our mind and our philosophical terminology. In fact, in a tradition that goes back to Duns Scotus, Taurellus predicates *ens* univocally of all being, including God.[37]

By contrast, Descartes' first philosophy is of course no ontology in the strict sense of the word, as Descartes does not try to establish a definition of *ens qua ens*. Despite the profound differences between their respective reforms of metaphysics, it is, as we have already mentioned, conspicuous that that both felt the need to preface their metaphysics by an examination of the human mind. As for Taurellus, he feels that he needs to prove that our mind still works, despite original sin and our obvious cognitive limitations, and that it can still cognize the truth adequately.[38] Although Taurellus' psychology and epistemology are very much unlike Descartes' in their explicitly theological worry about the consequences of original sin, they do resemble Descartes' inasmuch as Taurellus, too, stresses the sensory and mental inadequacy that we must overcome. He, too, feels that he must begin his philosophical project by removing skeptical doubts—although for him, these skeptical doubts are primarily of a theological, not of an epistemological, nature.[39]

Taurellus' eagerness to prove the powers of cognition of our soul has a conspicuous result. He ends up with a complete separation of the soul from the body! For him, the two are separate substances. In his eyes, the Aristotelian idea that the soul is the substantial form of the body is really quite mistaken: "Man is not one by himself (*unus per se*), constituted by two immutable forms. The composition of man is an accident, because its forms can subsist by themselves."[40]

This idea of an autonomy of both body and soul as two independent forms flies of course in the face of all scholastic schools of thought. But note that Taurellus does not speak of mind and body as two different substances, but as two different forms. As Andreas Blank has convincingly argued, Taurellus' ultimate and irreducible entities are forms.[41] We may safely conclude that Taurellus would have disliked the way in which Regius used his phrase *ens per accidens* to defend a Cartesian two-substance theory.

In contrast to Descartes' philosophy, moreover, Taurellus' entire ontology is composed of *entia*. Each *ens* is, unlike Aristotle's *ousia*, simply "that which is." For Taurellus, being and existence fall together. Whatever is an *ens*, actually exists and can have no potential being. For that reason, there cannot be any such thing as prime matter, which according to Aristotle is pure potentiality. To the extent that prime matter exists, it must be conceived either as the same as privation, or negation, or the 'nothing' of the theologians out of which God created the world.[42] But if the latter, it is a *prima substantia*. In this denial of potentiality and the image of the cosmological first substance, we may—rightly or wrongly—once again imagine a tenuous breeze of *Descartes avant Descartes*. Yet, in the next step he takes, Taurellus differs significantly from Descartes, for contrary to the latter, he is an ontological atomist, and worse yet, he is, following Andreas Blank, an immaterial atomist, in the sense that his ultimate components are forms. For him, *esse* necessarily means *esse unum*: whatever is, is one and only one.[43] A plurality of *entia* can never form a single new *ens*. Things can mix, of course, but this does not lead to a total merger, but to a *compositio*, just as the four elements don't perish in the *mixtio*, but only combine into a new form, which is, significantly enough, not a *substantial* form. Any such composition, whether a chemical mixture or the constellation of body and mind, is a mere *ens per accidens*, made up of *entia per se*.[44]

It is, then, from this combination of an ontological first philosophy and theology that the notorious definition of man as an *ens per accidens* follows once more. Rejecting the traditional meaning of this term, which designated something that is such-and-so thanks to the essence of something else, Taurellus reutilized the term *ens per accidens* to mean something that is made up of various individuals. For him, as later also for Gorlaeus and Regius, the human body and the human soul never merge into a single new entity, although they do of course interact, but they nevertheless remain distinct. Interestingly, whereas Voetius was to combat this view seventy years later also for the reason that it contradicted the resurrection of soul *and* body,

Taurellus is in fact eagerly combating those who believe that it is only the soul, but not the body, that will resurrect.[45]

Importantly, seventy years later, Regius was convinced that for Descartes, too, body and mind were separate things that did not merge. Whether he was right in this respect is of course a famous question. As is well known, in his Sixth Meditation and elsewhere, Descartes insists on the unity of body and mind. Significantly, however, both for his friend Regius and for such contemporary opponents as Voetius or Schoock (Schoockius), it seemed obvious that his two-substance dualism did not allow for such a unity.[46]

Regrettably, the development and later *fortuna* of Taurellus' work has never been investigated. Yet, it is clear that his philosophy was used by at least some Dutch professors of philosophy in the first years of the seventeenth century, alongside such other interesting German metaphysicians and ontologists as Otto Casmann or Rudolph Goclenius. We find Taurellus regularly quoted, for example, in disputations held at the University of Franeker, at which Descartes was briefly enrolled in 1629 and at which Gorlaeus took his first degree.[47]

Whatever interest there may have been in Taurellus' views, it grew considerably after 1610. We recall from above that according to Voetius' complex genealogical reconstruction of Regius' description of man as *ens per accidens*, the Heidelberg theologians had declared that Taurellus had negatively influenced the theological views of Conrad Vorstius.[48]

But who was Vorstius? In 1610, this German theologian, who had up to that year taught at the Academy of Bentheim alongside such luminaries as Otto Casmann and Clemens Timpler, was appointed as the successor of Jacob Arminius, the Leiden theologian who has given his name to the Calvinist movement that even today is know as "Arminianism." The theological positions on predestination and on the relation of state and church associated with Arminius and his followers had since about 1606 created ever-expanding public commotion, but the matter came to a first climax with Vorstius' appointment to Arminius' vacant Leiden chair in theology. While Arminius deviated only marginally from what with hindsight we tend to call "orthodox Calvinist" positions, Vorstius espoused theologically quite radical views and was on some issues close to the Socinians, an anti-Trinitarian movement that was maybe the most dreaded heterodoxy of the day. These theological battles, the details of which need not detain us here, led to an acute crisis within Dutch Protestantism, which would culminate in 1618/19, when the Synod of Dort (or Dordrecht), in the presence of representatives from France, Scotland, Germany, the Swiss Confederacy, Geneva, and elsewhere, condemned Arminian theology and banned Vorstius from Dutch soil.

Back in 1610, when the anti-Arminian theologians enquired with their Heidelberg colleagues about the credentials of Vorstius, of whom we have just heard that he had been chosen as Arminius' successor at Leiden, they received precisely the negative report they had hoped for. The letter from

Heidelberg not only listed the various aberrant views they had found in Vorstius' works, but it also mentioned Taurellus' damaging influence on Vorstius' mind—an allegation that, as we have seen above, Voetius was to repeat thirty years later, in 1641. Although the degree of this influence is not easy to verify, the alleged link between Taurellus and Vorstius led to a vivid interest, on the part of both friends and foes of Arminianism, in the contents of the allegedly influential *Philosophiae triumphus*. We know from the sources that in the years between Vorstius' appointment in 1610, and his condemnation in 1619, Taurellus' book was much read and discussed in the Netherlands. This interest explains of course the above-mentioned, otherwise inexplicable second, Arnhem edition of the *Triumphus* of 1617. Whether Descartes came across this book is not known; but while staying in the Netherlands as the dramatic Synod of Dordrecht was being convoked, it would have been impossible for him not to hear about the imminent confessional schism.

Gisbertus Voetius, an ardent anti-Arminian even as a young man, had not only been directly involved in the Synod of Dordrecht, but continued throughout his theological career to combat Arminian positions, which for him included Taurellus' philosophy, Gorlaeus' ontology and Conrad Vorstius' deviant theology.[49] This explains why, from his point of view, the combination of Descartes' philosophy, Regius' physiology, the latter's use of Gorlaeus, and the revival of Taurellus' tenet before an Utrecht audience looked like an Arminian resurgence. Apart from undermining the notion of substantial forms (to which, as Han van Ruler has shown in his exquisitely documented study, Voetius was for several reasons wedded), Descartes' and Regius' mind-body dualism also smacked generally of the heresy of the Arminians (or Remonstrants, as they were also called).[50] This association explains Voetius' violent reaction to Regius' philosophical positions in 1641. For Voetius, central theological positions were at stake. As Theo Verbeek explains:

> [i]t was the unity of faith, and hence the moral unity of the Academy, indeed of the entire country, which was called into question: by denying, even indirectly, the dogma of the resurrection of bodies, Regius showed his sympathies with the Remonstrants—or worse![51]

It has been suggested that Voetius' suspicion was well founded. According to Paul Mengal, who investigated the psychological authors, not the metaphysicians, Descartes was heavily indebted to the philosophical views of the Remonstrants (or Arminians).[52] But alas, as is so often the case with Descartes, one senses an intellectual nexus, but lacks conclusive proof for it. Descartes is notoriously silent about the authors he read, and the discussions to which he reacted.

Gorlaeus

But let us turn to the subsequent figure in the ancestral tree. It was in those agitated years around 1610, when the Arminian struggle first erupted, when

Vorstius was appointed to the Leiden chair and when Taurellus was indicated by the Heidelberg theologians as the evil genius behind the metaphysics contained in Vorstius' work *De Deo*, that a young student called David Gorlaeus (or, in Dutch, van Goorle) wrote two books.[53] These were published only posthumously, because the author died in 1612, aged 21, probably from malaria, after having recently enrolled as a theology student at Leiden University.

The longer and more detailed of Gorlaeus' treatises, the anti-Aristotelian *Exercitationes philosophicae*, was published in 1620. Although only a very limited number of former fellow students from Franeker and Leiden would have recognized the name of the author, the *Exercitationes* established Gorlaeus quickly in the canon of the so-called *novatores*. In his *Questions on Genesis* (1623), for example, Marin Mersenne named Gorlaeus alongside Campanella, Bruno, Telesio, Kepler, Galileo, Gilbert, Bacon, Fludd and Hill as a heretical author. In his subsequent *L'impiété des déistes* (1624), Mersenne announces an encyclopedia in which he was to refute various lies, notably those of "Gorlaeus, Carpenter, Basson, Hill, Campanella, Bruno, Vanini and others."[54] In the same city of Paris, and in the same period, Gabriel Naudé took a more favorable view on Gorlaeus. In his *Advising on Establishing a Library* (1627), he proposed a chronological set-up for an ideal philosophical library:

> In Philosophy, one has to start with the philosophy of Hermes Trismegistus, which is the oldest, continue with Plato's, Aristotle's, Ramon Lull's and Ramus', and finish with the *novatores* Telesio, Patrizi, Campanella, Bacon, Gilbert, Bruno, Gassendi, Basson, Gomez, Carpenter, and Gorlaeus, who are the most important authors in a flood of others.[55]

In Jean Bachout's 1651 introduction to his own translation of Jean d'Espagnet's *Enchyridion physicae restitutae* (originally of 1623), we find an enriched pedigree of *novatores* that is conspicuous for several reasons. Bachoud praises the *novatores* (or one should rather say, *renovatores*) for trying to recover the splendor of various ancient philosophical schools. The list begins in a standard way with the Italians, Telesio, Patrizi and Campanella, but then the epicenter moves north:

> Germany and England, too, have had several men who only followed Aristotle's opinions in those places where they found them at their most reasonable, as did Bacon, Fludd, Gorlaeus, Taurellus, Carpenter and others, of whom some have proposed new principles.[56]

As for France—Bachoud continues his reconstruction—Ramus had started the post-Aristotelian return to a genuine philosophy. He had been followed by d'Espagnet (whose French edition contains Bachoud's preface from which we are citing), who in turn had been followed by Descartes.[57]

What is conspicuous about this list is, first, its fairly exceptional coupling of Gorlaeus with Taurellus, and second, the fact that it ends with Descartes.

Incidentally, as one moves to later lists of *novatores*, one finds that to the extent that Descartes came to be widely recognized as the best and most successful *novator*, the earlier ones began to sink into oblivion. With respect to Gorlaeus, Charles Sorel provides us with a good example of this gradual process. While Sorel's prolix *Science des choses corporelles* (1634) had adopted Gorlaeus' atomistic two-element theory (the theory according to which earth and water were the only genuine elements, as they alone could mix), his attitude toward this Dutch author became dismissive after Descartes had appeared on the philosophical stage.[58] "Among the *novatores*," Sorel said of Descartes in his *De la perfection de l'homme* (1655), "one finds none that has removed himself father from common thought." And even though he found that many of Descartes' hypotheses looked fanciful and ill-founded, Sorel drew attention to their persuasive nature. Descartes' was a philosophy that came with "paintings" and "pictures . . . full of these little bodies, which are so little known, but which are represented there with such assurance as if he [*sc*. Descartes] had seen them clearly."[59] As for Gorlaeus, Sorel treats him in a short and critical section, in which he mentions not only the latter's two-element theory, but also the order in which his philosophy is presented:

> Among the moderns who deserved to be mentioned here, there is a certain David Gorlaeus, a Dutchman, who has written a book called *Exercitationes philosophicae*, where he engages in a fight against the entire theoretical philosophy of the Peripatetics. After dealing with metaphysics, he turns to physics [. . .][60]

These few references to a number of French responses to Gorlaeus, between Mersenne's of 1623 and Sorel's of 1655, should suffice to show that Gorlaeus was an author that in the few decades in which Descartes was active, enjoyed a certain standing in the canon of the *novatores*. Just as in the case of Taurellus, it cannot be established whether Descartes himself had ever held a copy of the *Exercitationes philosophicae* in his hands. In a list of *novatores* mentioned in a 1630 letter to Beeckman, Descartes referred to Telesio, Campanella, Bruno, Basson and Vanini, but not to Gorlaeus.[61] What can be established with certainty is, however, that in the Dutch and French circles in which Descartes moved, Gorlaeus was a well-known author.

But then, Descartes was not the only one to remain silent about the sources of his ideas. The same is true of Gorlaeus, who nowhere mentions Taurellus. In his two extant books, he in fact cites no living author (excluding an indirect reference to Galileo), and the only recent author he names (and quite obsessively so) is Julius Caesar Scaliger. But we know that at Franeker, where Gorlaeus had studied philosophy, Taurellus' writings were regularly cited by his teacher Henricus de Veno.[62] Moreover, the debt of Gorlaeus' ontological conception of philosophy to Taurellus is evident. Let us see why.

Gorlaeus' treatise opens with the typical question of ordinary philosophy textbooks: "What is philosophy?" The surprising character of his answer can only be gauged in comparison to the type of answer that ordinary philosophy textbooks around 1600 would have given in response. They would have sounded like this:

> Philosophy is an acquired power of the intellect, made up of wisdom and prudence, which contemplates everything and governs human actions in a congruent manner, enabling man to obtain the highest good.[63]

Gorleaus' answer contrasts starkly with the above. He writes: "It is the naked knowledge of entities."[64] For Gorlaeus, all philosophy is ultimately ontology. As for Taurellus, such a knowledge of entities has for Gorlaeus soteriological purposes: it perfects our souls in this life and helps us gain access to the heavens in the hereafter.[65] The reason why our soul is in need of such help is that as a consequence of the Fall, as it is limited by grave deficiencies. Our intellect, for example, suffers from the problem of referentiality: there is no correspondence between our notions (*notiones*) and the real things (*res*). A "first or universal philosophy" (*prima aut universalis philosophia*) will provide us with the necessary "knowledge/science of being" (*scientia entis*) to re-establish this correspondence between notions and things. This ontological *scientia* contains three branches: *theosophia*, which deals with God and His attributes; *angelographia*, which deals with supernatural beings; and *physica*, which deals with natural beings.[66] As for physics (or natural philosophy), contemporaries would have defined it as "a speculative science which studies the world of changing material things—celestial and terrestrial, animate and inanimate—culminating in the study of man."[67] But Gorlaeus once more begs to differ. For him, *physica* is not about a world of change, but instead about constellations of unchanging entities.

The ontological science that must be developed, Gorlaeus insists, must neither be confused with Aristotelian metaphysics nor with a universal logic, because it precedes all other disciplines:

> But this general enquiry (*tractatus*) does not require a particular name, and it cannot be called 'metaphysics', as it must be placed even before metaphysics. We may in the meantime call it 'first or universal philosophy', because it is allotted the first place and treats of the most universal.[68]

This position is obviously even more radical than the position of an author like the above-mentioned Théophraste Bouju, for whom the general part of metaphysics was inserted between logic and physics. For Gorlaeus, no discipline precedes his "first or universal philosophy," defined ontologically.

Right at the beginning of Gorlaeus' treatise, we encounter thus the same reversal of physics and metaphysics that we have found in Taurellus and that would later be taken to be characteristic of Descartes. Like Pierre de la Ramée and Francis Bacon, Gorlaeus insists that metaphysics as developed by Aristotle is no philosophical discipline, because it is too heterogeneous and contains too many different logical, physical and theological ingredients.

In his *prima aut universalis philosophia*, Gorlaeus begins by drawing a distinction between *entia realia* and *entia rationalia*, where "real entities" are those whose being (*esse*) is rooted in their independent existence (*existentia*), while "rational entities" depend for their *esse* entirely on the human intellect. Second, he distinguishes, following Taurellus, the *ens per se* from the *ens per accidens* (or *ens per aggregationem*). The *ens per se* is a fundamental, self-supporting unit, while the *ens per accidens* is an entity whose independence is but apparent, but which in truth is an aggregate of several *entia per se*. There is a whole typology of such accidental beings: [i] a heap of stones, whose entities "touch each other only confusedly"; [ii] the world (*mundus*), where the entities touch each other "in some order"; [iii] and water drops, which join up to form larger entities like rivers or lakes. Finally [iv], there exist cases of *entia per accidens* "where one [entity] is intimately in the other, penetrates it and acts through it: as the soul does in the body."[69] So, although soul and body are not spatially separated, as stones would be in a heap, but instead penetrate each other and interact, they are yet separate entities. Here, then, we have reached the view that would later serve as the basis of Regius' thesis that man is merely an *ens per accidens*:

> We gladly concede that there are composites, but we do not recognize any one being that should be called the composite. Instead, there are many things. For there is indeed the composing parts that we call composites, inasmuch as they are the things composed. We hold that each part has its essence before composition and also retains it afterwards, nor is any being made that is numerically one, or one being made from these parts. Instead, they are united and mixed so that one continuous thing is made, which is one being by aggregation and not by essence. Thus in a human being there is a soul and also a body, and these two are united in such a way that the body is made the soul's residence, vehicle, and instrument through which the soul exercises its operations. But these two are not made into one being, called a human being. Instead, each retains its complete and perfect essence, by which it is what it is. Still, the human being is not the same as the soul, nor the same as the body: rather, it is the same as the soul and the body taken together and aggregated. If, however, the human being is to be considered not as a being by aggregation, but as a thing *per se*, then it will be the same as the soul existing in the body.[70]

But what, then, is an *ens*? In the same Scotist fashion that we have already encountered in Taurellus, Gorlaeus takes *ens* as a term that equally and univocally applies to God, to souls, to angels and to material particles. It must therefore possess the following properties: "unity, truth, goodness, existence, locality, durability."[71]

All of this implies that Gorlaeus' philosophy, like Taurellus', implies a form of logical atomism, since being, actual existence and numerical oneness all fall together in the definition of *ens per se*. However, while for Taurellus, the ultimate entities were forms, and therefore ultimately immaterial, Gorlaeus is an outspoken physical atomist. His ontology, as Taurellus', excludes such Aristotelian entities as potential matter or substantial forms; however, unlike Taurellus, but like *novatores* of the same period such as Sébastien Basson, Gorlaeus was drawn to a physical atomism. These ultimate, solid entities were his answer to the Aristotelian forms, which to him seemed to come out of nowhere and disappear into nowhere, 'informing' a Protean *materia prima* in one way now, and in another way later.[72] Finally, his theory is explicitly dualistic. Thinking backwards from Descartes, we might be tempted to perceive him as foreshadowing the latter's mind-body dualism; but looking at it from his own period, the model must have seemed Platonic. Robert Pasnau has recently observed:

> Very occasionally, one finds post-scholastic authors denying that the mind-body composite is a *per se* unity. The most striking example from our period is David Gorlaeus, whose *Exercitationes philosophicae* (*circa* 1611) offers up a shockingly explicit version of Platonic dualism.[73]

But whether we may draw a straight line of influence from Gorlaeus' dualism to Descartes', as Helen Hattab has recently suggested, remains to be seen.[74] Staying in the world of Protestant metaphysics and anthropology, there would in fact be other routes to be explored, including one that goes from Melanchthon through Otto Casmann to Rudolph Snellius, as Davide Cellamare has recently suggested.[75] To assess the plausibility of either route, one would need to engage in a much closer terminological examination of the various Protestant metaphysicians than has hitherto been done.

Anticipating or Influencing Descartes?

But alas, despite Voetius' allegations, Taurellus and Gorlaeus have so far played no role in scholarly analyses of Descartes' conception of metaphysics and his reversal of the order of the philosophical disciplines, nor of his methodology, which begins with an epistemological exercise and which, by way of a proof of God as a warrant of certain knowledge, leads up to his substance dualism from which in turn physics and the other philosophical disciplines sprout as a tree does from its roots. But is it correct to exclude them? Is it obvious that the doctrinal link that Voetius thought to perceive between Descartes, Regius, Gorlaeus and ultimately Taurellus is only a figment of the imagination of a neurotic anti-Arminian watchdog of orthodoxy?

To be sure, there are a number of reasons for thinking the latter. Descartes, who claimed that he had never heard of either Taurellus or Gorlaeus, disapproved of the way in which Regius handled the issue, and advised him against using the formula of man as *ens per accidens*, although, as was already mentioned, he also stated that he could see the advantage of using that term. It is also evident that Descartes' own "first philosophy" does not contain an ontology in the proper sense of the word: he does not try to define *ens*, nor would he have dreamed of locating it in individual, atomic units, material or otherwise. Even though his visually alluring corpuscular theory as presented in his *Météores* and his *Principia* was taken by various contemporaries to constitute an illustrated elaboration of Democritean atomism, his matter theory was of course the pure contrary: his *res extensa* resembles a liquid in which material particles are merely temporary phenomena brought about by the displacement of adjacent chunks of material space.

And yet, it is also evident that our two Protestant metaphysicians did play a role in the reception of Descartes by his friends and foes alike. We have seen that Regius, in a disputation that was dedicated to expounding what to us looks like an otherwise standard Cartesian theory of substances, invoked a piece of Gorlaeus' philosophy as if it were a natural part of it. Voetius' irate reaction testifies furthermore to the structural similarities that could be perceived to hold between Descartes, Regius and Gorlaeus as expressions of a philosophico-theological program that had been condemned at the Synod of Dort more than twenty years earlier and now seemed to attempt a comeback. For Voetius, these philosophers all erred, as had the theologian Conrad Vorstius, in taking reason to precede faith, or even more strongly, to serve as the touchstone of theological positions: "Human reason is not prior, better known or more certain that faith; therefore it cannot be its principle."[76]

Whether as a consequence of Voetius' reaction or independently of it, it is certainly also interesting to notice that outside the Dutch Republic, and from a certain temporal distance, several early historiographers of philosophy opined that Gorlaeus had in certain ways anticipated Descartes. In the eighteenth century, Daniel Georg Morhof, in his *Polyhistor*, praises Gorlaeus as someone who "recognized before Descartes, what later Descartes claimed to be his own doctrines."[77] And Jacob Friedrich Reimann, in his *Historia literaria*, concludes his description of the *Querelle d'Utrecht* with the words: "The Cartesians, who were to integrate most of Gorlaeus' hypotheses into their philosophical system, managed to respond to Voetius with respect to all these objections."[78]

Are these statement correct? And what should we conclude with regard to Descartes himself? When in his polemical exchange with Voetius, he denied ever having heard of Taurellus or Gorlaeus, no one believed him, because, as Martin Schoock wrote in his anti-Cartesian *Admiranda methodus*, the books of these two authors were at the time in everybody's hands.[79] Moreover, we know that Descartes followed the composition of Regius' disputations closely, suggesting various improvements. Even though he disliked the definition of man as an *ens per accidens*—because his own philosophy didn't proceed along a definition of *ens* and *entia*, and also, because he thought it unnecessary to unnerve his opponents unduly with a novel terminology—it is quite obvious that he was acquainted with Regius' views. Whether he had actually read Regius' own sources, and more specifically, whether he had actually read Taurellus or Gorlaeus may, in this instance, not even be decisive. After all, one can be acquainted with, or even influenced by, the ideas of an author without having read his works. (There used to be lots of Marxists who had never read Marx).

What we do know for certain is that Descartes' closest Dutch allies and interlocutors such as Henricus Reneri or Henricus Regius possessed and cited the works of our Protestant metaphysicians. When Reneri moved to Deventer to teach there, Descartes settled, between May 1632 and February 1634, in the same town so as to be close to his friend. When Reneri subsequently moved on to Utrecht, where he was appointed to the Illustrious School (which was quickly upgraded to the status of full university), Descartes once again moved along with him.[80] Now, Reneri is known to have worshipped Descartes. "He is my light, my sun, and what Vergil said in his *Bucolics*, I can say about him: 'He will be for me forever a god,'" he wrote to Mersenne; and in Theo Verbeek's words, had become "the main herald of Descartes' glory."[81] And yet, from his few extant publications, Reneri appears by no means as a Cartesian propagandist, but as a teacher who combined a variety of sources. For our current purposes, it is relevant to know that he owned a copy of Gorlaeus' *Exercitationes* and that in his university disputations, he defended views that look like a blend of Descartes, Gassendi, Gorlaeus and other ancient and modern authors.[82] To give just one example to buttress this claim: in his disputation *On the Elements*, which a student defended in 1635, Reneri rejects Aristotle's four-element theory and proposes instead a corpuscular conception of elementary particles. Gorlaeus' influence there is quite evident. Elements do not transmute, Reneri has his student say, but they have been generated directly and entire by God (theses 17–18). Each element has a specific figure (thesis 14), and sensory experiences are caused by the figures of the particles impinging on our organs. Although the best-known ancient source for the mapping of particular shapes and sensory experiences is Plato's *Timaeus*, Reneri rejects Plato's specific explanation of the burning (or lacerating) quality of pyramidal fire particles. And he does so because he follows Gorlaeus' view that neither fire nor air are elements:

> Water and earth, considered in their purity, are the only elements properly so called, because they are simple and because one has to recur to

> them to explain the generation of all mixts and the resolution of anybody whatsoever into them."[83]

A similar story can be told about Henricus Regius, the other early Dutch champion of Cartesian physics. Regius, who had almost exactly the same age as Descartes, had read up on a certain number of *novatores*, including Gorlaeus, long before he read anything written by Descartes, with whom he only entered in contact in 1638. Although it is obvious that he was much impressed by Descartes' corpuscular physics, we find in his correspondence with Descartes various ideas that are closer to Gorlaeus than to Descartes.[84] The same holds true for that notorious series of disputations of 1641 that triggered the *Querelle d'Utrecht* and the subsequent ban on Descartes' philosophy at Utrecht. Take, for example, the following physiological thesis, in which Regius refutes the traditional explanation of the temperament as a mixture of elements into a new homogeneous substances and proposes an alternative model:

> Therefore we define a good temperament as follows: [it is] the location, figure, quantity and motion or quiet of the insensible parts which constitute the sensible parts in such a way that it fits the actions that have to be performed. From this temperament, or from the first qualities from which it is made up, all other qualities of the human body as well as all other homogeneous or heterogeneous bodies derive their origin.[85]

The insistence, in the first sentence, on the "figure, quantity and motion or quiet of the insensible parts" sounds Cartesian; the explanation of the emergent properties of the "temperament" and its "first qualities" is however much closer to Gorlaeus. Similarly, when Regius explains heat as "the various agitation of insensible particles" and cold as "their quiet," we may point again to Gorlaeus, who had explained that heat was "produced by the motion and friction of bigger particles."[86] But whereas Gorlaeus had added, "how this [transformation of motion into heat] happens, escapes me, and I marvel at it, just as I marvel at many other things," Descartes, in his *Principia* of 1644, was to include a woodcut which illustrated how corpuscles could be brought into agitation by the pressure of sun rays.[87] Regius, caught half-way between the two authors, did not provide a mechanical explanation of heat in his 1641 disputation, provoking Schoock's laughter over the vacuity of his claims.[88] Indeed, when one studies the way Regius argued in those years, one cannot but wonder how many of his corpuscular explanations were of Cartesian extraction, and how many were instead indebted to earlier, atomist authors, among whom we must obviously count Gorlaeus.

In sum, then, that Gorlaeus was in the hands of Descartes' intellectual allies and foes is evident; that they approved—or disapproved—of Gorlaeus and Descartes at the same time has just been documented; and that Regius found it logical to link the metaphysical principles of the two authors led, as we have now repeatedly heard, to the well-known condemnation. Leaving all the undeniable differences between the authors intact, we may, I think, nevertheless conclude that what Taurellus, Gorlaeus and Descartes shared was the conviction that any natural philosophy had to be erected on a metaphysical base, which in turn had to contain a theory of the mind and of cognition—in brief, on an epistemology. Equally importantly, but more specifically, all three of them viewed the body and the soul as things that could interact without losing their identity, and they all rejected substantial forms. In the way that they defined body and soul or mind, and their interaction, the three differed very strongly; on that score, all parallels end.

All in all, it might well be that to the recognized Dutch influences on Descartes' thought—Isaac Beeckman, the Franeker and Leiden mathematicians as well as optical experts—we should now wish to add Gorlaeus, whose elaborations on Taurellus' ontological first philosophy shares a number of features with Descartes—features that at least in Descartes' own time were taken, by friend and foe alike, to be anything but coincidental. While there is no proof of a direct debt, there is at least solid evidence to show that Descartes' Dutch friends did not need Suárez to agree with Descartes over the primacy of metaphysics in the order of learning.

Acknowledgment

I would like to thank Sophie Roux and Gary Hatfield for precious critical comments on the first version of this paper.

Notes

1 Daniel Garber, *Descartes' Metaphysical Physics* (Chicago: Chicago University Press, 1992), 61. Cf. similarly Gary Hatfield, "First Philosophy and Natural Philosophy in Descartes," in A.J. Holland, ed., *Philosophy, Its History of Historiography,* 149–164 (Dordrecht and Boston: D. Reidel, 1985).

2 Notably in Aristotle, *Metaphysics*, book XI, part I, *passim.*

3 This standard account has been questioned, however, notably by Immanuel Kant, who argued that the word "metaphysics" fits the theory too well, and that "meta" might refer to "beyond," in the sense of referring to the first mover. See Hans Reiner, "Die Entstehung und ursprüngliche Bedeutung des Namens Metaphysik," *Zeitschrift für philosophische Forschung* 8 (1954), 210–237; Takatura Ando, *Metaphysics: A Critical Survey of its Meaning*, 2nd enlarged ed. (The Hague: Nijhoff, 1974).

4 Hatfield, "First Philosophy and Natural Philosophy," 151. Gary Hatfield, "Metaphysics and the New Science," in David C. Linberg and Robert S. Westman, eds., *Reappraisals of the Scientific Revolution* (Cambridge: Cambridge University Press, 1990), 93–166, on 97n9, refers to Aristotle's *Physics* I.1–2 and *Metaphysics* III.2 as places where Aristotle appears to deny the ability of metaphysics to provide the principles of physics.

5 Théophraste Bouju, *Corps de toute la philosophie* (Paris: Charles Chastellain, 1614), "Sommaire de ce qui est contenu en ce corps de toute la philosophie." Cited in, and translated by, Roger Ariew, *Descartes and the First Cartesians* (Oxford: Oxford University Press, 2014), 97.

6 See, for example, Descartes' letter to Clerselier, of 12 January 1646, AT IX-I, 293–217, at 205: "l'erreur qui est ici la plus considérable est que cet auteur [Gassendi] suppose que la connaissance des propositions particulières doit toujours pouvoir être déduite des universelles, suivant l'ordre des syllogismes de la dialectique; en quoi il montre savoir bien peu de quelle façon la vérité se doit chercher; car il est certain que pour la trouver on doit toujours commencer par les notions particulières, pour venir après aux générales, bien qu'on puisse aussi, réciproquement, ayant trouvé les générales, en déduire d'autres particulières."

7 See, for example, Stephen Gaukroger, *Descartes. An Intellectual Biography* (Oxford: Oxford University Press, 1995), 204–5.

8 See, e.g., Marko J. Fuchs, "Univozität und Distinktion. Metaphysische Grundstrukturen bei Duns Scotus, Suárez, Descartes und Spinoza," in Lukás Novák, ed., *Suárez's Metaphysics in Its Historical And Systematic Context* (Berlin, Walter de Gruyter, 2014), 105–116, esp. 109–111; or Asa Goudriaan, *Philosophische Gotteserkenntis bei Suárez und Descartes im Zusammenhang mit der niederländischen reformierten Theologie und Philosophie des 17. Jahrhunderts* (Leiden: Brill, 1999), 'Einleitung' and the literature quoted in footnotes 6 and 7.

9 But see the passing remark by Édouard Mehl, *Descartes en Allemagne, 1619–1620* (Strasbourg: Presses Universitaires de Strasbourg, 2001), 141: "Il n'est pas certain que Descartes soit clairement conscient de répéter exactement le tournant qui a été opéré par le trio Keckermann-Trimpler-Alsted, mais c'est pourtant avec eux et avec Descartes que la philosophie première s'arrache à la 'pragmatie' aristotélicienne et se constitue en une discipline autonome."

10 See, above all: Theo Verbeek, *René Descartes et Martin Schoock: La Querelle d'Utrecht* (Paris: Les Impressions Nouvelles, 1988); idem, "*Ens per accidens*. Le origini della *Querelle* di Utrecht," *Giornale critico della filosofia italiana*, 6th series, 71 (1992), 276–88.

11 Henricus Regius, *Physiologia, sive Cognitio sanitatis, tribus disputationibus in Academia Ultrajectina publice proposita* (Utrecht: Aegidius Roman, 1641), 5, thesis 14 (transcribed in Erik-Jan Bos, *The Correspondence between Descartes and Henricus Regius* [Utrecht: Zeno, 2002], 202): "Mens, mensura, quies, motus, positura, figura, / Sunt cum materia cunctarum exordia rerum." On the fortuna of this didactic verse, see Erik-Jan Bos, "Een kleine geschiedenis van een cartesiaans versje," in Marco van Egmond, Bart Jaski & Hans Mulder, eds., *Bijzonder onderzoek: Een ontdekkingsreis door de Bijzondere Collecties van de Universiteitsbibliotheek Utrecht* (Utrecht: Universiteitsbibliotheek Utrecht/ Waanders, 2009), 244–51.

12 Henricus Regius, disputation of 8 December 1641: "VIII. Forma specialis est mens humana, quia per eam cum forma generali in materia corporea homo est id quod est. Haec ad formam generalem seu materialem nullo modo potest referri: quoniam ipsa (utpote substantia incorporea) nec est corpus, nec ex motu aut quiete, magnitudine, situ aut figura partium oriri potest. IX. Ex hac et corpore non fit unum per se, sed per accidens, cum singula sint substantiae perfectae seu completae. X. Cum autem dicuntur incompletae, hoc intelligendum est ratione compositi, quod ex harum unione oritur." Reprinted in: *Testimonium Academiae Ultrajectinae, et narratio historica quae defensae, qua exterminatae novae philosophiae* (Utrecht: W. Strickius, 1643).

13 See Verbeek, "*Ens per accidens*."

14 Cf. Pierre Bayle, *Dictionnaire historique et critique*, 5th ed. (Amsterdam, Leiden, The Hague, Utrecht: P. Brunel et al, 1740; repr. Geneva: Slatkine Reprints, 1995), vol. 7, s.v. Gorlaeus: "Régius, disciple de M. Descartes, se voyant harcelé pour une thèse qui concernait l'union de l'âme et du corps, allégua qu'il s'était

servi de propres termes de Gorlæus. Cela ne lui servit de rien, et fut cause que Voëtius, professeur en théologie, flétrit autant qu'il lui fut possible les sentiments de Gorlæus." See also Verbeek, "*Ens per accidens*" and idem, *Querelle*.

15 "Assertio παραδοξόλογος Taurelli (quem Atheum Medicum vocabant Theologi Heidelbergenses, in judicio suo de Vorstii tractatu, de Deo, perscripto ad deputatos Synodi Hollandicae anno 1610), in Triumpho Philosophiae, in praemissis axiom. D.4 & D.5, quam imprudentia juvenili ex illo adoptare voluit popularis noster David Gorlaeus . . . incurrit in veritatem non tantum Physicam . . . sed et Metaphysicam, pneumatologicam, & theologicam." This is the extended version of the *Corollaria* of the disputation defended on 18 December 1641 under Voetius, as reprinted in the *Testimonium Academiae Ultrajectinae*, 28.

16 The evidence given for the link from Taurellus to Gorlaeus and from Gorlaeus to Regius is provided in greater detail in my own *David Gorlaeus (1591–1612). An Enigmatic Figure in the History of Philosophy and Science* (Amsterdam: Amsterdam University Press, 2012).

17 Letter by Descartes to Regius, January 1642, AT III: 491–510; the explanations concerning the thesis of man as an *ens per accidens* are contained in Descartes' observations *Ad septimam*, 507–8.

18 Ulrich Leinsle, *Das Ding und die Methode: Methodische Konstitution und Gegenstand der frühen protestantischen Metaphysik*, 2 vols. (Augsburg: Maro, 1985), 1: 147.

19 See Nicolaus Taurellus, *Philosophiae Triumphus, hoc est, Metaphysica Philosophandi Methodus, qua divinitus inditis menti notitijs, humanæ rationes eò deducuntur, ut firmissimis indè constructis demonstrationibus, apertè rei veritas elucescat, & quæ diu Philosophorum sepulta fuit Philosophia victrix erumpat: quaestionibvs enim vel sexcentis, ea quibus cum revelata nobis veritate Philosophia pugnare videnbatur, adeo vere conciliantur, ut non fidei solum servire dicenda sit, sed eius esse fundamentum* (Basel: Henricpetri, 1573; 2nd ed. Arnhem: Jansonius, 1617), Introduction to theses, a1r: ". . . doctissimus aetatis nostrae Philosophus Iacobus Scheckius praeceptor noster charissimus, qui nobis in suis praelectionibus aliquoties inculcaverat"; idem, *Alpes caesae*, 2r: "Scheckianae philosophiae perpetuo fui studiosissimus."

20 Id., Introduction to "Theses quibus libri summam lectori paucis exponibus" [s.p.]: "Propositiones aliquot hac de re scriptas, ante quinquennium edidimus. . . ." I have recently located a copy of that set of theses, which I plan to publish in due time.

21 On Taurellus' life, see Johann Jacob Baier, *Biographiae professorum medicinae qui in Academia Altdorfina unquam vixerunt* (Nuremberg: Johann Daniel Tauber, 1728); Jacob Wilhelm Feuerlin, *Taurellus defensus. Dissertatio apologetica pro Nic. Taurello philosopho Altdorfino atheismi et deismi iniuste accusato, et ipsius Taurelli Synopsis Aristotelis Metaphysices ob raritatem recusa cum annotationibus editoris* (Nuremberg: Schmidius, 1734); Johann Heinrich Zedler, *Grosses vollständiges Universal-Lexicon aller Wissenschaften und Künste* (Lipzig & Halle: Carl Günther Ludovici, 1732–1754; repr. Graz: Akademische Druck- und Verlagsanstalt, 1961–1964), vol. 42, col. 401–2; Xaver Schmid, *Nikolaus Taurellus, der erste deutsche Philosoph* (Erlangen: Theodor Blaesing, 1860); Hans-Christian Mayer, *Nikolaus Taurellus, der erste Philosoph im Luthertum. Ein Beitrag zum Problem Vernunft und Offenbarung* (PhD thesis in theology, University of Göttingen, 1959).

22 For the full title, see above, footnote 19.

23 Descartes, "Lettre à Dinet," AT VII: 563–603, at 583.

24 See Alan Gabbey, "Philosophia Cartesiana Triumphata: Henry More (1646–1671)," in *Problems of Cartesianism*, edited by Thomas M. Lennon, John M.

Nichols, and John W. Davis (Kingston and Montreal: McGill-Queen's University Press, 1982), 171–250.

25 To my knowledge, John Henry has made to date the strongest argument to the effect that Descartes' theologically laden metaphysics was merely a tactical move to render his mechanical explanations acceptable; see John Henry, "Metaphysics and the Origins of Modern Science: Descartes and the Importance of the Laws of Nature," *Early Science and Medicine*, 9 (2004), 73–114.

26 See, e.g., Nicolaus Taurellus, *Emblemata physica ethica* (Nuremberg: Lochner, 1602), "Ad Lectorem," s.p., [i]: "Aristotelicae philosophiae et medicinae Galenicae professor." By contrast, he introduces himself as 'Philosophus Christianus' in the 'Praefatio' to his *Medicae praedictionis methodus, hoc est recta brevisque ratio coram aegris praeterita, praesentia futuraque praedicendi, morbos scilicet morborumque causas, mortem, sanitatem, recidivam aliqua symptomata. . .* (Frankfurt: Johann Feierabend, 1581).

27 Nicolaus Taurellus, *Synopsis Aristotelis metaphysices ad normam Christianae religionis explicatae, emendatae, et completae* (Hanau: Guilielmus Antonius, 1596).

28 Feuerlin, *Taurellus defensus*.

29 For the circumstances of this second edition, see Lüthy, *David Gorlaeus*, 101–2.

30 Peter Ramus' criticism of Aristotle's metaphysics is most forcefully expressed in his *Scholarum metaphysicarum libri quatuordecim, in totidem metaphysicos libros Aristotelis. Recens emendati per Joan. Piscatorem* (Frankfurt: Heirs of A. Wechel, 1583), 5: 'Aristotle has thus written fourteen other works, as I just said, about logic, although he gives them various titles, once "wisdom," once "philosophy," once "first philosophy," then again "theology," and then again "metaphysics," as they are commonly called in the schools, being "after the books on Physics." About theology, however, you will find only a few lines; the whole rest is about logic, either officially, as when he speaks of causes, opposites, comparison, and other such decrees of logic; or under the name of mathematical ideas, in which logical matters are being disputed in terms of causes, genus and species' (translation mine).

31 Taurellus, *Philosophiae triumphus*, 87: "Christo non credit, qui Deum nescit."

32 Ibid., "Introduction," 42th thesis: "Porro cum mens voluntatis sit subiectum: Ut ante lapsum perfecta Dei contemplatio foelicitatis erat, sic nostram nunc Philosophiam, fidei, laudisque Dei qualiscunque fundamentum esse statuimus." That theologians rejected his book precisely for this reason is mentioned by Taurellus himself in the preface to his *Alpes caesae* [s.p.]: "Prodiit ergo in multorum conspectum Philosophiae Triumphus, qui plerosque omnes offendit graviter. Et Theologos quidem propterea, quod Philosophiae tribueremus ea, quae solius viderentur esse Theologiae: . . . ."

33 Theology and philosophy must agree: "Haec ipsa Deo volente demonstrabimus, ut facile quivis ea quibus fides fundatur, ratione possit assequi. Hoc ergo posito, falsum esse quicquid Theologiae repugnant, Philosophorum errores examinabimus, ut rationum vitia depraehendantur, post demonstrationis extruemus, ut veritatem Philosophiae restituamus, sicque dignitatem eius tuebitur Philosophice" (*Ibid.*, "Introduction," 47th thesis). This agreement must be brought about by redefining the principles of philosophy: "At quoniam dato uno absurdo consequuntur infinita, principia primum examinanda sunt, ut expurgata scaturigine, pura tandem veritas" (*Ibid.*, 48th thesis).

34 Costantino Esposito, "Suárez and the Baroque Matrix of Modern Thought," in Victor M. Salas and Robert L. Fastiggi, eds., *A Companion to Francisco Suárez* (Leiden: Brill, 2015), 124–47, 128.

35 *Ibid.*, 43th thesis: 'Eadem enim mente credimus et intelligimus.'
36 Taurellus, *Triumphus*, "Conclusio," 374: ". . . de Philosophia dici possit, ipsam fundamentum esse Theologiae."
37 For Scotus' position, see Tobias Hoffmann, ed., *Johannes Duns Scotus: Die Univozität des Seienden. Texte zur Metaphysik* (Göttingen: Vandenoek & Ruprecht, 2002).
38 Taurellus, *Triumphus*, 6, [Title of *Quaestio*]: "Eandem per se humanam mentem esse, quae ante lapsus fuerit."
39 The theses prefixed to the *Triumphus*, which run under the title *De philosphia seu viribus humanae mentis tractatus*, have in fact the aim of deciding what the post-Adamitic man can still claim to understand despite of his impairment.
40 Glossary, s.v. "Homo, Corpus et Anima": "Homo non est unum per se, quod duabus immutatis constituatur formis. Accidens est hominis compositio, quod eius formae per se subsistere possint."
41 Andreas Blank, "Nicolaus Taurellus on Forms and Elements," *Science in Context*, 27 (2014), 659–82.
42 *Ibid.*, 171 ("Materia prima nihil est"), 302 ("Materia prima Physicorum NIHILUM est Theologorum").
43 *Ibid.*, Glossary, s.v. "Singularia, et Universalia": "Quicquid existit est singulare. In intellectu solum sunt universalia. Singularia sunt universalium causae, non e contrario."
44 *Ibid.*, e.g., 122: "Rem enim quae facta est per se non causarum respectu consyderamus, cum per se multa sint simplicia quae per accidens composita sunt." Taurellus' matter theory is not fully worked out. For diverging attempts to come to grips with it, see Christoph Lüthy, "Entia & Sphaerae. Due aspetti dell'atomismo bruniano," in *La filosofia di Giordano Bruno. Problemi ermeneutici e storiografici*, ed. Eugenio Canone (Florence: Leo S. Olschki, 2003), 165–98, part I; and Blank, "Nicolaus Taurellus."
45 Taurellus, *Triumphus*, thesis 159: "Animas equidem Philosophi nunquam morituras esse iudicarunt, sed non satis probe rem sunt assequuti. . . . Postquam hic ergo mundus suum fuerit finem consequutus, corpora Deum credimus esse suscitaturum, aliumque mundum hoc sublato constructurum, qui sufficiat omnibus et integris hominibus excipiendis."
46 See, e.g., Schoock, *Admiranda methodus*.
47 See Christoph Lüthy and Leen Spruit, "The Doctrine, Life, and Roman Trial of the Frisian Philosopher Henricus de Veno (1574–1613)," *Renaissance Quarterly*, 56 (2003), 1112–51.
48 Incidentally, the Heidelberg theologians also accused him of defending some Scotist theses; what they referred to was most likely the univocal way of applying the term ens to God and to His creatures alike.
49 On Voetius' repeated rebuttals of Gorlaeus' atomism, see J.A. van Ruler, *The Crisis of Causality. Voetius and Descartes on God, Nature and Change* (Leiden: E.J. Brill, 1995), esp. 298–300.
50 Van Ruler, *Crisis of Causality*, esp. 188–90.
51 Verbeek, *Querelle*, 44.
52 Paul Mengal, *La Naissance de la psychologie* (Paris: Editions L'Harmattan, 2005), 355: "Entre 1618 et 1650, il n'était pas très prudent, au Pays-Bas, d'affirmer ses sympathies pour le courant arminien. C'est donc en vain que l'on chercherait dans l'œuvre de Desscartes les noms de Snellius ou Bertius. Pourtant la psychologie cartésienne porte l'empreinte de ces auteurs que Descartes a très vraisemblablement lus."
53 David Gorlaeus, *Exercitationes philosophicae post mortem auctoris editae, quibus universa fere discutitur philosophia theoretica, et plurima ac praecipua*

Peripateticorum dogmata evertuntur (Leiden: Widow of Johannes Commelin, 1620). Idem, *Idea physicae, cui adjuncta est Epistola cuiusdam Anonymi [= Georg Joachim Rheticus] de terrae motu* (Utrecht: Johannes a Waesberge, 1651).

54 Marin Mersenne, *Quaestiones celeberrimae in Genesim, cum accurata textus explicatione. . .* (Paris: Sebastien Cramoisy, 1623), col. 1838; idem, *L'impiété des déistes, athées et libertins de ce temps. . .* (Paris: P. Bilaine, 1624), I: 237–38. Translations from Roger Ariew, *Descartes and the Last Scholastics* (Ithaca: University of Cornell Press, 1999), 126.

55 Gabriel Naudé, *Advis pour dresser une bibliothèque* (Paris: François Targa, 1627), 135: "En Philosophie, commencer par celle de Trismegiste qui est la plus antienne, poursuivre par celle de Platon, d'Aristote, de Raymond Lulle, Ramus, & achever par les Novateurs Telesius, Patrice, Campanella, Verulam, Gilbert, Iordan Bruno, Gassend, Basson, Gomesius, Charpentier, Gorlee, qui sont les principaux d'entre une milliace d'autres."

56 Jean d'Espagnet, *La Philosophie naturelle restablie en sa pureté*, edited and translated by Jean Bachout (Paris: Edme Pepingué, 1651), Preface by Bachout: "Discours a la recommendation de la Philosophie ancienne restablie en sa pureté. Et sur le nom de son premier Autheur," s.p. [viii]: "L'Allemagne & l'Angleterre ont eu aussi plusieurs Autheurs qui n'onst suivy les opinions d'Aristote qu'aux endroits où il les ont treuvées les plus raisonnables, comme ont fait Bacon, Flud, Gorlaues, Taurellus, Carpentarius & autres, dont quelques-uns ont escrit sur de nouveaux principles."

57 The establishment of a pedigree of the novatores has recently attracted the attention it deserves, as this category predates such labels as 'Baconianism', 'Cartesianism' or 'mechanical philosophy'; see Sophie Roux, "An Empire Divded: French Natural Philosophy (1670–1690)," in Daniel Garber and Sophie Roux, *The Mechanization of Natural Philosophy* (Dordrecht: Springer, 2013), 55–95, at 58–61; and Daniel Garber, "Remarks on the Pre-history of the Mechanical Philosophy," *Ibid.*, 3–26, at 20–21.

58 Charles Sorel, *La Science des choses corporelles. Première partie de la science humaine, où l'on connoist la Verité de toutes les choses du Monde par les forces de la Raison et l'on trouve la refutation de Erreurs de la Philosophie vulgaire* (Paris: Pierre Billaine, 1634), 342–3.

59 Charles Sorel, *De la perfection de l'homme, où les vrays biens sont considérez, et spécialement ceux de l'âme, avec les méthodes des sciences* (Paris: Robert Le Nain, 1655), 255–6: "Entre tous les Novateurs on n'en voit point qui s'esloigne d'avantage des Pensées communes. Les Peintures de ses Tourbillons imaginaires. . ., ses Figures & quantité d'autres, sont pleines de ces petits Corps si peu connûs, qui y sont representez avec autant d'asseurance que s'il les avoit veûs clairement."

60 *Ibid.*, 248: "Entre les Modernes de qui l'on peut faire quelque cas, il y a un David Gorlaeus Hollandois, qui a fait un livre appellé, Exercitationes Philosophicae, où il entreprend de combatre toute la Philosophie Theoretique des Peripaticiens. Ayant parlé de la Metaphysique, il vient à la Physique; Il traicte de toutes les qualitez des Corps. . . ."

61 Descartes' letter to Beeckman, 17 October 1630, AT I: 159. See Roux, "Philosophy 1670–1690", in Garber-Roux, *Mechanization of Natural Philosophy*, 59–60.

62 Lüthy and Spruit, "The Doctrine, Life, and Roman"; also Lüthy, *David Gorlaeus*, 83–93.

63 This definition is from Daniel Sennert, *Epitome naturalis scientiae* (Wittenberg: Gronenbergius, 1599/1600), "Disputatio prima, de natura philosophiae," tenet XI: "Philosophia est habitus intellectus, Sapientia et prudentia constans, omnia scilicet contemplans, atque actiones humanas congruenter gubernans, ut hinc homo summum bonum adipiscatur." Sennert's definition of philosophy follows the general pattern found in philosophy textbooks; cf. Patricia Reif, 'Natural Philosophy in Some Early Seventeenth-Century Scholastic Textbooks'. PhD thesis, St. Louis University, 1962.

64 Gorlaeus, *Exercitationes philosophicae*, ex.I, sect. i, p. 4: "nuda entium cognitio."
65 *Ibid.*, ex. I, sect. i. p. 5: "Philosophia est doctrina de perfectione animae humanae in hac vita."
66 *Ibid.*, ex. I, sect. iii, pp. 12–16. Gorlaeus' (underdeveloped) division of philosophy appears to echo ideas expressed in Bartholomaeus Keckermann's *Scientiae metaphysicae compendiosum systema* (Hanau: Guilielmus Antonius, 1609) and Otto Casmann's *Angelographia, sive Commentationum disceptationumque physicarum prodromus problematicus* . . . (Frankfurt: Palthen, 1605).
67 Patricia Reiff, "The Textbook Tradition in Natural Philosophy 1600–1650," *Journal for the History of Ideas*, 30 (1969), 17–32, 20.
68 Gorlaeus, *Exercitationes*, ex. I, sect. iii, p. 13: "Ille autem generalis tractatus non vendicat peculiare sibi nomen, neque dici potest metaphysica, sed etiam ante metaphysicam instituendus est. Potest interim prima aut universalis appellari Philosophia, eo quod primum sortiatur locum aut universalissimum tractet."
69 *Ibid.*, II.i.25: Definition of ens per accidens sive per aggregationem: "Sed varium est pro horum entium entium unione. Nonnunqueam haec se invicem confuse contingunt; ut in acervo lapidum: aliquando ordine; ut in mundo: quandoquoque habent easdem qualitates, ut etiam ejusmodi unio sit inter illa; sicut inter hanc et illam aquae guttam: sunt quoque, ubi unum est in alio intime, illudque penetrat, et per illud agit; sicut anima in corpore. Quomodocumque plura entia inter se jungantur, totum illud semper erit ens per aggregationem. Neque enim unio mutat rerum essentiam, ut ex duabus rebus possit fieri una res numero."
70 *Ibid.*, XII.i.222–23. Translation taken with some modifications, from Robert Pasnau, *Metaphysical Themes, 1274–1671* (Oxford: Oxford University Press, 2011), 598.
71 *Ibid.*, II.iv.54: "unitas, veritas, bonitas, existentia, localitas, durabilitas."
72 See Gorlaeus, *Exercitationes*, ex. 14 ("De materia et forma"), and specifically his outburst against the forma mixti, p. 268–9. His physical atomism is explained in exercise XIII, "On atoms."
73 Pasnau, *Metaphysical Themes*, 597.
74 Helen Hattab, *Descartes on Forms and Qualities* (Cambridge: Cambridge University Press, 2009), 159: ". . . even though Descartes never publically affirmed Gorlaeus' view that the human mind and body form an accidental union, the substance/mode ontology Descartes adopts from the Meditationes onwards resembles Gorlaeus' metaphysics in key respects. In both cases, these new metaphysical foundations imply the elimination of substantial form from the metaphysical as well as the physical realm."
75 Davide Cellamare, "Psychology in the Age of Confessionalisation. A Case Study on the Interaction between Psychology and Theology c. 1517–c.1640," PhD thesis, Radboud University, Nijmegen, 2015.
76 See Gijsbert Voetius, *Selectarum disputationum theologicarum pars prima* (Utrecht: Johannes a Waesberge, 1648), 4: "Ratio humana non est prior, notior, certior fide; ergo non est ejus principium."
77 Daniel Georg Morhof, *Polyhistor literarius, philosophicus et practicus*, with additions by Johannes Frick and Johannes Moller, 2nd ed., 3 vols. (Lübeck: Böckmann, 1732), part 2, lib. ii. cap. i.3, p. 273: "Laudem certe meretur, quod ante Cartesium ista videret, quae postea Cartesius dogmata sua esse voluit."
78 Jacob Friedrich Reimmann, *Versuch einer Einleitung in die Historiam literariam insgemein und derer Teutschen insbesonderheit*, 3 vols. (Halle: Renger, 1710–1734), s.v. 'Gorlaeus', vol III, p. 563: Voetius' attacks against Gorlæus turned out to be invalid: "Denn alle diese Einwürfe haben die Cartesianer, welch die meisten Hypoteses Gorlæanas hernachmals in ihr Systema philosophicum aufgenommen, dem Voëtio beantwortet."
79 Schoock, *Admiranda methodus*, Preface, s.p. [p. xxxiv]: ". . . Gorlaeum et Taurellum . . . authores in vulgus notissimos. . . . An vero Voetius ignoret Gorlaeum ac Taurellum discat ex studiosis, qui authores eos, quotidie ab eo commodato

accipere solent, aut si iis fidem deroget in consilium adhibeat suum Medicum." This is a reply to Descartes' allegation that Voetius himself was not acquainted with these two authors.

80 See Robert Buning, "Henricus Reneri (1593–1639): Descartes' Quartermaster in Aristotelian Territory." PhD thesis, University of Utrecht, 2013.

81 Letter by Reneri to Mersenne, probably of March 1638: "Is est mea lux, meus sol, et quod Virgilius in Bucolicis dixit, idem possum de ipso dicere: Erit ille mihi semper Deus . . ."; quoted from *Correspondence du P. Marin Mersenne, religieux minime*, eds. Marie Tannery-Prisset, Cornelis de Waard, Jr., and René Pintard, 17 vols, (Paris: Beauchene / Presses Universitaires de France, 1932–1988). Vol. 7: 115. Theo Verbeek, *Descartes and the Dutch. Early Reactions to Cartesian Philosophy 1637–1650* (Carbondale/Edwardville: Southern Illinois University Press, 199), 96–7.

82 That Reneri owned a copy of Gorlaeus' *Exercitationes* can be concluded from *Catalogus librorum Reneri* (Utrecht: Aegidius Romanus, 1639), C3: "Gorlei Exercitationes philos." Interestingly, Gorlaeus is listed right after Gassendi's *Exercitationes* of 1624 and his attack on Fludd of 1630. I owe this precious reference to Robin Buning. On Reneri's use of Gorlaeus, see Robin Buning, "Henricus Reneri (1593–1639): Descartes' Quartermaster in Aristotelian Territory," PhD thesis, University of Utrecht, 2013.

83 Reneri, *De elementis*, 31: "Aqua & terra in sua puritate considerata, sunt elementa proprie dicta: quia & simplica sunt, & ad ea mixtorum omnium generationem concurrere arguit vel resolutio quorundum corporum in ea . . .". Reneri almost never cites the sources for the theses that he has his students defend. In this case, too, there is no reference to Gorlaeus. However, the combination of a two-element theory and a corpuscular view of matter was in those days generally ascribed to Gorlaeus. (I owe the reference to this disputation once more to Robert Buning).

84 See Geneviève Rodis-Lewis, "Problèmes discutés entre Descartes et Regius: L'âme et le corps," in *Descartes et Regius. Autour de l'explication de l'esprit humain*, ed. Theo Verbeek (Amsterdam: Rodopi, 1993), 35–46, esp. 36–38.

85 Henricus Regius, *Physiologia*, 5–6 (= *Disputationum medicarum primae, de sanitate, pars prior*, transcribed in Bos, *Correspondence between Descartes and Regius*, 202), theses 14–15: "Idcirco bona temperies a nobis definitur: situs, figura, quantitas, et motus vel quies particularum insensibilium partes sensibiles constituentium, actionibus perficiendis conveniens. A temperie, sive a primis qualitatibus ex quibus constat, omnes aliae corporis humani atque etiam reliquorum omnium tam homogeneorum, quam heterogeneorum corporum qualitates originem ducunt."

86 *Ibid.*, 6, thesis 17 (transcribed in Bos, *Correspondence between Descartes and Regius*, 202), "Calor actualis est varia agitatio insensibilium particularum: frigus autem est earum quies." Gorlaeus, *Exercitationes*, VII.iii.115: "Motu enim corporum crassorum producitur, et attritu nonnunquam."

87 René Descartes, *Principia philosophiae* (Amsterdam: Elzevier, 1644), part IV, §198. This image is discussed in Christoph Lüthy "Where Logical Necessity Turns into Visual Persuasion: Descartes' Clear and Distinct Illustrations," in *Transmitting Knowledge: Words, Images and Instruments in Early Modern Europe*, edited by Ian Maclean and Sachiko Kusukawa (Oxford: Oxford University Press, 2006), 97–133, at 119.

88 Schoock, *Admiranda methodus*.

5 Strange Bedfellows

Physics and Metaphysics in Descartes

Emanuela Scribano

In a famous passage in the Sixth Set of Replies, Descartes theorizes three grades of sensory knowledge.[1] The first grade is a pure material event that concerns the brain and the nervous system; Descartes specifies that this level of sensation is common to men and animals:

> When I see a stick . . ., [the] rays of light are reflected off the stick and set up certain movements in the optic nerve and, via the optic nerve, in the brain, as I have explained at some length in the Optics. This movement in the brain, which is common to us and the brutes, is the first grade of sensory response.

The second grade of sensation requires the presence of a mind and is thus restricted to men; it consists of the conscious perception of the material events that take place in the brain:

> This leads to the second grade, which extends to the mere perception of the color and light reflected from the stick; it arises from the fact that the mind is so intimately conjoined with the body that it is affected by the movements which occur in it.

The realm of sensation in the strict sense comprises nothing more than these two grades, since the third grade always implies the presence of judgment, and hence the activity of the intellect:

> But suppose that, as a result of being affected by this sensation of color, I judge that a stick, located outside me, is colored; and suppose that on the basis of the extension of the color and its boundaries together with its position in relation to the parts of the brain, I make a rational calculation about the size, shape and distance of the stick: although such reasoning is commonly assigned to the senses (which is why I have here referred it to the third grade of sensory response), it is clear that it depends solely on the intellect. I demonstrated in the Optics how size,

> distance and shape can be perceived by reasoning alone, which works out any one feature from the other features.

This passage in the Sixth Set of Replies poses several questions that are hard to answer.[2] Below I shall explain what, in my opinion, are the chief difficulties.

The first and most serious problem is that these lines appear to contradict an important part of the physiological research developed in *l'Homme*. In the Sixth Set of Replies, Descartes' intention was to reply to those who called into question his theory that certainty always comes from the intellect and not from the senses. The objectors had countered Descartes with a theory of empiricist stamp: in actual fact the certainty of the intellect depends on the senses, and an error induced by one sensation can be corrected only by another sensation. Descartes' response aims to turn the tables on empiricism, pursuing the approach he had explored in the *Meditationes* and later in the reply to Gassendi.[3] What are mistaken for mere sense perceptions are indeed judgments, and hence depend upon the intellect. So that when we believe we are correcting one sensation with another we are in actual fact correcting one judgment with another judgment.

The strong point of Descartes' reply is the dependence of sensation on judgment, so that in his defense strategy the most interesting grade of sensation is the third. Apropos this third grade of sensory response, he claims that it is through judgment that—starting from the two distinct perceptions of color and of a stick—we arrive at the perception of that stick as a colored object. The sense perceptions are punctiform and disorganic: only the judgment can bring the mosaic of perceptions together and build what we strictly call "experience." Indeed, in experience what we see is what we judge we see, because of the action of the intellect.

The interpretation of sensory experience in the third grade of perception is perfectly consistent with the outcomes of the *Meditationes*. In the Second Meditation, Descartes had sought to refute empiricism by maintaining that experience is the product of an active intervention by the mind. This theory was argued through the famous experiment of the wax and illustrated by the example of the hats and cloaks. The wax remains the same even when it changes from solid to liquid because the intellect judges it to be the same. When we look out the window, we say that what we see are men passing in the street, even though in fact all we really see are hats and cloaks, because the judgment bypasses the perceptive data and shows more than what the visual perception actually contains:

> We say that we see the wax itself, if it is there before us, not that we judge it to be there from its color or shape; and this might lead me to conclude without more ado that knowledge of the wax comes from what the eye sees, and not from the scrutiny of the mind alone. But then if I look out of the window and see men crossing the square, as

> I just happen to have done, I normally say that I see the men themselves, just as I say that I see the wax. Yet do I see any more than hats and coats which could conceal automatons? I judge that they are men. And so something which I thought I was seeing with my eyes is in fact grasped solely by the faculty of judgment which is in my mind.[4]

The Sixth Set of Replies simply restates what was already argued in the Second Meditation: the perception of the external world is a mental construct. However, this theory, employed by Descartes again to combat the empiricism of the objectors, appears difficult to reconcile with the physiological research developed in *l'Homme*. Here Descartes had proposed a completely different scenario regarding the construction of sensory experience. Faithful to his program of explaining even cognitive functions such as perception, imagination, and memory purely through neurocerebral events, Descartes had identified a way of integrating the perceptive data of sensation in the association of ideas made by the material memory.

The material memory—we read in *l'Homme*—consists in the disposal toward reopening of the cerebral traces. When two cerebral traces have been produced together several times, then the opening of one will cause the other to reopen too, and in this way past experience reintegrates present experience: "For example, if I see two eyes with a nose, I immediately imagine a forehead and a mouth and all the other parts of a face, because I am unaccustomed to seeing the former without the latter."[5] Here Descartes speaks of "imagining" and not exactly of "seeing" them, although just a few pages earlier in *l'Homme* he had shown how imagination and sensation can be indistinguishable,[6] and in the *Dioptrique*, too, imagination and sensation had been strictly united.[7]

In the physiological writings, memory and the association of cerebral traces play the role that in the *Meditationes* is entrusted to thought alone: it is the memory that fills the gaps of sensitive experience and makes it possible to see a face where, in fact, we see only two eyes and a nose.

There is an evident discord between the Sixth Set of Replies and the text of *l'Homme*. Consequently, it is scarcely surprising that the passage in question represents an obstacle for the interpreters who stressed the significance of the physiological research contained in *l'Homme*, in which Descartes had sought to reduce cognitive mental phenomena to purely physical phenomena.[8]

Another difficulty consists in the interrogative opened in the Sixth Set of Replies regarding the mechanisms presiding over animals' reactions to the external environment. Animals as such are never studied in *l'Homme*, but it is legitimate to derive from it that animals, since they have no soul, find their way through the world by filling the gaps of current experience with the cerebral traces of the past, that is with material memory. This renders plausible the theory that makes its first appearance in the *Discours*:

animals' reactions to the outer world do not allow them to be distinguished from machines. But if what Descartes states in the Sixth Set of Replies is true, namely that animals do not possess the faculty of judgment, then they would also be lacking the capacity to organize cerebral changes into an experience.[9] Moreover, Descartes claims that already in the *Dioptrique* it had been argued that the third grade of sensory response was the *only* way of perceiving distance: "I demonstrated in the Optics how size, distance and shape can be perceived by reasoning *alone*, which works out any one feature from the other features." If it is true that these characteristics of the objects are perceived through judgment alone, then this further exacerbates the difficulty stated above: animals would not perceive the distance of an object based on its size, or rather, they are indeed incapable of perceiving distance. And so how can their appropriate reactions to the external environment be explained?

In the face of these difficulties, the strategy of recent interpreters has split into two main approaches:

1. The first proposes an interpretation of the Sixth Set of Replies in the light of *l'Homme*. The judgment that Descartes speaks of regarding the third grade of sensory perception is "a psycho-physiological process"—to borrow the term used by Stephen Gaukroger to indicate the purely neurophysiological aspect of sensory perception. Consequently, this level of sensation is common to men and animals, perception in both cases being the result of a material interference made up of links between cerebral traces.[10] The problem of this interpretative strategy is that it unarguably clashes with the text of the Sixth Set of Replies. Descartes clearly states that the first grade of perception alone is common to men and animals, since the second grade implies the presence of a mind and is hence restricted to man. Finally, Descartes states explicitly that the third grade "depends solely on the intellect" and hence to an even greater degree cannot possibly be shared by animals.
2. The second interprets *l'Homme* in the light of the Sixth Set of Replies. One of the proponents of this approach is Thomas Vinci,[11] according to whom in *l'Homme* there is already a mental component implied in human perception and it is this that accounts for the unification of the multiple sensory perceptions into an experience. On the contrary, the world perceived by animals—and in general the world perceived through physical instruments alone—is "a fragmented world." The main problem with this approach is that it actually renders the program inherent to the research of *l'Homme* incomprehensible. Indeed the essence of this program was to describe the set of cognitive processes connected with the body without any mental intervention.[12]

What the two strategies do share is the "charitable principle": namely a prejudice in favor of the consistency of Descartes' writings. Nevertheless, it is clear that the application of this principle inevitably saves one or other of the texts—*l'Homme* or the Sixth Set of Replies—at the expense of the other. I myself intend to pursue a different path and call into question the shared assumption of the interpreters, namely the consistency of Descartes' writings.

In order to address this question it is as well first to clarify, at least briefly, the relationship between the two aspects of Cartesian philosophy that are at issue in these texts: metaphysics and physics, in other words the role of the mind in sensory knowledge on the one hand and physiology on the other. At the same time, we need to clarify that what we are dealing with here is not exactly the relation between metaphysics and physics (both singular) but rather between metaphysics (plural) and physics (singular).

We know that the elaboration of metaphysics was a long and difficult operation for Descartes, and that it went through at least three successive stages of development. In the years between 1628 and 1629, he was working on a "short treatise" on metaphysics,[13] addressing the two subjects which for him had always represented the quintessence of metaphysics: God and the soul. The short treatise has been lost, or was possibly destroyed: indeed, in a letter Descartes had declared that he was tempted to set fire to it as soon as it was completed.[14]

The famous letter to Mersenne of 15 April 1630, in which Descartes stated for the first time that the rules of logic were created freely by God, leads one to think that this theory was also included in the short treatise. However this theory—generally indicated as the "free creation of eternal truths"—does not appear in the *Monde*, where Descartes had told Mersenne he wished to include it, and is also absent from the first public exposition of the metaphysics, that is in the fourth section of the *Discours de la méthode*, and in the second and more elaborate exposition contained in the *Meditationes de philosophia prima*.

The first public presentation of metaphysics explicitly set itself a foundational task: to prove that the clear and distinct ideas of science are true. But Descartes was not satisfied with it. Above all, he was dissatisfied with the arguments supporting doubt.[15] In the *Discours*, Descartes had decided not to advance the argument of divine deception, since he considered it too daring for a work intended for wide distribution such as the *Discours*. However, the presentation of metaphysics in the *Meditationes*, which was conceived to resolve the defects of the *Discours*, reveals changes and novelties that go well beyond the theory of divine deception as the apex of doubt in the First Meditation. We owe to Geneviève Rodis-Lewis the entirely convincing theory that the text of the *Meditationes* presents two completely new additions to the substance of the pre-existing short treatise on metaphysics: the Fourth and the Sixth Meditations.

Particularly cogent in the reconstruction proposed by Rodis-Lewis is the theory that the Third Meditation was directly connected with the Fifth, whereas the Fourth Meditation is a novel element not only in comparison to the short treatise, but also in relation to the exposition of metaphysics that took up the legacy of the treatise, that is the fourth section of the *Discours de la méthode*.[16] Now, both the Fourth and Sixth Meditations contain a justification of God in the face of human error. This is therefore a strategy introduced into the *Meditationes* to bolster divine truthfulness and as a better guarantee of the science that Descartes founded on this very truthfulness of God from the *Discours* on. There are, however, other novelties in the *Meditationes*, including one that goes in the same direction as the Fourth and Sixth Meditations. We are speaking here of the role of thought and judgment in the construction of experience, which was the subject of attention in the Second Meditation. The analysis of the piece of wax and the example of the hats and cloaks are aimed at demonstrating that experience is always the result of an intellectual construction, and that what we think we see is, in reality, what we judge we see. If this strategy is successful, another obstacle to the foundation of science is removed. The Second Meditation would indeed have vanquished empiricism on its own ground, demonstrating that experience is not at the origin of intellectual knowledge but rather that the one is the condition for the other. In this way, the innatism underpinning the certainty of mathematics would be fully grounded. In a word, some of the most significant novelties introduced into the *Meditationes* as compared to the *Discours* relate to consolidating the metaphysical foundations of science.

As against this mobility in the metaphysical project, we have to observe that the elaboration of science in both physics and physiology does not undergo any significant changes. The *Dioptrique* offered a detailed exposé of the observations on vision already briefly anticipated in *l'Homme*, while the fifth part of the *Discours* offers a summary of *l'Homme*, expanding the part dealing with the circulation of the blood while reducing to the richness of analysis featured in the unpublished work to its bare bones. The metaphysics developed in the fourth section of the *Discours* posed no problem to the physiology expounded in the fifth section, or even to the physiology of *l'Homme*. The absence of the theory that perceptions are isolated data that can be unified only by the judgment, and also of the theory that experience is conditional on thought, ensured that the foundation of science as presented in the fourth section of the *Discours* featured no significant conflicts with the physiology of *l'Homme*—summarized in the fifth section of the *Discours*—or with the parts of the *Dioptrique* dealing with perception.

The same cannot be said of the metaphysics expounded in the *Meditationes*. Although the analysis of the piece of wax and the example of the cloaks and hats pulls the rug from beneath the feet of empiricism, it nevertheless moves in the opposite direction to that of the physiology. The foundation of science demands that only the active intervention of intellectual judgment can generate an authentic experience starting from the atomic perception transmitted by the action of the external senses. On the contrary, physiology had shown that the association of ideas was able to fill the gaps of sensory perception and thus build a real experience. If science is now served by a much more potent guarantee than the version of metaphysics elaborated in the *Discours*, the price to be paid for this result is very high.

If this is so, it is possible that Descartes set himself to make the two halves of his thought—science and its foundation—compatible, or at least to conceal their profound discord as much as possible. This task was facilitated by an important fact: *l'Homme* was still unpublished when Descartes wrote the *Meditationes*, and shortly afterwards he decided to rewrite the former work entirely in the light of revolutionary new discoveries.[17] Therefore, in order to attempt an operation of homogenization between science and its foundation, he would have had to bear in mind above all the *Dioptrique*, that is the work in which physiology had been used for the material explanation of cognitive phenomena such as sensation in general and vision in particular.

But let us return to the Sixth Set of Replies. Here, Descartes resumed the toppling of empiricism of the Second Meditation, hypothesizing a third level of experience made possible by the judgments accumulated in childhood which later became habitual and were hence easily mistaken for perceptions. In addition, he referred back to the *Dioptrique* to prove that he had always supported such a theory: "I demonstrated in the Optics how size, distance, and shape can be perceived by reasoning *alone*, which works out any one feature from the other features".[18]

In the *Dioptrique*, Descartes had identified four modes of perceiving distance. Of these only the fourth implied an inference—and hence a judgment—more specifically that of inferring the distance from the size and vice versa. The other three modes of perceiving distance, namely the modification of the eye, binocular convergence, and the greater or lesser clarity of the image, were cases of psychic registration of physical alterations.[19] Indeed, it was precisely these that represented the original contribution of Descartes, since the perception of distance via inference of the perception of size was the most traditional approach of the four and extensively present in previous studies of optics.[20]

Although it is true that in the *Dioptrique* Descartes had maintained that size, distance, and shape can be perceived *also* by reasoning, it is definitely not true that he maintained that size, distance, and shape can be perceived by reasoning *alone*, deriving any one from the others.[21] And thus we have to explain why Descartes inserted an obvious falsehood in the text of the Replies. My theory is that Descartes took advantage of the presence in the *Dioptrique* of a means of perception of distance that implies the presence of a judgment to demonstrate that he had always maintained what, in actual fact, he had sustained only in the *Meditationes*: namely that the sense perception *always* requires the presence of an intellectual judgment. Because only if sense perception always requires the presence of a judgment does physiology speak the same language as metaphysics. In this way the science of the *Dioptrique* is subjugated to the progress and the changes introduced to Descartes' metaphysical foundations in the interim, albeit at the price of an "author's fake."

Descartes' need to harmonize the new results affecting the foundation of science with science itself was not restricted to the Sixth Set of Replies. Another example of this strategy can be found in the different destiny of men and animals as it appears in the replies to Gassendi. Being unfamiliar with the unpublished *l'Homme*, and hence unable to charge Descartes with inconsistency, Gassendi focuses on the theory developed in the Second Meditation, namely that without the active intervention of the intellect there can be no experience in the strict sense of the term. The questions that Gassendi brings up in relation to this theory address the difficulty posed by animal behavior. It is sufficient to believe—as everyone does, first and foremost the Aristotelians—that animals do not possess intellect but only sense and imagination, to consider untenable the notion that intellect alone can permit recognition of the fact that wax in the solid state is the same thing as wax in the liquid state. Indeed, it is patently obvious that a dog despite lacking intellect recognizes his own master whatever clothes he may be wearing, so why on earth would man need intellect to recognize wax in its changing forms?

> A dog, which you will not allow to possess a mind like yours, certainly makes a similar kind of judgment when it sees not its master but simply his hat or clothes. Indeed, even if the master is standing or sitting or lying down or reclining or crouching down or stretched out, the dog still always recognizes the master who can exist under all these forms, even though like the wax, he does not keep the same proportions or always appear under one form rather than another.[22]

In *l'Homme*, Descartes had had no difficulty in admitting that the dog recognizes its own master. This would be a "material" recognition, in other words the reopening of cerebral traces impressed upon the brain by past experiences that are similar in certain respects to the present experience. He had also added that man can do the same, with the difference being that the human mind produces a conscious recognition and a conscious memory.[23] Possibly the fifth section of the *Discours* ought also to have forced him to a physiological response to the question of recognition of the external world despite its changes. It is indeed well known that in the *Discours* Descartes had maintained that it is only language and versatility of conduct that distinguish men from animals. In all other behavior men and animals are indistinguishable. However, at this stage Descartes can no longer give such an answer to Gassendi any more than he can concede that both men and animals can recognize the same object in its changing forms through purely cerebral means. This would totally undermine his anti-empiricist offensive. And so Descartes' reply represents an elusive attempt to separate human cognitive acts from those of animals:

> I do not see what argument you are relying on when you lay it down as certain that a dog makes discriminating judgments *in the same way as we do*. Seeing that a dog is made of flesh you perhaps think that everything which is in you also exists in the dog. But I observe no mind at all in the dog, and hence believe there is nothing to be found in a dog that resembles the things I recognize in a mind.[24]

Descartes does not deny that the dog recognizes his master even when he has changed his clothes or when he assumes unusual postures: what he argues is that although the dog recognizes its master, it does not do so *in the same way* that the human mind can recognize wax in its changes. But how the dog does actually recognize his master remains unsaid. Gassendi's mistake lies in explaining the similar behavior of the man and the animal in the same way. It is unnecessary to recall that this was precisely Descartes' own physiological program: to explain all the behavior common to men and animals without the intervention of the mind.[25]

If the elusive reply to Gassendi is a strategic ploy to sidestep the problems caused by the developments of metaphysics, we could argue that this strategic path had already been paved in the Second Meditation. In the analysis of the piece of wax, having asserted that neither the senses nor the imagination can justify the fact that the wax is perceived as the same wax despite the changes that it undergoes, Descartes asks,

> Was there anything in it which an animal could not possess? But when I distinguish the wax from its outward forms—take the clothes off, as it were, and consider it naked—then although my judgment may still contain errors, at least my perception now requires a human mind.[26]

In accordance with the meditative style, the question had been formulated within a conceptual framework still of Aristotelian stamp, which attributed to animals sensibility and imagination but not intellect.[27] Following the analysis of the piece of wax, even an Aristotelian would have had to attribute to the intellect what he had previously ascribed to the imagination. In this way, the Aristotelian would have been induced to distinguish human sense perception from that of animals. Seen in the light of the replies to Gassendi, this elusive notation on the difference between human and animal sensory perception perhaps paves the way for a conciliation between the new theories about the organization of experience elaborated in the *Meditationes* and the text of *l'Homme*. The manner in which animals have experience of the world could continue to be described through the material cognitive actions studied in *l'Homme*, whereas the presence of a human soul would permit the organization of the data of experience through intellectual ideas and judgments. The paths of human experience are not the same as those pursued by the brains of animals. The human mind alone resorts to judgment to construct experience, and without the mind there would be no human experience in the true sense.

There is a significant consequence of the strategy just mentioned which relates to animals. If it is true that human experience is different from animal experience, and that only human experience is a mental construct, it is no longer necessary for the world of animals, devoid of minds, to be "a fragmented one." Even without the use of judgment, the animals could employ the unifying effect of memory and the association of ideas and would behave like men, albeit for different reasons. In short, the intellect is not the only instrument for the unification of sensory data.

Nevertheless, such a strategy would have a paradoxical consequence. In her classic study on animal-machines, Leonora Cohen Rosenfield had already observed the fact that animals were not mentioned in *l'Homme*.[28] Although Descartes claimed that he was summarizing *l'Homme* even as regards the animals, the pages devoted to this subject in the *Discours sur la méthode* were in actual fact a new chapter, included possibly through the desire to reap the fruits and enhance the apologetic potential of physiology. In the *Discours de la méthode*, Descartes introduced *l'Homme* as a text in which he would address the human body and *also* deal with animals. After the publication of the *Meditationes*, despite its title *l'Homme* proved to deal *only* with animals.

Regardless of Descartes' intentions, metaphysics became the keystone supporting a new study of human perception. At the same time, the divergence between his mature metaphysics and the scientific results that had already been achieved in 1630 represents a shining example of the obstacles encountered in the attempts at a metaphysical justification of science. If Descartes was the first philosopher-scientist of the modern age to have consciously undertaken a foundational project of ample scope, he was also the first philosopher who had to tackle all its inherent difficulties.

Translation by Helen Cleary

Notes

1 AT VII 436–8, CSM II 294–5.
2 See Alison Simmons, "Descartes on the Cognitive Structure of Sensory Experience," *Philosophy and Phenomenological Research* 67 (2003): 549–79.
3 See in particular the Fifth Set of Replies, AT VII 381–2, CSM II 262.
4 AT VII 32, CSM II 21.
5 René Descartes, *Traité de l'homme*, AT XI 179.
6 *Id.*, 177.
7 AT VI 138: "cela suffira pour faire que la grandeur de la ligne Ss et des deux angles XSs et XsS se trouvent ensemble en nostre fantaisie, et nous facent apercevoir la distance du point X: et ce, par une action de la pensée, qui, n'estant qu'une imagination toute simple . . ." *Id.*: "cela nous peut servir, non pas proprement a voir, mais a imaginer sa (d'un objet) distance."

8 Stephen Gaukroger speaks of an "anomalous" statement, adding that if we were to take it literally, "then Descartes' detailed account of the psycho-physiology of automata, in chapter 3 of *l'Homme* . . . would either be completely mysterious, or we should have to say that, by the time of the *Meditationes*, Descartes had abandoned this potentially fruitful approach . . . in favor of the view that only human beings are capable of genuine cognition." Stephen Gaukroger, *Descartes' System of Natural Philosophy* (Oxford and New York: Oxford University Press, 1997), 203. The divergence between the Sixth Set of Replies and *l'Homme* is also underscored by Gary Hatfield, "Descartes' physiology and its relation to his psychology," in *The Cambridge Companion to Descartes*, ed. John Cottingham (Cambridge and New York: Cambridge University Press, 1992), 356–8. See also Wilson, Margaret, "Descartes on the Perception of Primary Qualities," In *Ideas and Mechanism: Essays in Early Modern Philosophy* (Princeton: Princeton University Press, 1999), 26–40.

9 S. Gaukroger admits that the statement in the replies to the Sixth Objections clashes with his interpretation whereby animals feel. Gaukroger, *Descartes' System*, 203.

10 Gaukroger, *Descartes' System*, 213.

11 Thomas C. Vinci, "Reason, Imagination, and Mechanism in Descartes' Theory of Perception," in *Oxford Studies in Early Modern Philosophy*, eds. Daniel Garber and Steven Nadler (Oxford: Clarendon Press, 2005), 35–73.

12 AT XI 201–2, CSM I 108: "I should like you to consider, after this, all the functions I have ascribed to this machine, such as the digestion of food, the beating of the heart and arteries, the nourishment and growth of the limbs, respiration, waking, and sleeping, the reception by the external sense organs of light, sounds, smells, tastes, heat, and other such qualities, the imprinting of the ideas of these qualities in the organ of the 'common' sense and the imagination, the retention or stamping of these ideas in the memory, the internal movements of the appetites and passions, and finally the external movements of all the limbs (movements which are so appropriate not only to the actions of objects presented to the senses, but also to the passions and the impressions found in the memory, that they imitate perfectly the movements of a real man). I should like you to consider that these functions follow from the mere arrangement of the machine's organs every bit as naturally as the movements of a clock or other automaton follow from the arrangement of its counter-weights and wheels."

13 Descartes to Mersenne, 25 November 1630, AT I 182.

14 Descartes to P. Gibieuf, 18 July 1629, AT I 17.

15 Descartes to Mersenne, March 1637, AT I 349–50.

16 Geneviève Rodis-Lewis, "Hypothèses sur l'élaboration progressive des Méditations de Descartes," *Archives de philosophie* 50 (1987); 109–23, and by the same author, "On the Complementarity of Meditations III and V: From the 'General Rule' of Evidence to 'Certain Science'", in *Essays on Descartes' Meditations*, ed. A. Oksenberg Rorty (Berkeley-Los Angeles-London, University of California Press), 271–95.

17 Descartes to***, 1648 or 1649 (?), AT V 261. Moreover Regius had expressed his doubts about the publication of *l'Homme*, recalling that it was a text not explicitly approved by the author. Roberto Bordoli, *René Descartes: Henricus Regius. Il carteggio. Le polemiche* (Napoli: Cronopio, 1997), 64–5.

18 AT VII 438, CSM II 295.

19 Binocular convergence calls for a "natural geometry" that makes it possible to perceive distance through the angle of the eyes. Natural geometry has given rise to highly divergent interpretations: from an extreme intellectualist reading in Nancy Maull, "Cartesian Optics and the Geometrization of Nature," in *René*

Descartes: Critical Assessments, ed. Georges J. D. Moyal, vol. 4 (London and New York: Routledge, 1991), 263–79, to a reading that assigns natural geometry to a case of "natural institution" that envisages no calculation on the part of either the human or the divine mind. Particularly convincing in this sense is Elisa Angelini, *Le idee e le cose: La teoria della percezione di Descartes* (Pisa: ETS, 2007), 88–106. But see also Gaukroger, *Descartes' System*, 275 ff.

20 See Angelini, *Le idee e le cose*, 92 and Gaukroger, *Descartes' System*, 210.

21 Falsehood was nothing new to Descartes. We could mention the famous case of the discussion of God as *causa sui*. Against Caterus, Descartes argues the legitimacy of using the notion of *causa sui* in the case of God and of interpreting it as analogous to efficient causality. AT VII 108: "Denique non dixi impossibile esse ut aliquid sit causa efficiens sui ipsius." Later, with Arnauld, Descartes transforms the *causa sui* from efficient causality to formal causality, AT VII 239–43, while also reasserting his claim that if we do not grant that God is *positively* self-caused it is impossible to prove his existence, AT VII 239. Finally, in the *Notae in Programma*, he states with shameless emphasis: "moneo libellorum istorum authorem, me nunquam scripsisse *Deum non modo negative, sed et positive, sui causam efficientem dici debere. . .* Quaerat, legat, evolvat mea scripta: nihil unquam simile in illis reperiet, sed omnino contrarium"! AT VIII-2, 368–9.

22 AT VII 272–3, CSM II 190.

23 The study that has most profoundly explored the role of the material memory in cognitive phenomena in Descartes is that of John Sutton, *Philosophy and Memory Traces: Descartes to Connectionism* (Cambridge: Cambridge University Press, 1998).

24 AT VII 359, CSM II 248. Emphasis mine.

25 AT VI 46, CSM I 134: "And when I looked to see what functions would occur in such a body I found precisely those which may occur in us without our thinking of them, and hence without any contribution from our soul (that is, from that part of us, distinct from the body, whose nature, as I have said previously, is simply to think). These functions are just the ones in which animals without reason may be said to resemble us. But I could find none of the functions which, depending on thought, are the only ones that belong to us as men; though I found all these later on, once I had supposed that God created a rational soul and joined it to this body in a particular way which I described."

26 AT VII 32, CSM II 22.

27 On the rules of the meditative style I refer to Harry G. Frankfurt, *Demons, Dreamers and Madmen: The Defense of Reason in Descartes' Meditations* (New York: The Bobbs-Merrill Company, 1970) and to Emanuela Scribano, *Angeli e beati: Modelli di conoscenza da Tommaso a Spinoza* (Roma, Bari: Laterza, 1996).

28 See Leonora Cohen Rosenfield, *From Beast-Machine to Man-Machine* (New York: Octagon Books, 1968), 5–6.

6 Descartes, *A Priori* Knowledge, and Metaphysics

Desmond M. Clarke

In the history of philosophy, Descartes' name has been linked so frequently with a range of familiar claims about eternal truths and innate ideas, and with the foundationalism that he defended in the *Meditations*, that he appears to be a characteristic defender of *a priori* knowledge and of metaphysics as a foundation on which all other knowledge is based. Despite this familiar interpretation, however, Jean Laporte suggested that Descartes should be understood as an empiricist: "If we wish to insist on characterizing the philosophy of Descartes by a single name, then the description that would best fit his philosophy, without paradox, would be 'empiricism,' a comprehensive and radical empiricism."[1] Even if one accepts that authors may change their epistemology during the course of their lifetime, it seems implausible that Descartes could have adopted both positions. This chapter examines the extent to which Descartes endorsed *a priori* knowledge claims or, alternatively, that he relied on various kinds of experience as the ultimate warrant for all knowledge claims. While it is not possible in the space available to examine all his publications, some of the most significant texts that might be expected to reveal Descartes as a supporter of *a priori* knowledge demonstrate that he conformed throughout his life to Laporte's description.

Before examining some of these Cartesian texts, one of the fundamental motivations for defending *a priori* knowledge can be made explicit by Frege's question about the epistemic status of logical truths, which are often characterized as paradigms of such knowledge. Frege formulated the question as follows:

> The question why and with what right we acknowledge a law of logic to be true, logic can answer only by reducing it to another law of logic. Where that is not possible, logic can give no answer.[2]

If logic cannot supply an answer, and if an infinite regress of logical laws supporting other laws is futile, the question is, can anything else apart from logic provide an answer? In broad terms, there are two philosophical traditions in response to this question.[3] One approach, which may be called transcendentalism, assumes that philosophy has a distinctive method by which it examines the conditions for the possibility of human thought and discovers some fundamental truths that are independent of experience. The products of such a philosophical analysis or investigation are necessary truths, such as axioms of logic or mathematics, and they are assigned the value "true" independently of any experience. In other words, no matter what we may ever experience, we either cannot or will not reconsider the truth-value of such basic principles.

The other alternative, naturalism, conceives of the role of philosophy in a different way. For a naturalist, no sources of knowledge are ultimately independent of experience. Even the human activity of acquiring knowledge is merely a particularly interesting and complex natural phenomenon, so that one's theory of knowledge shares the same fundamental dependence on experience as any other theory we may hold. Quine adopted the full implications of this naturalist position. For him, there is no claim or axiom that is not subject to possible rejection or revision if it fails to cohere with a web or network of beliefs that are compatible with our experience of reality.[4] Quine's conclusion was that while we may treat some propositions as analytic—that is, as necessarily true simply by virtue of the words or symbols used to express them—there is no guarantee that the conceptual scheme in which the relevant words or symbols function is suitable for achieving our epistemic goals (whatever they may be). Thus, some propositions are contingently necessary; they are always true on condition that one adopts a particular language or conceptual system, but the adoption of that language remains contingent.

I shall use the term "*a priori*" in the same way as Quine, and then examine if Descartes endorsed the thesis that some beliefs are necessarily true or true regardless of what we experience about the world in which we live. I shall argue that, with the exception of the ontological argument for God's existence, Descartes took a decisive step in the direction of epistemological naturalism in the *Regulae* and that, subsequently, he never resiled from that position.

The *Regulae*

One of the first indications of naturalism occurs in Rule 12 of the *Regulae*, when Descartes summarizes everything that was said in the previous eleven rules. "Where knowledge of things is concerned, only two factors need to be considered: ourselves, the knowing subjects, and the things which are the objects of knowledge" (AT X 411, CSM I 39). He then offers some comments, in turn, about each of these two factors. For the subject, he suggests a hypothesis about how sensation works, how images are stored in the imagination, and how judgments are made in the intellect.[5] Once he had outlined this hypothetical account of the subject of knowledge, he needed to say something about the objects of knowledge. Here again he says that "certain assumptions must be made in this context" (AT X 417, CSM I 43), which may not be accepted by everyone.

> But even if they are thought to be no more real than the imaginary circles which astronomers use to describe the phenomena they study, this matters little, provided they help us to pick out the kind of apprehension of any given thing that may be true and to distinguish it from the kind that may be false.
>
> (AT X 417, CSM I 43–4)

However, just when we are expecting him to say something about the *objects* of knowledge, as he had promised, Descartes retreats from discussing those objects to offer suggestions about how we conceive of them.

> First, when we consider things in the order that corresponds to our knowledge of them, our view of them must be different from what it would be if we were speaking of them in accordance with how they exist in reality. . . . That is why, since we are concerned here with things only in so far as they are perceived by the intellect, we term "simple" only those things which we know so clearly and distinctly that they cannot be divided by the mind into others which are more distinctly known. Shape, extension, and motion, etc. are of this sort.
>
> (AT X 418, CSM I 44)

Our simple conceptions may be distinguished further into those that are "purely intellectual or purely material, or common to both," although here again Descartes was talking about "things that are said to be simple with respect to our intellect" (AT X 419, CSM I 44).

It remains to be seen whether Descartes subsequently abandoned this explicit acknowledgement that, when he writes about the objects of knowledge, he was actually writing about the ways in which he conceives of them. It is difficult to understand what other perspective one could adopt. The very recognition of this subjective perspective implies the possibility of alternative conceptual frameworks (either within a given discipline over a period of time, or even simultaneously for the same knower) and the need to give reasons for choosing between such alternatives when they are available. This is precisely what Descartes did when confronted with the choice between the scholastic and mechanical conceptions of natural phenomena. He argued that one should choose the way of conceiving the objects of knowledge that he believed would deliver a better explanation.

Le Monde and the *Essais*

When Descartes rejected scholastic forms and qualities in chapter 2 of *Le Monde*, the reason he gave for doing so was that they were redundant to his explanatory program.

> When a flame burns wood or some other similar material . . . others may, if they wish, imagine the form of fire, the quality of heat, and the

> process that burns it to be completely different things in the wood. For my part, I am afraid of mistakenly supposing there is anything more in the wood than what I see must necessarily be in it, and so I am content to limit my conceptions to the motions of its parts.
>
> (AT XI 7, CSM I 83)

The claimed relationship between the kinds of entity that are included in a Cartesian universe and their explanatory success is repeated in many places in the *Essais* of 1637. Rather than review the more familiar *Discours* and the *Dioptrique*, one might consult the *Météores*, where Descartes writes:

> in order to keep my peace with the philosophers, I have no desire to deny that which they imagine to be in bodies in addition to what I have given, such as their *substantial forms*, their *real qualities* and the like; but it seems to me that my explanations ought to be approved all the more because I shall make them depend on fewer things.
>
> (AT VI 239)[6]

The meteorology is explicitly structured as a hypothetical enterprise:

> It is true that since knowledge of these matters depends on general principles of nature which have not yet, to my knowledge, been accurately explained [*expliqués*], I shall have to use certain hypotheses at the outset, as I did in the *Optics*. But I shall try to render them so simple and easy that perhaps you will have no difficulty in accepting them.
>
> (AT VI 233)[7]

The *Meteorology* includes explanations of a wide range of natural phenomena.[8] Descartes claims that his explanations agree so well with experience that, in the case of the rainbow, it would be impossible to doubt that the matter was as he had explained it (AT VI 334). Of course, one might object that Descartes' logic of confirmation was so unreliable that he was claiming much more for his speculative hypotheses than the supporting evidence warranted. But that is beside the point here. Whether Descartes' hypotheses are plausible or otherwise, when deciding which kinds of entity exist in the universe, he confines himself to those that are alleged to be successful as explanations of observed phenomena. Even if his resulting beliefs are false or unwarranted, the reason given for adopting one conception rather than another of the objects of knowledge is their explanatory success.

The few examples mentioned so far are, perhaps, easy cases for a thesis about the lack of *a priori* knowledge in Descartes, or they may represent only one phase of the Cartesian project that was prominent in his early work. One also needs to examine what type of knowledge he claimed to have acquired in the *Meditations on First Philosophy*. This provides an opportunity to comment on Descartes' use of the term "metaphysics."

The *Meditations*

Descartes invariably defines metaphysics as the study of God and the separated soul or mind, which he describes in the Preface to the French edition of the *Principles* as "immaterial or metaphysical things" (AT IX-2 10, CSM I 184).[9] Although this seems to have been what Descartes consistently meant by the term "metaphysics," we are not limited to that usage when examining the relationship between metaphysics and other kinds of knowledge in his work. We might wish, for example, to include in his metaphysics the most general assumptions of his natural philosophy. However, since *a priori* claims are most likely to be made about God and the separated soul, the discussion here is confined to them. If the scope of the term "metaphysical" is extended to include fundamental principles of natural philosophy, the latter are unlikely to rely on *a priori* truths if Descartes' claims about God and the soul fail to do so.

With one exception—namely, the ontological argument for God's existence in the Fifth Meditation, to which I return below—Descartes claims that we acquire from experience all the concepts required to make claims about the mind and God, and that we confirm the truth of those claims by experience. If that were true, it suggests that Descartes may use a variety of terms, in Latin and French, for different kinds of experience and that "experience" may extend much wider than the observation of external natural phenomena. This, in fact, is what one discovers by examining the Cartesian concept of experience. Descartes' vocabulary includes "*experientia*" and "*experimentum*" in Latin, and "*expérience*," "*sentiment*," and "*épreuve*" in French, and the corresponding verb forms.[10] Without reviewing the full range of Cartesian experiences, however, it suffices to notice that he also includes a type of experience that Locke described as follows:

> The other Fountain, from which Experience furnisheth the Understanding with *Ideas*, is the *Perception of the Operations of our own Minds* within us . . . which Operations, when the Soul comes to reflect on, and consider, do furnish the Understanding with another set of *Ideas*, which could not be had from things without: . . . This Source of *Ideas*, every Man has wholly in himself: . . . so I call this *Reflection*, the *Ideas* it affords being such only, as the Mind gets by reflecting on its own Operations within it self (emphaisis in original).[11]

Hume likewise assumes that all ideas are "deriv'd from sensation and reflection."[12] If Locke and Hume do not compromise their empiricism by using this type of experience, then neither does Descartes.

Descartes acknowledges the psychological source of his concepts of God and the soul in the *Conversation with Burman*. Burman queried the suggestion he had made, in the First Meditation, that everything that he had "accepted as most true" had been acquired "either from the senses or through the senses" (AT VII 18, CSM II 12), because it seemed to Burman to overlook the concepts of the mind and of God. However, Descartes replied,

> The objection cannot be made here that this leaves out the common principles and ideas of God and of ourselves. . . . For, firstly, I acquired these in the same way, through the senses, that is to say, through hearing. (AT V 146, CSMK 332)

This corresponds to the language used in the arguments about the self and God in the *Meditations*, in which Descartes recommends that we look for the appropriate concepts in experience.

The concept of a thinking thing emerges, in a confused manner, in the Second Meditation. Descartes initially clarifies the extension of the term "to think" by reflecting on the psychological experience of various kinds of thinking, so that a thinking thing is one that "doubts, understands, affirms, denies . . . and also imagines and has sensory perceptions" (AT VII 28, CSM II 19). He had not yet decided, in the Second Meditation, if thinking was a bodily activity or something else, and he therefore left open the possibility that the bodily features that he had doubted in the previous meditation may not be distinct from the subject which is responsible for thinking.[13] Once he recognized different kinds of thinking—by reflection on his experience of thinking—he had to choose a concept whose extension coincided at least with the range of activities just mentioned. It is not clear what Descartes thinks is going on at this stage in the argument. His exercise in conceptual clarification seems to oscillate between simply naming activities (as in: I wish to call those activities "thinking") and making claims about the nature of the activities that are being named (as in: those activities are operations of a spiritual substance). The interplay of conceptual analysis and confirmation of truth-claims, and the precise point in the *Meditations* where one changes into the other, is equally unclear in the summary provided in the "Synopsis." Here Descartes contrasts the foundations of the sciences (*scientiarum fundamenta*) that he will offer with those that were provided by earlier philosophers. However, on the next page he claims that "the premises which lead to the conclusion that the soul [*mens*] is immortal depend on an account of the whole of physics" (AT VII 13, CSM II 10).[14] In other words, the conceptual clarification of the term "mind" is inextricably linked with a reciprocal clarification of the term "matter," and the claim that the human mind is immortal rests on a foundation in natural philosophy. Physics precedes metaphysics and provides its evidential basis.

Descartes is more explicit and clear about how he came to have the concept of God (assuming that the concept of a mind had been adequately clarified first). This concept is generated by extrapolating the concept of a human mind, or by denying the limitations that are associated with that concept. That is how he explained the acquisition of the concept to Mersenne, in replies to the *Second Objections*:

> I readily and freely confess that the idea which we have of the divine intellect, for example, does not differ from that which we have of our own intellect, except in so far as the idea of an infinite number differs from the idea of a number raised to the second or fourth power. And the same applies to the individual attributes of God of which we recognize some trace in ourselves.
>
> (AT VII 137, CSM II 98)

He gave a similar reply to Hobbes, who had claimed not to have had any concept of God:

> Now everyone surely perceives that there are things he understands. Hence everyone has the form or idea of understanding; and by indefinitely extending this he can form the idea of God's understanding. And a similar procedure applies to the other attributes of God.
>
> (AT VII 188, CSM II 132)

This hypothesis about the origin of the idea of God coincides exactly with Hume's account of the same idea, which originates in an impression of reflection:

> The idea of God, as meaning *an infinitely intelligent, wise, and good Being*, arises from reflecting on the operations of our own mind, and augmenting, without limit, those qualities of good and wisdom.[15]

It is well known that Descartes describes both concepts—the concept of a human mind and the concept of God—as innate, but they are innate only in the weak sense that (as he explained in reply to Regius) they originate by reflection on his faculty of thinking:

> I have never written or taken the view that the mind requires innate ideas which are something distinct from its own faculty of thinking. I did, however, observe that there were certain thoughts within me which . . . came solely from the power of thinking within me.
>
> (AT VIII-2 358, CSM I 303)

In summary, Descartes acquired the concepts of mind and God by reflecting on his own psychological experience of thinking. The metaphysical arguments for the distinction of the mind from the body and for the existence of God depend, therefore, on the experiential origins of those concepts.

Laws of Logic

As indicated above, one of the central claims in Quine's naturalistic epistemology is that even the most fundamental laws of logic are, in principle, open to revision. Descartes also concedes that the laws of logic merely reflect our way of thinking rather than truths that can be known *a priori*, and he made this explicit when reflecting on his knowledge of God.

Henry More wrote to Descartes in 1649 and challenged his denial of the possibility of a vacuum. More asked whether the concept of extension applies to God and whether God's extension might fill a vacuum. In reply, Descartes made a distinction between what God is capable of doing and what he (Descartes) was capable of conceiving or, as in the *Regulae* text quoted above, between how things may be in reality and how we conceive of them.

> For my part, I know that my intellect is finite and God's power is infinite, and so I set no limits to it; I consider only what I am capable of perceiving and what not, and I take great pains that my judgement should accord with my perception. And so I boldly assert that God can do everything which I perceive to be possible, but I am not so bold as to assert the converse, namely that he cannot do what conflicts with my conception of things—I merely say that it involves a contradiction.
>
> (AT V 272, CSMK 363)

That suggests that the laws of logic apply to our ways of perceiving or understanding things; although our beliefs are constrained by those logical rules, the realities to which they are applied may be otherwise. This is even more explicit later in the same letter when Descartes tells More: "our mind is not the measure of reality or of truth; but certainly it should be the measure of what we affirm or deny" (AT V 274, CSMK 364). He had given a similar commentary on God's powers to Mesland, in 1644, when he conceded that God could make contradictory propositions true at the same time (AT IV 118–19, CSMK 235).

This understanding of God's power is consistent with Descartes' well-known claim that God creates eternal truths. Descartes had distinguished, in reply to Mesland, between God creating necessary truths and God creating them necessarily.

> And even if God has willed that some truths should be necessary, this does not mean that he willed them necessarily; for it is one thing to will that they be necessary, and quite another to will this necessarily, or to be necessitated to will it. I agree that there are contradictions which are so evident that we cannot put them before our minds without judging them entirely impossible.
>
> (AT IV 118, CSMK 235)

God freely created necessary truths, as he had explained to Mersenne in 1630: "the mathematical truths which you call eternal have been laid down by God and depend on him entirely no less than the rest of his creatures" (AT I 145, CSMK 23). In other words, necessary truths are contingent on God's free choice and, in principle, they could have been otherwise. Our minds are also created to perceive them as necessary, but that is simply because God is the creator of both the truths and our minds, and he matched the latter to the former. We should not, however, conceive of God as being constrained by our logic or reason.[16] Since God creates eternal truths, the arbitrariness of God's decrees dilutes the apparent necessity of the resulting truths. Thus, eternal truths are necessary only in the sense that God created us in such a way that we perceive them as necessary. Descartes' consistently held concept of God did not limit anything in God, including his creative decisions, to what would seem to be rational to human beings.

Descartes had come close to making a similar distinction in reply to one of the Second Objections. Mersenne was rightly skeptical of the argument that, if some proposition is perceived clearly and distinctly, then it must be true. He asked Descartes: "Why should it not be in your nature to be subject to constant . . . deception?" (AT VII 126, CSM II 90). He thereby raised the possibility that God might have given human beings a nature such that they are often deceived even about matters of which they have a clear and distinct perception, and that God may even have had benevolent reasons for doing so, in the same way that doctors or parents sometimes deceive children for their own benefit.

Descartes' reply is less than convincing, but it includes a distinction between the kind of certainty that it is possible for human beings to acquire and some other kind of absolute truth that is available only to God or an angel.

> What is it to us that someone may make out that the perception whose truth we are so firmly convinced of may appear false to God or an angel, so that it is, absolutely speaking, false? Why should this alleged "absolute falsity" bother us, since we neither believe in it nor have even the smallest suspicion of it?
>
> (AT VII 145, CSM II 103)

Although this is more dismissive of Mersenne's objection than is warranted by the strength of the reply, it at least seems to concede a distinction between (1) the limits within which human cognitive efforts operate and (2) some other ideal of which we have no conception. In effect, Descartes is claiming that when we use our cognitive faculties as best we can, we have reached the limits of what is possible epistemically.

Despite these explicit concessions about God's freedom to create alternative eternal truths and that reality may not coincide with the limits of human logic, Descartes' use of the ontological argument in the Fifth Meditation seems to be an exception to what is otherwise a constant theme in the Cartesian project. Descartes argued, in Meditation V, as follows: "But from the fact that I cannot think of God except as existing, it follows that existence is inseparable from God, and hence that he really exists" (AT VII 67, CSM II 46). In this context, Descartes anticipates the kind of justification that he offered eight years later in reply to More, about the kind of mind he had. In the case of things that are known clearly, "the nature of my mind is such that I cannot but assent to these things, at least so long as I clearly perceive them" (AT VII 65, CSM II 45). Therefore, given the concept with which Descartes thinks about God, he cannot avoid thinking that such a God exists. One might understand that conclusion in line with Descartes' description of how his mind functions, as follows: although he cannot avoid thinking that God exists, it may still be possible that God does not exist. However, there is a simpler solution to this apparent aberration on Descartes' part, which is to suggest that he made a logical mistake.

Regius wrote his famously disparaging letter about Cartesian metaphysics in July 1645, in the course of which he suggested that many readers did not believe that Descartes actually approved the metaphysical views that he published, especially since he expressed them in scholastic language. Regius wrote:

> Many honourable and intelligent people have often told me that they think too highly of your intelligence not to believe that, in the depths of your soul, you hold opinions that are the opposite of those that appear publicly under your name . . . many of them are convinced that you have discredited your philosophy very much by publishing your metaphysics. (AT IV 255)

When Regius published his own account of how the human mind works and how we acquire knowledge, he rejected innate ideas as superfluous—a position that Descartes accepted retrospectively as being identical with his own; they disagreed, Descartes said, merely verbally. Regius also rejected the concept of a pure intellect, in the second edition of his *Philosophia naturalis*.[17] It was not surprising, then, that Regius also rejected the ontological argument in the same book. He argued that the fact that necessary and actual existence is included in the idea or concept of God does not imply that God exists.

> For it does not follow from that concept that God exists necessarily and actually, but only that, if he existed, he would exist necessarily and not contingently or that it would then be impossible for him not to exist.[18]

In the course of presenting this argument in the Fifth Meditation, Descartes had acknowledged the general principle that the mere fact of having ideas of things could not possibly confirm the existence of anything corresponding to those ideas.[19] We can only conclude that he made a mistake in logic by confusing the scope of the modal qualifier "necessary." In the concept of God, "necessary" qualifies the kind of non-contingent existence that God would have if God existed. In the conclusion of his argument, however, "necessary" qualifies the truth of the claim that God exists.

Apart from this exception, however, it is plausible to read Descartes' project as Regius had understood it, as follows.

Epistemology Naturalized

Descartes summarized the task of epistemology in the *Regulae*, in two questions:

> We ought once in our life carefully to inquire as to what sort of knowledge human reason is capable of attaining, before we set about acquiring knowledge of things in particular. . . . But the most useful inquiry we can make at this stage is to ask: what is human knowledge and what is its scope?
>
> (AT X 396–7, CSM I 30–1)[20]

Whether Descartes assumed the possibility of *a priori* knowledge depends on how he answered these questions.

In Rule 8, he compared the acquisition of knowledge with the activity of artisans who exercise their skills in two stages. For example, a blacksmith initially uses some naturally occurring thing (e.g., a rock) as a hammer to fabricate appropriate tools that are subsequently used to make swords or helmets. Descartes suggested that the exercise of our cognitive powers involves a similar two-stage process: we initially use certain rules that "*appear* [*videntur*] to be innate in our minds" and, just like the blacksmith, we use those to refine the methods by which we subsequently acquire reliable knowledge. In the blacksmith case, there is an empirical test by which the two-stage process can be evaluated; if it produces good swords that are effective in battle, then the blacksmith has succeeded both in producing tools and in producing swords. Is there any similar criterion of success for someone who wishes to acquire knowledge, by which to evaluate the methods used to achieve one's epistemic goals? Alternatively, does Descartes ignore this type of evaluation and prescribe *a priori* the rules that must be followed in order to acquire knowledge?

Recent discussions in theory of knowledge make explicit the unavoidable circularity involved in making normative judgments about the success of the principles, methods, or procedures that should be used to achieve one's epistemic goals. If such norms are specified *a priori*, we cannot avoid the question: what evidence have we for believing that *a priori* prescriptions or procedures that appear to coincide with our concept of rationality will lead creatures like us to achieve our epistemic goals?[21] If, on the contrary, we try to justify our choice of epistemic tools by their successful application in the past—for example, if we find that a particular instance or kind of knowledge is somewhat successful and if then endorse the methods that are used there—we cannot avoid the question that emerged in history of philosophy of science, especially in the work of Thomas Kuhn: how could the scientific methods used in one period of history guarantee against the possibility of radical conceptual or methodological changes at a subsequent time? Why should the (limited?) success of the past be normative for the future? It seems as if a blacksmith must repeatedly refashion his tools in response to new objectives for his mechanical art; by analogy, it seems as if our attempts to acquire knowledge must invent new methods of discovery and justification as the scope and objectives of our epistemic goals vary.

It should also be evident that however we come to have rules or methods for acquiring of knowledge, we need to specify achievable epistemic goals for any cognitive enterprise. Descartes assumes that the goal of knowledge is the truth, but that fails to provide any criterion by which we could measure success. In the *Discourse* (Part I), however, he said that he was not prescribing a definite epistemic goal or a specific method for realizing it; he was merely describing "a method whereby, it seems to me, I can increase my knowledge gradually and raise it little by little to the highest point allowed by the mediocrity of my mind and the short duration of my life" (AT VI 3, CSM I 112). That seems closer to the iterations of the blacksmith who is open to new tools and novel products than one who has decided what to make with tools that are definitively adopted. A comprehensive review of Descartes' account of different kinds of knowledge (in mathematics, metaphysics, ethics, and natural philosophy), of the epistemic goals that are appropriate to each discipline, and of the methods by which those goals are likely to be realized is outside the scope of this short chapter. A few suggestions must suffice.

The well-known choice of mathematics by Descartes as a paradigm of certain knowledge should be understood in the context of his claims that the imagination, rather than the intellect, is the most appropriate faculty for doing mathematics,[22] and that the imagination provides the most reliable ideas of physical things. Even if mathematics provided a model of certainty, it does not follow that every type of knowledge should be evaluated by that standard or, as Descartes acknowledged, that it is even possible to realize mathematical certainty in every discipline. The epistemic goal in natural philosophy was to explain the appearances of natural phenomena in terms of what are, for the most part, unobservable parts of matter in motion, and this could be done only hypothetically. Descartes replied to Morin, in 1638, that if one compared the explanations of the scholastic tradition with his, one had to choose the latter because of their intelligibility, their simplicity, and their explanatory scope.

> Compare my assumptions with the assumptions of others. Compare all their *real qualities*, their *substantial forms*, their *elements* and countless

> other such things with my single assumption that all bodies are composed of parts. This is something which is visible to the naked eye in many cases and can be proved by countless reasons in others. . . . Compare the deductions I have made from my assumptions—about vision, salt, winds, clouds . . . I hope this will be enough to convince anyone unbiased that the effects which I explain have no other causes than the ones from which I have deduced them (emphasis in original).
>
> (AT II 200, CSMK 107)

He answered similar objections from Mersenne in the same year: "to require me to give geometrical demonstrations on a topic that depends on physics is to ask me to do the impossible" (AT II 142, CSMK 103).

Of course these concessions to the hypothetical character of our knowledge claims were made in the context of defending explanations in natural philosophy that seemed to many readers and correspondents to be implausible, and one might suspect Descartes of pretending to concede (because he lacked the kind of proof required) what he does not actually believe. There is a simpler explanation, however. The fundamental epistemological theses adopted in the *Regulae* and the *Discourse* were based on relatively detailed, hypothetical accounts of how perceptual experiences occur. This is particularly clear in the Sixth Discourse of the *Dioptrics*, in which Descartes explained and defended the thesis that our powers of perceptual discrimination are a function not only of the signals that come from objects to our senses, but also of the construction and sensitivity of our perceptual organs. The physical limits of our perceptual organs set the limits of what we can perceive. Since Descartes used his discussion of visual perception as a model for the perception of clear and distinct ideas, it is very plausible that he drew the same conclusion about conceptual perception as he had drawn about its corporeal model. The limits of what we can understand are set by the limits of the sensory and intellectual equipment with which we are endowed. If there are any *a priori* truths, therefore, they are empirically known facts about our cognitive capacities which are necessarily true only in the sense that they set limits to our epistemic ambitions.

Descartes' sympathy for a naturalized epistemology was overshadowed by the attempt, in the *Meditations*, to use God as a guarantor of our cognitive faculties. Despite this unsuccessful response to radical skepticism, Regius was correct to have noticed that Descartes did change the scholastic narrative about theories of knowledge by examining successful cases of knowledge and, for the most part, by endorsing an "empirical turn" which avoided *a priori* knowledge claims. He reflected on some kinds of knowledge that seemed to be successful or, at least, more successful than alternatives that were available in the seventeenth century, and he conditionally endorsed the methods used in those disciplines. This is reflected in the critical comment, in the *Regulae*, about philosophers who neglect experience and believe "that the truth will spring from their brains like Minerva from the head of Jupiter" (AT X 380, CSM I 21).

Notes

1 Jean Laporte, *Le Rationalisme de Descartes* (Paris: Presses universitaires de France, 1945), 477.
2 G. Frege, *The Basic Laws of Arithmetic*, trans. M. Furth (Berkeley: University of California Press, 1964), 15.
3 I borrow this summary analysis from Philip Kitcher, "The Naturalists Return," *The Philosophical Review* 101 (1992): 53–114, 74; "Mill, Mathematics, and the Naturalist Tradition," in *The Cambridge Companion to Mill*, ed. John Skorupski (Cambridge: Cambridge University Press, 1998), 57–8.
4 W. V. O. Quine, "Two Dogmas of Empiricism," *Philosophical Review* 60 (1951): 20–43. For different versions of naturalism in epistemology, see Ram Neta, "Quine, Goldman, and Two Ways of Naturalizing Epistemology," in *Epistemology: The Key Thinkers*, ed. Stephen Hetherington (London: Continuum, 2012), 193–213.
5 The summary in Rule 12 is offered as "the most useful way of conceiving everything within us which contributes to our knowledge of things" (AT X 412, CSM I 40). Descartes claimed merely that it helpful to conceive of our cognitive faculties in that way ["*haec omnia ita concipere multum juvat*"; AT X 413] because it is very simple.
6 English translation from Paul J. Olscamp, *Discourse on Method, Optics, Geometry, and Meteorology*, rev. ed. (Indianapolis: Hackett, 2001), 268. The hypotheses are called "suppositions" (AT VI 233, 238).
7 Olscamp, *Discourse on Method*, 264.
8 Descartes uses the terms "*raison*" (AT VI 253, 257, 264, 273) and "*explication*" (AT VI 324, 348) and its verbal form "*expliquer*" (AT VI 255, 323).
9 His first reference to "a little treatise of metaphysics," in a letter to Mersenne (25 November 1630), describes its main objectives as "*to prove the existence of God and of our souls* when they are separate from the body" (emphasis in original, AT I, 182, CSMK). In March 1636, Descartes told Mersenne that, *in the Discourse*, he will "try to prove the existence of God and of the soul apart from the body" (AT I, 339, CSMK 51); and in the letter to the Sorbonne, he claims that he had always thought that two questions "*de Deo et de Anima, praecipuas esse ex iis quae Philosophiae potius quam Theologiae ope sunt demonstrandae*" (AT VII 1, CSM II 3).
10 See Clarke, "The Concept of Experience in Descartes' Theory of Knowledge," *Studia Leibnitiana* 8 (1976): 18–39.
11 John Locke, *An Essay Concerning Human Understanding*, ed. Peter H. Nidditch (Oxford: Clarendon Press, 1975), II.i.4.
12 David Hume, *A Treatise of Human Nature*, eds. D. F. Norton and M. J. Norton (Oxford: Clarendon Press, 2007), vol. I, 10.
13 "*Fortassis vero contingit, ut haec ipsa, quae suppono nihil esse, quia mihi sunt ignota, tamen in rei veritate non differant ab eo me quam novi?*" (AT VII 27, CSM II 18).
14 "*Praemissae, ex quibus ipsa mentis immortalitas concludi potest, ex totius Physicae explicatione dependent.*" It is significant that Pascal, in *De l'esprit géometrique*, refers to Descartes as establishing the distinction of matter and mind as a firm principle which is supported by "*une physique entière*" (*Œuvres*, ed. Le Guern, II, 179).

15 David Hume, *An Enquiry Concerning Human Understanding and Other Writings*, ed. S. Buckle (Cambridge: Cambridge University Press, 2007), 2.6.
16 See Steven Nadler, "Conceptions of God," in *The Oxford Handbook of Philosophy in Early Modern Europe*, eds. D. Clarke and C. Wilson (Oxford: Oxford University Press, 2011), 525.
17 "Since all of its operations can be explained by means of the faculties of the intellect that were already explained above, there is no need to add to them a pure intellect or any other faculty as if it were distinct from them," *Philosophia naturalis*, 2nd ed. (Amsterdam: Louis Elzevier, 1654), 404. This explicit rejection of Descartes' language about a pure intellect is not found in the first edition of 1646.
18 *Philosophia naturalis*, 357.
19 "From the fact that I cannot think of a mountain without a valley, it does not follow that a mountain and valley exist anywhere, but simply that a mountain and a valley, whether they exist or not, are mutually inseparable" (AT VII 66, CSM II 46).
20 This is very similar to the objective that Locke chose for his *Essay*, when he asked about "the Original, Certainty, and Extent of humane Knowledge," *Essay*, I. i. 2.
21 Kitcher, "The Naturalists Return," 63.
22 See the discussion in *Descartes' Theory of Mind* (Oxford: Clarendon Press, 2003), 81–93.

7 A Deflationist Solution to the Problem of Force in Descartes

Sophie Roux

As Desmond M. Clarke has noted, the ease with which Descartes uses the term *vis* suggests that it was for him a spontaneous way of explaining the motion of bodies, rather than a term denoting a meticulously crafted concept.[1] In that sense, it is not clear that, for Descartes himself, there ever was a problem of force. But for Descartes scholars, there is a question about the ontological status of forces. This question is embedded in another more general one, namely to determine which causal activity should be attributed to God, who is the primary cause, and which causal activity should be attributed to the secondary causes, that is, bodies with their various capacities—motions, forces, laws. Let us call this other more general question the metaphysical question of causal agency. Three positions were recently attributed to Descartes in this respect: (1) he was an occasionalist, who attributed no causal power to bodies, so that forces, if they exist at all, are in God, (2) he was a pure conservationist, who attributed causal power to forces conceived as physical entities distinct from matter and motion, and (3) he was a concurrentist, who attributed causal activity both to God and to second causes in the physical world, whether these second causes are forces or laws.

At first glance, these three interpretations of Descartes' position exhaust the possibilities, whether the causal activity is attributed to God (occasionalism), to physical forces (conservationism), or to both God and bodily forces (concurrentism). In this chapter, however, I would like to defend a fourth interpretation of Descartes, according to which God intervenes in this world only to conserve it by his ordinary concurrence (in this I agree with the conservationist interpretation), without, for all that, forces or laws being specific entities (in this I agree with the occasionalist interpretation). This interpretation is deflationist: from a causal point of view, it reduces God to a first but distant cause; from an ontological point of view, it insists that, in the physical world, there is no room for real forces distinct from the matter and the motions of its parts. To put in bluntly, forces have indeed a causal role for Descartes, but, for all that, they are specific entities distinct from matter and motion. As we will see, such an interpretation leads to attribute a causal power to bodies which is distinct from the causal power of God; thereby, it contributes to suggest that the strong relationship which is assumed to exist between Descartes' physics and his metaphysics should be downplayed.

I shall proceed in three steps. I begin by highlighting the main arguments that lead me to reject certain aspects of the existing interpretations. Returning next to the *Principia philosophiae*, I show that two notions of causation should be distinguished, and that physical forces, without being additional ontological entities with respect to matter and motion, nevertheless have a real causal role when bodies exchange motion. In a third brief and conclusive part, I suggest that a common point to the three usual interpretations is that they put Descartes' physics and its metaphysics on the same plane; on the contrary, the deflationist interpretation I suggest insists that the physical discourse and the metaphysical discourse are indeed articulated, but that they developed on two distinct planes.

The Three Existing Interpretations

In the final chapter of *Descartes' Metaphysical Physics*, Daniel Garber defends two theses about the Cartesian physical world. The first thesis is "fictionalist," "nullibiquitous," or "nominalist" as far as forces are concerned: even if bodies behave as if there were forces, forces are actually nowhere, neither in God nor in bodies, so that they are just names.[2] Garber is here more radical than Gary Hatfield, who indeed banished forces from Cartesian bodies, but nevertheless attributed to them some kind of reality by putting them in the hands of God. Garber's second thesis, concerning the metaphysical question of causal agency, amounts to attributing to Descartes a restricted occasionalism: Descartes' bodies do not have the force to move each other, only God and finite spirits can do it.[3] Garber here agrees with Hatfield that, according to Descartes, God possesses a power to causally act on bodies, which they lack by themselves.[4] But contrary to Hatfield, for whom the truth of Descartes' world is Malebranche's entirely occasionalist world, he insists that Descartes' and Malebranche's occasionalisms have different foundations: the first is based on the passivity of bodies, while the second relies on their finite character. The difference between Descartes' doctrine and "standard" occasionalism can be made more explicit by distinguishing three cases, whether a body acts on another body, a mind on a body, or a body on a mind: Descartes defended occasionalism in the case of body-body interaction, he thought that finite minds are able to cause motion in bodies, and he should have thought that bodies cannot act on minds.[5] As for Garber's occasionalist interpretation of Descartes' physical interactions, it relies on two arguments, the first concerning God as a substantial cause of the existence of bodies and the second concerning God as a modal cause of their motion. The first argument comes from what Garber calls "the doctrine of divine sustenance," that is the doctrine of creation as continuous creation: since a finite substance cannot conserve itself and since the same action is necessary from the part of God to conserve a substance as to create it, God must not only bring bodies into existence, but sustain them.[6] In itself, the doctrine of divine sustenance is no argument for occasionalism: as Descartes reminds us, this doctrine was "commonly accepted among theologians," and even "a manifest truth" for all metaphysicians, including the majority who rejected occasionalism.[7] To lead to occasionalism, it should be combined with a second argument that comes from what Garber calls the "divine impulse view." For the scholastics, substantial forms used to explain what bodies are and what they do: after their elimination, God was needed to fulfill their tasks, and this is exactly what he does as the direct cause of their motion. He does not recreate bodies at different places from one instant to the other, but gives an impulse or a shove to them.[8]

Garber's occasionalist interpretation set the terms of the debate with particular clarity, but I am not sure that it is correct. Let us begin by noting that whenever Descartes evokes the way in which God acts in the world, he explicitly uses restrictions to emphasize that the *only* thing God does in the world is to sustain things as he created them. In *The World*, he notes that many changes happen "from the *mere* fact [*de cela seul*] that he [God] continues thus to conserve it [matter]"[9]; while

> the two [first] rules follow manifestly from the *mere* fact [*de cela seul*] that God is immutable and that, acting always in the same way, . . . supposing that God placed a certain quantity of motion in matter in general at the first instant he created it, . . . he always conserves the same amount of motion in it,

the third and last rule "depends *solely* on [*ne* dépend que *de*] God's conserving each thing by a continuous action."[10] Similarly, when he summarizes *The World* in the *Discourse on the Method*, he puts forth the hypothesis that, after creating matter and imparting some motion to it, God "did *nothing but* [il *ne* fist autre chose *que*] lend his ordinary concurrence to nature."[11] Last, but not least, in the *Principia*, he introduces the discussion on the general cause of motion by the following:

> as far as the general cause [of motion] is concerned, it seems clear to me that this is no other than God himself, who in the beginning created matter alongside with motion and rest, and who now, *merely* by his ordinary concurrence [*per solum concursum ordinarium*], conserves as much motion and rest in the whole of matter as he first introduced.[12]

None of these texts say exactly the same thing, but the general picture can be summarized by the three following propositions:

i. In order to account for the changes that we observe in nature, we don't need *more* from God than the continuation of conservation.
ii. In order to account for the changes that we observe in nature, we don't need *more* from God than his ordinary concurrence.[13]
iii. God's ordinary concurrence is the action through which he conserves bodies in their existence, which in turn is the action through which he created them.

We have now enough material to formulate an argument against Garber's occasionalist interpretation. In the absolute, iii/ can be understood in two ways, whether we say that no less is needed for conservation than for creation, or that we say that no more is needed for conservation than for creation. Garber thinks that it must be understood in the first way. It seems to me that it must rather be understood in the second way. The restrictions that I emphasized in the texts I quoted indicate namely that Descartes opposed theses that would attribute to God *more* than he himself did: against these theses, he argued that, once matter and motion are created, there is no need to attribute to God *more* than the action by which he conserved what he created; rather, his *mere* ordinary concurrence is sufficient to explain the changes that can be seen in the world. Now, occasionalism is the strongest thesis that can be defended concerning the action of God in the world since it amounts to attributing no causal agency to bodies and to placing causal agency involved in physical interactions entirely in the hands of God. Consequently, if Descartes had been an occasionalist, he would not have been able to oppose theses that attribute more to God than he did. In these circumstances, it seems reasonable to conclude that Descartes was not an occasionalist, but, rather, a conservationist.

In chapter 3 of *Descartes on Causation*, Tad Schmaltz defends two theses on the Cartesian physical world that are diametrically opposed to those of Garber. In the first place, he argues with different arguments than those that I just presented that, far from being an occasionalist, Descartes was a pure conservationist for whom God does nothing more in the world that he created than conserve it through his ordinary concurrence—which, in the range of doctrines conceivable at the time, amounts to attributing as little as possible to God. Schmaltz's second thesis is more problematic. It consists in defending a form of causal realism concerning forces by describing them as tendencies of bodies to persist in their motion, durational tendencies or, still, modes of duration.[14] Schmaltz reaches this conclusion by developing some suggestions from Martial Gueroult and Alan Gabbey. Starting with the remark that, when Descartes says for example that a body persists in its state of rest or motion *quantum in se est*, this "quantum in se est" cannot be reduced to purely geometric aspects of bodies, Schmaltz distinguishes matter and motion as considered abstractly, from a purely geometrical perspective, and matter and motion as concretely instantiated in a substance that exists and endures. This remark leads him to quote Gueroult, who suggested that, according to Descartes,

> force, duration and existence are one and the same thing (conatus) under three different aspects and the three notions are identified in the instantaneous action in virtue of which corporeal substance exists and endures, that is possesses the force which puts it into existence and duration.[15]

According to Gueroult's suggestion, force is, like duration and existence, an attribute that existing bodies possess, contrary to geometric bodies. This poses, however, a problem: substance, existence, and duration remain unmodified, while the forces of a body vary.[16] Gabbey proposed a solution to this problem by drawing on the scholastic distinction between causes *secundum esse* and causes *secundum fieri* that Descartes uses in *Quintae responsiones* and the distinction between invariable attributes and variable modes that he made in *Principia philosophiae*, to add on forces that are variable modes to the invariable forces that are identical to existence and duration.[17] Schmaltz endorses Gabbey's solution, but wants to go farther by determining the kind of modes that forces exactly are. This he does by applying the distinction between attributes and modes to duration: true enough, duration as such is an invariable attribute, but since there are real parts in substance, it should be possible to identify in duration various modal parts.[18] In these conditions, Schmaltz continues, *pace* Gabbey, it is only natural to identify variable forces with various modal parts of the duration of bodies, or, more simply, with various modes of the duration of bodies.[19]

Schmaltz's interpretation is quite subtle, but perhaps excessively so. As I will explain in the second part of this chapter, I do agree that forces are causes *secundum fieri* of the exchange of motion between bodies, but I am skeptical that it is meaningful to consider that they are modal parts of the duration of bodies. A first argument is that we don't gain anything by introducing duration. Since, as Schmaltz reminds us, there is only a distinction of reason between duration and substance that endures and exists, rather than saying that forces are modes of duration, we can as well say that forces are modes of bodies that endure and exist. Indeed, nothing that evokes duration or its parts intervenes when Descartes proposes to "calculate how much force there exists in each body [*calculo subducere, quantum in unoquoque sit virium*]"[20]—this is all the more true that, as we will see in the second part of this chapter, this calculation can be made only at the instant of collision. The dilemma is actually the following: if duration is only distinct by reason from the enduring and existing body, the force of the body is determined by its extension and motion; if duration is something else than extension and motion, it has no place in Descartes' physical ontology. Another objection against Schmaltz's emphasis on duration is that the principle on which the existence of modes of duration depends does not lead to the desired conclusion. This principle is, according to Schmaltz, the analogy that exists between the extended substance and its real parts on the one hand and, on the other hand, the duration and its real parts. But if we follow this analogy, we must conclude that, just as the parts of a substance are real substances,[21] so the parts of duration are durations—and not, as Schmaltz wants, *modes* of duration.

My general impression is that Descartes worried less about the ontological status of force than his commentators did. More precisely, I have the feeling that Schmaltz, while permuting the role of force and the role of God, nevertheless retained Garber's ideas that the causal efficacy that is not attributed to God should be attributed to forces and that what has a causal efficacy should have an ontological status. In the very systematicity of the opposition between Garber's interpretation and Schmaltz's interpretation, there is indeed something constant, which is the application of a principle of compensation, according to which, what God does not do, the physical forces do and, conversely, what the physical forces do not do, God does. Another presupposition on which I will come back later is that only something with a specific ontological reality can be a cause. As we will now see, to the disjunction underlying the principle of compensation (either God or the forces), historians who proposed a concurrentist interpretation of Descartes substituted a conjunction (both God and the forces).

Concurrentism designates in general a mixed position between occasionalism and mere conservationism, according to which both God (the first cause) and bodies (the second causes) concur in the production of the effect.[22] As historians of medieval metaphysics pointed out repeatedly, this does not mean that the two causes collaborate to produce the effect at stake, but rather that both of them produce it completely, though in different ways, the finite cause being subordinated to the infinite cause. This is for example the case if God causes the existence of a being, while other bodies cause its determinations—its being a body of this species rather than a body of that species particularly.

In the wake of these historians of medieval philosophy, several scholars—most notably Kenneth Clatterbaugh, Andrew Pessin, and Helen Hattab—attributed to Descartes some form or another of concurrentism.[23] For that, they put forward two main arguments. First, there are a number of occurrences of the term "concurrence [*concursus*]" in Descartes' works. This first argument is weak if one believes, as I do, that Descartes assimilates the ordinary concurrence of God with his action of creation and of conservation. The second argument is stronger. Since Descartes describes God as a universal cause of everything, while he presents bodies as genuine causes, attributing to him concurrentism would explain why he does not choose between ascribing causation to God and ascribing it to bodies—he can have both. But there is a general problem with concurrentism, which is to determine how God and secondary causes produce an effect without being redundant. To find an answer to this problem, Kenneth Clatterbaugh defended a deductive concurrentism while Helen Hattab and Andrew Pessin espoused a nomic concurrentism, according to which, while God is the first cause of motion, laws of motion are the secondary causes that specify how particular motions are conserved and exchanged.[24]

Here, I shall discuss only Hattab, whose position is, in my opinion, the most fully developed. As noted by Dennis Des Chene, a first argument against her interpretation is that, except in one occasion, Descartes says that the motion is conserved and exchanged *according to* laws, not that laws *cause* the exchange of motion.[25] For Descartes as for his predecessors, causes being necessarily substances endowed with causal powers, it is indeed difficult to understand how laws could be real causes. Another argument against Hattab's interpretation, also pointed out by Des Chene, is that saying that laws are secondary causes distinct from the first cause neglects the fact that Descartes presents them as manifestations of the immutability of God, in other words as inseparable from the nature of God. In fact, Hattab argues that laws not only depend of the immutability of God, but constitute also constraints on God's action that stem from the existence of the material world. The idea that they are constraints on God's action is however problematic: all causes being subordinated to God, nothing can truly prevent God from performing the actions that he wants.[26]

Beyond the details of Hattab's interpretation, the problem that concurrentism faces in general is to find room both for the first cause and for the secondary causes. Such a problem reveals which presupposition concurrentism shares with occasionalism and with conservationism. As we have seen, Schmaltz and Garber apply a principle of compensation according to which what God does not do, the forces do, and vice versa. The application of such a principle presupposes that metaphysics and physics are on the same plane—things that are on different planes cannot compensate each other. If we think about it a little, the concurrentists, even if they do not apply the principle of compensation, share as well this presupposition: if their problem is to determine how the primary cause and the secondary causes can concur in producing an effect without provoking overdetermination, it is precisely because they consider that the causal activities of the first cause and of the secondary causes are on the same plane. Before developing the implications of refusing this presupposition in the third and last part of this chapter, in the next part, I would like to return to the text of the *Principia philosophiae* to better grasp the nature and the function of forces in Descartes' world.[27]

Descartes' Two Notions Of Force

It is quite possible that, as Garber noted, "there may not be an altogether satisfactory view of the ontology of force in Descartes, one that is coherent and sensible, and is consistent with what he says about force in all of his writings and what he commits himself to in other contexts."[28] But I think that the essential step forwards amounts to recognizing that there are two notions of causation in the second book of *Principia philosophiae*.[29] I am not the only one to suggest that we should take such a step. For example, Gueroult distinguishes from the outset forces as causes, which being identical with divine conserving force, are not modes of corporeal substances, from forces as effects, which are modes of corporeal substances and, as such, belong to the conserved world.[30] Gabbey highlights the description that Gueroult gives of the ambiguous situation of forces with respect to God and the created world through the scholastic distinction between causes *secundum esse* and causes *secundum fieri* and through the Cartesian distinction between attributes and modes: by way of a conclusion, he underlines the distinction between the practical level of physical investigation and the true level of metaphysical enquiry in the following way: "Strictly speaking God is the ultimate real cause and the only true substance, but speaking at the 'practical' level of physical investigation, forces—whether of motion or of rest—are real causes in their own right and distinct from motion and rest."[31] Last, Des Chene maintains that forces present two distinct and uncorrelated aspects: as active powers that move bodies, they pertain to God; as measures of the quantity of motion by extension and speed, they concern bodies.[32] Gueroult, Gabbey, and Des Chene have different positions, and I may have another position still. But I agree with them that we should be very attentive to the distinctions that Descartes makes.

Two notions of causation are introduced in *Principia philosophiae*, II 36: on the one hand, "the universal and primary cause, the general cause of all the motions in the world," and on the other, "the peculiar cause through which individual pieces of matter acquire some motion which they did not have before."[33] The rest of the article 36 indicates that "the universal and primary cause" is God who, by his ordinary concurrence, conserves as much motion and rest as he first created. An elementary but crucial remark is that these two causes are not causes in the same way. In saying that the universal cause is the general cause of all motions, Descartes does not intend to derive a series of statements associating a particular cause to a particular motion, as he would if, for example, he used the claim that all rainbows have for their cause the rain in general to derive a series of statements associating each particular rainbow to the particular rain that is its cause. Rather, his point is to say that when we ask the question of what causes motion, we can answer this question in two different ways, according to the kind of causation that we refer to. Either we look for the efficient cause that produces the totality of the motions (all the motions in as much as they constitute a certain quantity of motion), or we look for the "peculiar cause" of a change of motion:

i. God, as the first cause, creates and conserves all the existing motions. The apparent redundancy between the terms "universal," "general," "all [the motions]" that are used to qualify this kind of causation points toward the idea that such a cause is a total cause, contrary to the sun, which may well be called "the universal cause of all flowers," but which is not their total cause, because other particular causes, which are not subordinated to the sun, intervene in explaining the differences between roses and tulips.[34]

ii. The second kind of cause specifies how certain parts of matter acquire motions that they did not have before, although these motions already existed in other parts of matter. In II 36, Descartes is not very explicit about the kind of causation that is at stake in this peculiar cause. Before making it as explicit as possible, I would like to insist on the difference between the two kinds of causation that are at stake here.

As the *Principia philosophiae* is a school textbook, one could say that this division of labor is a scholastic garment ill adapted to Descartes' thought.[35] This does not seem to be the case here because these two kinds of causation correspond to a division of labor that *The World* already introduced between, on the one hand, God who creates and conserves a quantity of motion (the general cause of the *Principia*), and, on the other hand, the nature which is responsible for the particular changes (the particular cause of the *Principia*):

> it follows of necessity from the mere fact that he [God] continues thus to conserve it [matter], that there must be many changes in its parts which cannot, it seems to me, be properly attributed to the action of God (because that action never changes), and which therefore I attribute to nature.

The World is all the more interesting for my point that, a few lines later, Descartes mentions the laws of motion "according to which it must be thought that God makes nature act."[36] Thus, two relations of causation intervene here: nature causes the changes of motion; God causes nature to act. If these two kinds of cause could be treated transitively, we should conclude that God is the cause of change, which is explicitly proscribed. This is another argument that shows, in my opinion, that we cannot treat these two causes transitively, but that we must on the contrary distinguish them strongly and say that they are two kinds of causation that do not operate on the same plane. It is in two different ways that nature causes changes of motion and that God causes nature to act. The only difference between *Principia philosophiae* and *The World* here is that, compared with the later, the former gained in clarity: by distinguishing article II 36, which deals with the primary cause, from article II 43, which deals with the peculiar cause of changes of motion, *Principia philosophiae* marked clearly the difference between the two kinds of causation. At this point, the question is to describe them as precisely as possible.

As we have already seen, the first and total cause, to which all the other causes are subordinated, is God.[37] In other more metaphysical texts, Descartes associates:

i. being a total cause and being an efficient cause,[38]
ii. being an efficient cause and conserving (in the sense of the doctrine of conservation as a continuous creation),[39]
iii. conserving (in the same sense) and being a cause *secundum esse*.[40]

Above all, it seems that, according to Descartes, only God can be properly called an efficient cause—or, more exactly, that it is not possible to be an efficient cause without being a total cause that creates and conserves, that is to say without being God. The most convincing text in this regard is an excerpt from *Sextae responsiones*, in which Descartes asks what name could be given to the kind of causation that links God and the eternal truths that he created (for example, goodness), and concludes that such a name has existed since a long time:

> There is no need to ask what kind of causation [*genus causae*] is applicable to the dependence of this goodness upon God, or to the dependence on him of other truths, both mathematical and metaphysical. For since the various kinds of causes [*causarum genera*] were enumerated by thinkers who did not, perhaps, attend to this type of causation [*causandi ratio*], it would be hardly surprising if they had not given a name to it. But in fact they did give it a name, for it can be called efficient, in the sense that a king is the efficient cause of a law [*eadem ratione quo Rex est legis effector*], although the law itself is not a thing which has physical existence, but is merely what they call a "moral entity."[41]

This text is interesting because Descartes inquires explicitly about types of causation and reaches the conclusion that there is a name for the kind of causation that God exerts, namely "efficient." More generally, there are several texts where God appears as the efficient cause of all the created or finite things.[42]

In contrast with what was just said, it is extremely probable that the second kind of causation is not efficient. In fact, apart from a few rare places where Descartes takes over the term "efficient cause" already used by a correspondent, he himself never uses it in relation to a finite creature.[43] Besides, once said that God is the efficient cause of all things, how could we find room for another efficient cause? However, as I will now emphasize, that the specific causation of II 36 is not efficient does not mean that it is not a cause.

It must be noted that, in general, Descartes is far from having reduced all causes to a single category. On the contrary, he did not hesitate to use the rich arsenal of scholastic distinctions that he had at his disposal to apprehend the different causal relations he wanted to identify. In the *Secundae*

Responsiones, he noted that more is needed to create and conserve a substance than for creating and conserving the attributes or properties of a mode, so that Garber introduced a distinction between being a substantial cause and being a modal cause.[44] Commenting on the *Notes on a Certain Broadsheet*, Steven Nadler established that, to understand how bodies cause ideas, Descartes introduced an occasional causation, in which *A* is the indirect cause of the effect *e*: *A* occasions or elicits *B* to engage in the activity of efficient causation in producing *e*. Occasional causation defined in this way is distinct both from standard transeunt efficient causation and from occasionalism: contrary to efficient causation, occasional causation does not require that something literally passes from the cause to the effect and is not grounded in some ontically real power; contrary to occasionalism, occasional causation is "a real causal relation, albeit an inferior or secondary variety if efficient causation is taken to be the standard."[45]

The more interesting distinction for me is, however, the distinction that Descartes introduced between two kinds of causation in his letters to Elisabeth concerning free will. Elisabeth could not understand how the existence of our free will can be compatible with the assertion that God is the universal and total cause: according to her, either our free will exists and we are independent from God, or God is the cause of our actions and we are dependent on him. Descartes answered her that "the independence that we experience and feel in ourselves . . . is not incompatible with a dependence of quite another kind [*qui est d'autre nature*], whereby all things are subject to God."[46] In other words, if we do not have to choose between the affirmation that, having free will, we are causes of our acts, and the affirmation that God, on which we are dependent, is the cause of our acts, it is because we and God are causes of our actions in two different ways. Elisabeth confessing that she still had difficulties adjusting the dependence of our will with its liberty, Descartes proposed a comparison: if a king, who has forbidden duels, orders two gentlemen to be at the same place on the same day, while he knows that they will duel once in presence, he determines causally their duel, although they engage freely in breaking the law by the king who forbade duels.[47] One can easily understand Elisabeth's difficulties; still, Descartes' general meaning is reasonably clear: that God is the total and efficient cause of all our actions does not prevent men from being responsible for their actions before the law. Descartes makes explicit that it amounts to attributing two kinds of volitions to the king of his story, "one according to which that these gentlemen should fight, since he causes them to meet, and the other according to which he did not so will, since he forbade duels."[48] But it also suggests a distinction, which Descartes does not make explicit, between two different discourses: on the one hand, the discourse of the royal historiographer, who reconstructs chains of events flowing from the decisions of the king, and, on the other hand, the discourse of the judge, whose duty is to condemn the infractions to the law, has to do more than saying that everything depends of the will of the king.

Indeed, human beings are in question in these letters, while I am examining physical bodies in this chapter: in the first case, the question is how our free will can cause our actions; in the second case, the question is how a change of motion of a given body can be caused by other bodies. Still, Descartes insists that the difference is not that important once we consider not a human king, but God:

> all the reasons that prove that God . . . is the first and immutable cause of all effects that do not depend on human free will prove similarly, I think, that he is also the cause of all those effects which do so depend.[49]

For this reason, it seems to me that it is possible to transpose to physical bodies what Descartes says about human affairs in these letters:

i. They are two kinds of causation, the first one being an efficient causation exerted by God, the second one having no name. Once it has been said that God is a total and efficient cause, it is tempting to omit this second kind of causation, at least if overdetermination is to be avoided, because it is impossible to pin it down, not to speak of describing it.
ii. Still, Descartes does not omit this second kind of causation. If one admits the distinction that I suggested between the discourses of the historiographer and of the judge in the case of human affairs, the second kind of causation falls within another kind of discourse than the discourse about God.

Before making the relationship between different discourses more explicit, I would like to come back to *Principia philosophia*, part II, and show that physical forces correspond to this second kind of causation. As we have seen, article II 36 says nothing about the "specific cause," except that it intervenes when a part of matter acquires a motion that it did not have. One can therefore think that this cause has something to do with the third law of nature, which says how bodies that collide exchange motion. In fact, things are a bit more complicated than that because article II 37, which introduces the very notion of laws of nature, present them somewhat mysteriously as "secondary and particular causes of the various motions we see in every specific body," as if all three laws of nature were indeed "secondary and particular causes."[50] Still, it is only when the third law is enunciated in article II 40 that something explicit is said about the secondary and particular causes: according to the third law, "all the particular causes of the changes which bodies undergo [*omnes causae particulares mutationum, quae corporibus accident, in hac lege continentur*]" depend on the relative forces of the bodies that meet.[51]

Articles II 43 and II 45 describe more precisely what happens when two bodies collide. They will help me to complete what has been said until now about the second kind of causes by determining the ontological status of forces, the causal role they play in the exchange of motions, and how to evaluate them at the moment of collision:

i. The ontological status of forces

Descartes warns us from the outset that, from an ontological point of view, the force of a moving body does not constitute a kind of ontological supplement that should be added to matter and motion.

> We must be very careful to note that the force of any given body to act on, or to resist the action of another body . . . consists simply in the fact that everything tends, so far as it can, to persist in the same state [*unaquaeque res tendat, quantum in se est, ad permanendum in eodem statu in quo est*], according to the first law.[52]

Descartes does not want us to imagine that there are forces and tendencies hidden somewhere in the physical world: he insists that in moving bodies, there is nothing else than motion, force being always the force of a moving body.[53] In this sense, physical forces are nothing more than matter and motion; in a counterfactual world where, *caeteris paribus*, bodies would move without meeting other bodies, we would not speak of forces. In our world, if A moves alone, its force is nothing but the fact that it perseveres in its state of motion; when it meets B, we say that it possesses a certain force from the point of view of B, because it could modify the motion of B. The Latin version of article II 40 of *Principia philosophiae* does not even mention force, unlike *The World* and the French version of *Principes de la philosophie*.[54]

ii. The causal role of forces in the exchange of motion

Although they have no ontological consistency, forces play a causal role when two bodies meet. As we have already recalled, the third law specifies the manner in which the exchange of motion between two bodies is realized, depending on the stronger one at the moment they meet. Article II 45, which introduces the seven rules of collision, insists that the quantitative ratio of forces determines the outcome of this meeting.

> To enable us to determine . . . how individual bodies [*singula corpora*] increase or diminish their motions or change direction as a result of a collision with other bodies [*ob aliorum corporum occursus*], all that is necessary is to calculate how much force there is to move or to resist motion [*quantum in unoquoque sit virium, sive ad movendum, sive ad motui resistendum*].[55]

In other words, although physical forces are not entities distinct from moving bodies and although they are not efficient causes, the instantaneous ratio of the forces present when two bodies collide causally determines how these bodies exchange motion.

iii. The evaluation of forces at the moment of collision

The causal role played by the forces of two bodies that collide depends on their respective quantities at the moment of collision. The evaluation of these quantities must therefore be related to this moment. Descartes actually sets up an evaluation of the forces at the moment of the collision, in which four parameters (the respective magnitude of the bodies, their contact surface, their respective velocity, their respective direction) intervene:

> An estimate of this force must depend firstly on the size of the body in question and the size of the surface which separates it from the other body, and secondly on the speed of the motion, and on the nature and contrariety, in which [these] different bodies collide.[56]

From this examination of articles II 36–45 of the *Principia philosophiae*, two points stand out. In the first place, the tendency of bodies to persist in their state is the supreme law of the Cartesian world. It is according to this law that a body tends to stay in its motion, but also resists the changes that other bodies try to impose upon it when they collide. The first law explains not only the persistence of motions, but also their changes; it explains not only the individual behavior of a body, but also its causal interaction with other bodies. Second, these articles rest on the distinction between two kinds of causation: on the one hand, the efficient causation of God, who creates and conserves bodies in motion, and on the other hand, an elusive specific causation, which seems to be described only in a negative way: it is not universal, it does not conserve, it is not efficient. This second type of causation corresponds to physical forces that manifest themselves at the moment bodies collide. Physical forces have no specific ontological reality but are reduced to a set of parameters; still, they have a causal role in the exchange of motions between bodies. It is quite counterintuitive to say that forces have a causal function without having any ontological reality, but it is what Descartes said.

Some Separation Between Physics and Metaphysics

More generally, the deflationist interpretation that I propose is based on the introduction of some separation between physics and metaphysics.

I do not think that it has ever been noticed how the title of Garber's seminal work, *Descartes' Metaphysical Physics*, is both surprising and revealing of Garber's perspective, which set the tone for other interpreters. Descartes himself never speaks of his "metaphysical physics," but of foundations [*fundamenta*] or principles [*principia*] of his philosophy or of its physics.[57] Unlike his Aristotelian professors, he thought indeed not only that metaphysics should precede physics, but also that his physics was founded or supported by his metaphysics.[58] But this does not make Descartes' physics a "metaphysical physics." Even in part II of the *Principia philosophiae*, one can wonder what exactly the idea that physics is founded on metaphysics actually implies. For sure, the existence of a non-deceiving God allows us to be certain that we know something when we think we know; the distinction of mind and body helps us to establish that the essence of matter is extension; the laws of nature flow from the immutability of God. But for all that, are physics and metaphysics discourses that develop on the same plane?

Let us consider for example the notion of conservation, which seems at first sight to entangle both physical and metaphysical elements, since we say that God retains as much motion as he created, that the laws of nature are laws of conservation, and that a body conserves its motion. But precisely Descartes makes here a distinction, by using the verb "conserve [*conservare*]" to describe the action of God, though he never uses it when he refers to the motion of bodies. Rather, he writes that "everything, in so far as it can, always continues in the same state [*unaquaeque res, quantum in se est, semper in eodem statu perseveret*]," that "everything, in so far as it can, always remains in the same state [*unaqu[ae]que res . . . manet, quantum in se est, in eodem semper statu*]," and that "everything tends, in so far as it can, to persist in the same state [*unaquaeque res tendat, quantum in se est, ad permanendum in eodem statu in quo est*]."[59] In a word, the action of conserving is reserved to God, who actually *conserves* the quantity of motion that he created in the world, while bodies only *tend* to stay in their state of motion.

In this chapter, I suggested that metaphysical causation and physical causation are of two different kinds. It may more generally be argued, at least as a plausible hypothesis, that Descartes did not want to place all the discourses on the same plane and that he was rather concerned to determine what is the proper plane to which each thing belongs, and in addition what is the arrangement of these planes with respect to each other. What Descartes says about the action of God who creates and conserves is not canceled when he begins to write physics—it remains preserved, but on a metaphysical plane that is not the plane where physics is practiced.[60]

If it is true that Descartes instituted some separation between the metaphysical discourse and the physical discourse, there was here a singular equilibrium, which his heirs had difficulty in preserving, whether they were contemporaries of Descartes or are today's historians. Descartes had managed to make opposite positions coexist precisely because he did not situate them all on the same plane; his heirs, because they situated these positions on the same plane, had to make some choices between them. The occasionalist interpretation of Descartes, starting with Claude Clerselier, Géraud de Cordemoy, Louis de la Forge, Malebranche, and ending with Garber, is the best known, probably because the canonical history of philosophy favor grandiose systems, even if they go against common sense. The conservationist interpretation adopted by Schmaltz also existed among the heirs of Descartes, as another paper could have shown.

Acknowledgments

I wish to warmly thank Jean-Pascal Anfray and Tad Schmaltz for their thoughtful comments on an earlier version of this chapter, which forced me to clarify my interpretation.

Notes

1 Desmond Clarke, "The Concept of Vis in Part III of the Principia," in *Descartes: Principia philosophiae (1644–1994): Atti del Convegno per il 350° anniversario della pubblicazione dell'opera (Parigi, 5–6 maggio 1994; Lecce, 10–12 novembre 1994)*, eds. Jean-Robert Armogathe and Giulia Belgioioso (Napoli: Vivarium, 1996), 321–39.
2 Daniel Garber, *Descartes' Metaphysical Physics* (Chicago: University of Chicago Press, 1992), 293–9. Garber's position is described as "nominalist" by Dennis Des Chene, *Physiologia: Natural Philosophy in Late Aristotelian and Cartesian Thought* (Ithaca and London: Cornell University Press, 1996), 313–4, as "nullibiquituous" by Clarke, "The Concept of Vis," 125, and as "fictionalist" by Tad Schmaltz, *Descartes on Causation* (New York: Oxford University Press, 2008), 105, 116.
3 Garber, *Descartes' Metaphysical Physics*, 299–305.
4 Gary Hatfield, "Force (God) in Descartes Physics," *Studies in History and Philosophy of Science* 10 (1979): 113–40.
5 Daniel Garber, "Descartes and Occasionalism," in *Causation in Early Modern Philosophy*, ed. Steven Nadler (University Park: Pennsylvania State University Press, 1993), 9–26.
6 Garber, *Descartes' Metaphysical Physics*, 263–6. According to Garber, it was not important for Descartes to specify how God sustains the world, that is whether he simply continues his initial act of creation or if he literally re-creates each substance in different places at every instant. On the distinction between a weak and a strong readings of the doctrine of creation as continuous creation, see however Jean-Pascal Anfray in this volume.
7 Descartes, *Discourse on the Method*, AT VI 45, CSM I 133; *Quintae responsiones*, AT V 369, CSM II 253.
8 Garber, *Descartes' Metaphysical Physics*, 273–80.
9 Descartes, *The World*, AT XI 37, CSM I 92–3, modified.
10 *Ibid.*, AT XI 43 and 44, CSM I 96, modified.
11 Descartes, *Discourse on the Method*, AT VI 42, CSM I 132, modified.
12 Descartes, *Principia philosophiae*, II 36, AT VIII 61, CSM I 240, modified.
13 On God's *concursus*, see as well Descartes to Hyperaspistes, August 1641, AT III 429 and to Mersenne, 21 April 1641, AT III 360 referring to *Primae responsiones*, AT VII 109. For comments, see Geoffrey Gorham, "Cartesian Causation: Continuous, Instantaneous, Overdetermined," *Journal of the History of Philosophy* 42, no. 4 (2004): 389–423, here 407–8, and Schmaltz, *Descartes on Causation*, 99–105, 127–8. Des Chene, *Physiologia*, 334–6, first suggested that God's concurrence manifests itself in moving forces, while his capacity to conserve manifests itself in resisting forces. However, in his answer to Helen Hattab, "The Problem of Secondary Causation in Descartes: A Response to Des Chene," *Perspectives on Science* 8 (2000): 93–118, here 105–6, Des Chene seems to

retreat by renouncing this distinction (Dennis Des Chene, "On Laws and Ends. Reply to Hattab and Menn," *Perspectives on Science* 8 (2000): 144–63, here 147).

14 Schmaltz, *Descartes on Causation*, 116–21. Tad Schmaltz, *Early Modern Cartesianisms: Dutch and French Constructions* (Oxford: Oxford University Press, 2016), 167–75, is more concise and does not mention the hypothesis that forces are durational tendencies.

15 Martial Gueroult, "Métaphysique et physique de la force chez Descartes et chez Malebranche: Première partie: Descartes," *Revue de métaphysique et de morale* 59, no. 1 (1954): 1–37, here 3, quoted by Schmaltz, *Descartes on Causation*, 117.

16 Descartes, *Principia philosophiae*, I 56, AT VIII 26, CSM I 211–2. Descartes then argues that there is only a conceptual distinction between the substance and its duration (I 62, AT VIII 32, CSM I 214).

17 Alan Gabbey, "Force and Inertia in the Seventeenth Century: Descartes and Newton," in *Descartes: Philosophy, Mathematics and Physics*, ed. Steven Gaukroger (Brighton: The Harvester Press, Barnes and Noble Books, 1980), 230–319, here 234–8.

18 Schmaltz, *Descartes on Causation*, 78 and 118.

19 *Ibid.*, 118–21.

20 Descartes, *Principia philosophiae*, II 45, AT VIII 67, CSM I 244.

21 *Ibid.*, I 60, AT VIII 28, CSM I 213.

22 See, for example, Alfred Freddoso, "God's General Concurrence with Secondary Causes: Why Conservation Is Not Enough," *Philosophical Perspectives* 5 (1991): 553–85.

23 Kenneth Clatterbaugh, "Cartesian Causality, Explanation, and Divine Concurrence," *History of Philosophy Quarterly* 12, no. 2 (1995): 195–207; Andrew Pessin, "Descartes' Nomic Concurrentism: Finite Causation and Divine Concurrence," *Journal of the History of Philosophy* 41, no. 1 (2003): 25–50; Hattab, "The Problem of Secondary Causation." Although Gorham himself, in "Cartesian Causation," criticizes the concurrentist interpretation of Descartes and defends instead an "overdeterminist" interpretation, according to which God causes the effect E and causes the cause C to cause the effect E, his interpretation can be seen as a form of concurrentism (both God and C cause E) or even as an occasionalism (both God and C causing E, but God being necessary, while C is not, C can be overlooked).

24 I mention the two texts motivating Hattab's interpretation below, notes 36 and 50.

25 Hattab, "The Problem of Secondary Causation," 114, recognizes the problem. Des Chene, "On Laws and Ends," 152, insists that Descartes, far from developing the law analogy, which could conduce to say that laws are causes, treats them as mere consequences of God's immutability.

26 Hattab, "The Problem of Secondary Causation," 112–4. Des Chene, "On Laws and Ends," 151.

27 In the following part, I develop some indications already given in Sophie Roux, "Découvrir le principe d'inertie," *Recherches sur la philosophie et le langage* 24 (2006): 453–515. Gabbey, "Force and Inertia" and Jean-Pierre Séris, *Machine et communication* (Paris: Vrin, 1987) were at the time important to my thinking.

28 Garber, *Descartes' Metaphysical Physics*, 297.

29 There are obviously other physical forces than the one that appears in articles II 43–5 of *Principia philosophiae*; see Clarke, "The Concept of Vis"; Richard S. Westfall, *Force in Newton's Physics: The Science of Dynamics in the Seventeenth Century* (London, MacDonald and New York: American Elsevier, 1971), 529–33.

30 Gueroult, "Métaphysique et physique de la force," 5–9. See also 35–6.

31 Gabbey, "Force and Inertia," 238.

32 Des Chene, *Physiologia*, 334–41 and Des Chene, "On Laws and Ends," 146–8. Relying on what Des Chene wrote in *Physiologia*, 340, Schmaltz, *Descartes on Causation*, 105, identifies his position to Garber's position, which amounts to neglect one aspect of Des Chene's position.
33 Descartes, *Principia philosophiae*, II 36, AT VIII 61, CSM I 240, modified.
34 Descartes to Elisabeth, 6 October 1645, AT IV 314: "Dieu est tellement la cause universelle de tout, qu'il en est en mesme façon la cause totale; et ainsi rien ne peut arriver sans sa volonté."
35 The expression "division of labor" flows naturally from the pen in this context; not surprisingly, it has been used by those who contest the occasionalist interpretation of Descartes; see Clarke, "The Concept of Vis," 333; Hattab, "The Problem of Secondary Causation," 110; Schmaltz, *Descartes on Causation*, 23, 122–3, 218.
36 Descartes, *The World*, AT XI 37, CSM I 92–3, modified. This passage is one of the reasons why Hattab, "The Problem of Secondary Causation," 109, suggests to put aside the problem of forces and to find in laws of nature an answer to the question of causal agency. Des Chene, *Physiologia*, 316–17, and Gorham, "Cartesian Causation," 411, point out that, contrary to what happens in *The World*, Descartes makes paradoxically the mutation of things an argument for divine immutability in *Principia philosophiae*.
37 Descartes to Elisabeth, 6 October 1645, AT IV 314.
38 Descartes to Mersenne, 27 May 1630, AT I 152: "il [Dieu] a créé toutes choses. . . *ut efficiens et totalis causa*."
39 *Primae responsiones*, AT VII 109: "non dubitarem illam causam, quae me conservat, efficientem appellare."
40 *Quintae responsiones*, AT VII 369: "Deus est causa rerum creatarum non modo secundùm fieri, sed etiam secundùm esse, ideosque debet semper eodem modo influere in effectum, ut eundem conservet."
41 *Sextae responsiones*, 8, AT VII 436.
42 *Quartae responsiones*, AT VII 236: "causa efficien[s], sine quâ res finitae esse non possunt" and 244: "ii qui Deum nondum norunt, in causam aliarum rerum efficientem inquir[unt]"; *Principia philosophiae*, I 28, AT VIII 15–6: "ipsum [Deum] ut causam efficientem rerum omnium considerantes"; Descartes to Chanut, 6 June 1647, AT V 54: "c'est Dieu seul qui est la cause finale, aussi bien que la cause efficiente de l'univers."
43 Descartes to Morin, 13 July 1638, AT II 209, in answer to Morin to Descartes, 22 February 1638, AT I 548; Descartes to Hobbes, 31 March 1641, AT III 355, in answer to Hobbes to Mersenne for Descartes, 30 March 1641, AT III 343.
44 Garber, *Descartes Metaphysical Phyics*, 276–8.
45 Steven Nadler, "Descartes and Occasional Causation," *British Journal for the History of Philosophy* 2 (1994): 35–54, here especially 42–3.
46 Descartes to Princess Elisabeth, 3 November 1645, AT IV 332–3, CSMK 277.
47 Descartes to Princess Elisabeth, January 1646, AT IV 353–4, CSMK 282.
48 *Ibid.*
49 Descartes to Princess Elisabeth, 6 October 1645, AT IV 314, CSMK 272.
50 *Principia philosophiae*, II 37, AT VIII 62, CSM I 240. Hattab, "The Problem of Secondary Causation," 108–16, insists on II 37 in order to defend her thesis that laws are secondary causes, but neglects II 40.
51 *Principia philosophiae*, II 40, AT VIII 65, CSM I 242.
52 *Ibid.* It must be noted that "vis" and "actio" appear often as synonymous in Descartes' vocabulary, see for example "vis et actio" (AT VII 49), "vim, vel actionem" (*Principia philosophiae*, II 25, AT VIII 54), "eadem vis et actio" (*ibid.*, II 29, AT VIII 55), "vis sive actio" (*ibid.*, III 38, AT VIII 96).
53 *Principia philosophiae*, II 25, AT VIII 54: "*Et dico [motum] esse translationem, non vim, vel actionem quae transfert, ut ostendam illum semper esse in mobili,*

non in movente, quae haec duo non satis accurate solent distingui." When commenting on this chapter, Tad Schmaltz told me that it was what he had in mind when he tied the force of a body to its tendency to endure, in which case I cannot but agree. Still, I think that there is a tendency to ontologize these tendencies in his book.

54 *The World*, AT XI 38, CSM I 93: "each individual part of matter . . ., if it has once begun to move, it will always continue with an equal force [*avec une égale force*] until others stop or retard it." *Principes de la philosophie*, II 40, AT IX 84: "lorsqu'elle a commencé une fois de se mouvoir, nous n'avons aussi aucune raison de penser qu'elle doive jamais cesser de se mouvoir de mesme force; pendant qu'elle ne rencontre rien qui retarde ou arreste son mouvement."

55 *Principia philosophiae*, II 45, AT VIII 67, CSM I 244.

56 *Ibid.*, II 43, AT VIII 67, CSM I 244.

57 See, for example, Descartes to Mersenne, April 1634, AT I 287; to Plempius for Fromondus, 3 October 1637, AT I 413, 421, CSMK 63, 64; to Plempius, 15 February 1638, AT I 529, CSMK 83; to Mersenne, 11 November 1640, AT III 233, CSMK 156; to Mersenne, December 1640, AT III 258, CSMK 160; to Charlet, October 1644, AT IV 140–1, CSMK 238/

58 Descartes to Mersenne, 25 April 1630, AT I 144, CSMK 22: "I would not have been able to discover the foundations of physics, if I had not looked after them along that road [the road of metaphysics]"; Descartes to Gibieuf, 11 November 1640, AT III 233, CSMK 157: "the little book which I sent [the *Meditations*] you contain all the principles of my physics"; Descartes to Mersenne, 28 January 1641, AT III 298, CSMK 173: "these six Meditations contain all the foundations of my physics"; *Principes de la philosophie*, Préface, AT IX 19: "quelques veritez de Metaphysique, sur qui toute la Physique doit estre appuyée."

59 *Principia philosophiae*, II 37 and 43, AT VIII 62 and 66, CSM I 240–1 and 243.

60 Against the metaphor of chain of reasons and via the expression "the space [*espacement*] of truth" that he borrows from Jacques Derrida, Denis Kambouchner, *L'homme des passions: Commentaires sur Descartes* (Paris: Albin Michel, 1995), II, 354–5, reaches a similar conclusion concerning the relation of metaphysics and ethics.

Part III

European Receptions

8 Duplex Intellectus Et Sermo Duplex

Method and the Separation of Disciplines in Johannes De Raey

Antonella Del Prete

When Johannes de Raey's last work was printed in 1692, readers gained access to two letters, published at the end of the volume. The first letter, dated 1680, was addressed to his friend Christoph Wittich, at that time professor of theology in Leiden. Here, De Raey discussed the relationship between physics (which is supposed to be part of philosophy) and medicine. The second letter commented on a controversy which, shortly before in 1686, had taken place at the University of Franeker: some philosophers and theologians had appealed to Cartesian philosophy to justify the use of reason in theology, prompting vigorous replies from the supporters of the *philosophia recepta*.[1] De Raey openly criticized the Cartesian group in Franeker. In his view, philosophy should be separated from other disciplines, including theology, law, and medicine. He even argued that Descartes' legacy had been diverted from its original purpose and that some of Gijsbert Voetius's opinions should be rehabilitated. Voetius had been right to oppose Regius's *Cartesiomania* but had made the mistake of identifying the opinions of the master with those of the disciple.[2] Although he kept critical distance from the former opponent of Descartes, De Raey believed that "it is easier to reach a state of grace by following Voetius than by following Regius."[3]

However, far from being a follower of the scholastic philosophy, as these judgments might suggest, De Raey was in many ways a very typical example of the diffusion of Descartes' ideas in the Netherlands: first, because he had known him personally and participated in the publication of his *Opera philosophica* and the *Epistolae;*[4] second, by his *cursus studiorum*—he had studied under the direction of two well-known Cartesians, Henricus Regius and Adriaan Heereboord. Third, De Raey played a leading role during the heated controversies opposing the supporters of the new philosophy and the followers of the *philosophia recepta* in Utrecht and in Leiden.[5] The publication of his first and most renowned book, *Clavis philosophiae naturalis sive introductio ad naturae contemplationem Aristotelico-Cartesiana* (1654), took place when the controversy was in full swing.

De Raey's teaching at the University of Leiden and the Amsterdam Athenaeum,[6] while being less challenged than his Cartesian activities as a student, did give rise to controversy. His private lessons were interrupted in 1648, following the protests of the Leiden theologian Jacobus Revius. When he began the 1654 edition of the *Clavis* with a dedication to the Curators of the University of Leiden, they were forced to publish a decree ordering the elimination of any reference to Descartes on the title page of published copies. Even his lectures on medicine prompted protests from his colleagues.[7] In short, De Raey's biographical data indicate sympathy for modern philosophy and opposition to Scholasticism. We must then try to understand what brought him to pronounce these disavowals toward the end of his life.

Let us begin with the 1680 letter to Wittich. Here, De Raey develops his intellectual biography. His narrative is organized around a single leitmotif: the separation of philosophy from other disciplines.[8] He first recalls that already in 1651 he had distinguished philosophical knowledge from common knowledge. The latter coincided with Aristotelian philosophy, which De Raey did not appreciate.[9] He then discussed the study of logic, concluding that there were two kinds of logic, the logic of common knowledge and the logic specific to philosophical knowledge. Good examples of the latter type were the Platonic dialectics, Descartes' *Meditations*, and the first part of his *Principles*. These two approaches should remain separate. In this respect, De Raey believed that even his friend Johann Clauberg could not be exempted from criticism.[10] The arrival of Franciscus Sylvius in Leiden in 1658 as Professor of Medicine gave De Raey an opportunity to reflect on contemporary examples of the methodological mistakes made by Peripatetic philosophers. Like Aristotle, Sylvius believed that there was a single scientific method, namely the experimental method, which applied to both physics and medicine. De Raey, on the contrary, was becoming increasingly convinced that his initial intuitions had to evolve toward a practical separation of philosophical knowledge from the topics taught in the disciplines of the higher faculties, that is to say law, medicine, and theology.[11] The difference between the two methods was explained by the twofold nature of the human mind: philosophy derived from a higher understanding compared to the one producing the knowledge specific to the other disciplines. The publication of Lodewijk Meyer's *Philosophia S. Scripturae interpres* (1666) and Spinoza's *Tractatus theologico-politicus* (1670) made it even more evident that such a separation was urgently needed.

We now arrive at the time De Raey moved to Amsterdam and to the long gestation period of his *Cogitata de interpretatione*. His thinking took its final form: the human mind has two parts, and to these parts correspond two kinds of language and two methods: one language and method common to all arts, and another language and method specific to philosophy. Everything separates philosophy from the other disciplines. It is as if De Raey wanted to ban the application of any results from philosophy to other areas of knowledge. Based on similar assumptions, De Raey condemned the Franeker Cartesians, who, as we have seen, were guilty of not respecting the existing separation between philosophy and theology:

> The understanding is two-fold and so is language. One, as it seems, is prior in nature and common to all men, to all other arts and disciplines.

> The other is posterior in nature, and proper to Philosophy. We must therefore distinguish, as far as possible, this double understanding and language, and for the sake of this fundamental distinction, we must think that philosophy is by nature different from every other art and discipline.[12]

Now, generally, we should be wary of autobiographies and, as we shall see, this also applies to De Raey's. In order to understand the genesis of his ideas and verify whether the steps described in his 1680 letter truly match his intellectual development, we must begin by looking at what he wrote in the 1654 edition of *Clavis philosophiae aristotelico-cartesiana*. Moreover, we must try to grasp what he meant when speaking of "philosophy." In the first essay of the *Clavis*, originally delivered in 1651, De Raey began by distinguishing revealed knowledge from natural knowledge.[13] Next, natural knowledge was further divided into knowledge shared by all humanity, knowledge non-specific to the philosopher, and philosophical knowledge.[14] In this context, however, this distinction did not give rise to a complex classification of the disciplines and the arts, as it did in subsequent works. Hence, at the beginning of De Raey's academic career, the opposition between these two types of knowledge was not used to develop a tree of knowledge, based on a classification of disciplines involving a methodological differentiation. It was, it seems, rather aiming at discarding Aristotelian philosophy in favor of Descartes and the modern philosophy. As for the classification of disciplines, in 1654, De Raey separated theology from the other disciplines. It was not affected by the distinction between common knowledge and philosophical knowledge, because it had a revealed origin. As for other kinds of knowledge, they entertained relations that subsequently changed. Only medicine was explicitly described from early on as utilitarian knowledge, giving it an empirical and practical nature. The praise of philosophy as contemplative knowledge, along with the classification of medicine as art, should remind us of Jacopo Zabarella's attempt to defend the independence of philosophy in general, and natural philosophy in particular, against the interference by representatives of the higher faculties, i.e., physicians and theologians. The same battle that took place in Italy in the name of Aristotle now took place in the Netherlands in the name of Descartes.[15] The same method seemed to apply, according to De Raey, not only to all the objects that escaped the senses (God, spirits, and mathematical objects), but also to natural things. In all these cases, we should attempt to attain the first principles and see intuitively the *praecognita*, assumptions providing access to the knowledge of these objects.

In subsequent works, this classification radically changed. Without losing any of his polemic aim, the common/philosophical dichotomy would come to organize all kinds of knowledge.

Let us now focus on the *Disputatio philosophica de constitutione logicae*, defended first in Leiden in 1668, and then again in Amsterdam in 1684. Here, De Raey began by separating ordinary and common knowledge from the philosophical knowledge, organizing their respective qualities in two columns. Ordinary knowledge was based on common sense and looked for what is useful to our lives. Thus, it classified all things by examining their relationship to us. Philosophical knowledge, on the contrary, went beyond common sense and aimed at showing us things as they are in themselves, without regard for their usefulness or harmfulness. Ordinary and philosophical knowledge stemmed from two different kinds of logic: the first was dialectic and the abuse of it had led to false philosophy; the second was true logic.[16] In 1668, the text stopped here, providing no explicit example of the abuse in question. Everything suggests that De Raey aimed at Scholastic philosophy, just like in 1654. In a roughly contemporary dissertation (defended in 1669, but printed in the 1677 edition of the *Clavis*), the *Specimen logicae interpretationis*, he did, however, also argue that Cartesians were guilty of an error similar, or rather inverse, to the one committed by Aristotle's followers. Just like the Peripatetic philosophers, they did not respect the separation between philosophy and common knowledge. However, while Aristotle and the Aristotelians used common logic to establish a true philosophy, developing only common philosophy, Cartesian philosophers applied true philosophy to other disciplines, based on ordinary knowledge.[17] In 1692, when the *Disputatio philosophica* was reprinted in *Cogitata de interpretatione*, De Raey added a note explaining that he was aiming at both Aristotelian and Cartesian philosophers. However, if the former merely repeated their master's mistake, it was not the same for the Cartesian philosophers. In fact, Descartes himself had shown how to separate common and imperfect knowledge from the knowledge of the truth by the first causes, i.e., how to distinguish dialectics from true logic.[18]

The *Disputatio philosophica* also gives us some indications about the nature of true logic. The purpose of true logic was to detect fallacies related to hidden reasons, to avoid precipitate judgment, and to eliminate obscurity and confusion in our concepts. These precepts, established with reference to both Bacon and Descartes, was followed by an exposition, not very original, of the four methodological rules of the *Discours*. Like Clauberg, Arnauld, and Nicole, De Raey believed that these rules expressed the mathematical ideal of Cartesian science, introducing a new logic.[19] However, the objects to which this method was applied had changed in comparison to Descartes and the 1654 edition of the *Clavis*. Clear knowledge—distinct, ordered, and sufficient—could be reached, in De Raey's opinion, only in mathematics. Medicine, law, and theology studied composite and confused objects: they were not real sciences, but arts or practical disciplines, aiming at what is useful. They did not seek the truth, but relied on habits, examples, and authority. They did not rely on reason and judgment, but on experience, opinion, and anticipation.[20] In short, they were not philosophical knowledge, but common knowledge. A first displacement here becomes evident. In 1668, and contrary to what he held previously in 1654, De Raey generalized his classification of knowledge, based on methods used by different disciplines and on the faculties of the mind they appealed to. By focusing on the sensible/intellectual dichotomy, rather than on the natural/revealed dichotomy, he was now able to separate philosophy from all the subjects taught in the higher faculties. Theology, however, had lost its uniqueness and was grouped with law and medicine as practical and sense-based utilitarian knowledge.

Let us now turn to philosophy where De Raey reserved a surprise for us: physics did not fully correspond to the criteria necessary to be accepted into philosophy because it focuses on composite and difficult objects.[21] The status of physics was somewhat ambiguous: it was not an art nor practical discipline such as medicine, law, and theology. Therefore, it should not use common logic. The nature of the object of physics, i.e., the body, did however prevent the full application of rules of the method that we find in mathematics. Should we then conclude that mathematics was the only true philosophy? No, we should not, because in addition to method, there was also a science of logic, defined as the study of the principles of human knowledge. This logic could legitimately be identified with metaphysics or first philosophy.[22] When giving some examples of this true logic, De Raey unexpectedly referred, not to a contemporary, but to an ancient philosopher, Plato, whose dialectics he held in great esteem. More explicitly than in the 1654 edition of the *Clavis*, the criterion of science had become the use of the understanding: the science of logic or dialectic excluded recourse to the senses, the imagination, discursive reason (which derived conclusions from suppositions), and certainty arising from faith (which was based on experience, not on evidence.) An immediate consequence of this epistemology was that the arts and disciplines concerned with the useful could not be considered scientific, not only because they could not use the Cartesian rules of method, but also because they did not live up to this second criterion, to the extent that they were based on sense experience or authority.[23] The opposition between hypothetical-deductive reasoning and intellectual intuition, however, ended up being more restrictive than the rules of the *Discours*: mathematics, in fact, relied on a demonstrative and discursive process, and therefore could not be classified as a real science, although mathematics was not mere opinion either.[24]

Regarding physics, the discipline should be divided into two branches. In the first branch, we find natural history, which, in a Baconian vein, relied on observation, experience, sometimes even faith. It was not separated from medicine, chemistry, or other arts. In its second branch, physics as a science, it used a hypothetical-deductive method, like mathematics, and was therefore subject to the same problems, when compared to knowledge obtained through intuition. However, contrary to mathematics, physics also required the prior elimination of prejudice. Hence, the four rules of the method were not sufficient to confer scientific status on physics, but it also required a science of logic, namely the Platonic dialectics.[25] The four rules of Cartesian method therefore applied to physics, but the complexity of its object prevented it from making effective use of them, as mathematics does. However, scholars of physics could receive some assistance from the science of logic, namely from the Platonic dialectic, which provided real hypotheses that deductive reasoning could be based upon. Hence, physics could obtain at least a partially scientific status. A corollary here established a connection between the Platonic dialectics and Cartesian philosophy: "The first part of *Principles of Philosophy*, which is concerned with the principles of human knowledge, is the science of logic that I call Philosophy."[26]

We now better understand the internal division within philosophical knowledge, including the relationship between metaphysics and physics. The essay *De vera metaphysica*, published in the 1677 edition of the *Clavis*, provides us with another piece of the puzzle. According to this text, true metaphysics dealt with the principles of human knowledge:

> True metaphysics, which more correctly is called First Philosophy, is concerned with the principles of human knowledge and therefore teaches to philosophize in order. The other parts of philosophy, and physics first of all, discover here their own objects and principles, not to mention their certainty and evidence, which must be based on these.[27]

These definitions require some comments. Maybe De Raey was trying to build on the *Lettre-Préface* to the 1647 edition of the *Principles* to justify this series of equivalences between philosophy, the science of logic, first philosophy, and metaphysics. Descartes here prompted readers to study first the true logic in order to learn how to conduct his reason, and then to apply "the true philosophy," the first part of which is "metaphysics, which contains the principles of knowledge,"[28] and then finally move on to physics, which must first establish the principles of material things, and then analyze the universe. De Raey, however, distorted this famous Cartesian text in a small but significant manner. First, if the *Lettre-Préface* allowed him to describe Descartes' method as a "logic," the French philosopher himself did not identify his method with metaphysics in the way that De Raey does in his works. Descartes distinguished the principles of knowledge, which were the subject of first philosophy or metaphysics, from the principles of material things, which belonged to physics. The latter were deduced from the former, but they did not coincide with them: metaphysics was concerned with the existence of thought and of a truthful God, physics with the existence of extended bodies, insofar as they assume different shapes and movements. Instead, De Raey transformed metaphysics into the study of the principles and objects of physics. In this regard, it is useful to compare these passages to the text of Franco Burgersdijk, perhaps the most important teacher of philosophy in Leiden in the seventeenth century. Burgersdijk distinguished metaphysics from logic. He also argued that metaphysics was superior to other sciences, because it provided them with their objects and principles:

> it gives the object and the principles of all things: where metaphysics stops, the other sciences begin. And so, those things that are finally demonstrated by metaphysics are the principles of other sciences, and they are presupposed by these without demonstration.[29]

Combining Descartes and Burgersdijk, De Raey gave metaphysics a broader epistemological role to play than Descartes did. Taken on its own, not only physics could not establish its own certainty, but it also had to look elsewhere for its starting point and purpose. Metaphysics, however, lost its autonomy and was now conceived as the foundation of physics and other sciences: the study of the mind and God was only a premise for doing physics.

Without going into detail, it suffices to note here with regard to the specific content of metaphysics that De Raey distanced himself from Burgersdijk and the Calvinist tradition, and rather followed Descartes. Burgersdijk was one of the instigators of the Calvinist use of Suárez. For him, the object of metaphysics was the *ens reale*, which was studied both as *ens qua ens* and as an immaterial being. The contents of metaphysics were close to those of theology. The latter, however, did not rely on reason, but on faith.[30] Contrary to this, in his 1677 *Clavis*, De Raey contented himself with an exposition of Descartes: doubt, understood as a way to get rid of prejudice, was followed by the observation that we exist as thinking beings, the distinction of thought and body, and the idea of God. Note, however, that in the second part of the *De vera metaphysica*, De Raey engaged in a broad controversy against metaphysics as the study of *ens qua ens*. Since it began with universals, he argued, ordinary metaphysics could only be a purely verbal discipline, separating what should not be separated or inventing things that do not exist. It is evident from these considerations that his disagreement with Clauberg not only concerned the project to create an Aristotelian-Cartesian logic (which for De Raey amounted to a contradiction since it involved mixing common with philosophical logic), but also Clauberg's plan to develop a real ontology. According to De Raey, such a discipline would be useless verbiage. By rejecting the study of *ens qua ens*, De Raey was also opposed to the large group of Calvinist philosophers and theologians who applauded the emphasis that Suárez had put on this discipline. He was in particular opposed to the philosophical strategies of the Voetians, whose appreciation of Suarezian metaphysics has been amply documented.[31] When defining metaphysics as the principles of knowledge and proposing an exposition of the path leading from Cartesian doubt to ideas, through God, the *res extensa* and the *res cogitans*, De Raey made the same choices as his friend Wittich, who, in his *Theologia pacifica*, used the division of disciplines outlined by Descartes in *Notae in programma quoddam*. According to this division, theology was concerned with revelation and belonged to faith; physics dealt with what we can achieve by the natural light; metaphysics, or the study of God and the mind, occupied an intermediate field, accessible to both reason and faith.

Returning now to physics, in the *Disputatio de constitutione physicae*, De Raey criticized the common definition of physics as "knowledge of the natural body to the extent that it is natural." This was a very common definition in textbooks of the time, from Duns Scotus, to Zabarella, to Burgersdijk and Heereboord, the latter being the most likely source for the *Clavis*.[32] The traditional definition had the advantage of allowing us to demarcate physics immediately. De Raey, for his part, first delimited physics from theology and metaphysics: physics was not concerned with the first cause, but only with secondary causes; it had to remain focused on nature, without rising to the contemplation of God, as theology and metaphysics did. Nature also opposed art: man, being endowed with an incorporeal mind, was capable of altering nature, like when creating drugs. Finally, physics was different from mathematics, because it studied existing objects, not their abstractions.

As we have seen, when analyzing the essays devoted to philosophical logic, the scientific status of physics was somewhat problematic in De Raey, to the extent that he adopted the intellectual character of knowledge and the methodological rules established by Descartes as criteria of scientific knowledge. In the essay *De constitutione physicae*, however, he leaned toward granting physics scientific status. It could become certain and evident knowledge grasped by our natural light. As scientific knowledge, physics did not fall under the category of *fides*, based on the testimony of others, nor under opinion, that is to say the merely probable.[33] Physics was also something more than experience, otherwise it would be natural history.[34] De Raey went on to refer to two maxims that traditionally indicated the difference between physics and medicine: "where the physicist stops, the physician begins" and "where the physician stops, the physicist begins." De Raey relied on the second maxim to support his thesis: where the sense experience of the physician ended, the intellectual work of the physicist began. Medicine, therefore, was based solely on experience and close to natural history; but physics as a science was totally separate from it, as De Raey had argued since his youth.

Physics as a science should study sensible phenomena, connecting them to their intelligible causes.[35] It therefore belonged to intellectual knowledge. When a physicist grasped what matter or motion were, he grasped by intuition an object that is simple and unique, or a truth *per se nota*.[36] Some objects of physics were, however, composites: in this case, physics knew them through their causes, i.e., through deductive reasoning.[37] In addition to this, this causal knowledge not only excluded the use of all the notions that ordinary physics adopted from ordinary logic (natural powers, the substantial forms, occult qualities, etc.), but also the consideration of final causes.[38] Scientific physics had to limit itself to the study of efficient causes. But efficient causes could be searched out in three distinct ways. First, we can determine them through observing phenomena with our senses: this is the method of natural history. Next, we can deduce causes from effects, but often we do not exceed the level of opinions and conjectures. Finally, we can—and this is the only *consideratio potissima et legitima*—know the causes first and then seek confirmation in the effects.[39] De Raey was not explicit about this, but it could be supposed that the latter kind of knowledge, i.e., knowledge through the causes, ultimately required a logical regression from physics to metaphysics, the discipline that discovered the principles of knowledge.

Let me summarize De Raey's itinerary. After 1668, he radically changed his thoughts about method and outlined a picture that remained unchanged until 1692. This radical change involved giving a new structure and a new content to the distinction between common knowledge and philosophical knowledge, which was the subject of his first book. He initially used this distinction to challenge any claim to primacy made by Aristotelian philosophers. Afterwards, he used it to classify the different disciplines and describe their relationships. When classifying medicine, law, and theology together as common knowledge and philosophy as true knowledge, he had a dual purpose: he no longer did this only to establish the *libertas philosophandi* and allow Cartesian philosophy to spread freely in Dutch universities, as in his early years. He also did it to save other disciplines, including theology, from the kind of attacks that thinking appealing to clarity and distinction was open to.

Already in 1654, the external front was accompanied by an internal front. De Raey not only opposed fanatical Aristotelian philosophers, but also the "wrong" interpretations of Descartes' thought. He already decided to classify medicine among the arts and thus to remove it from the field of philosophical method around 1650, before the arrival of Sylvius in Leiden. On this point, De Raey was most likely responding to the potentially materialist reading of Cartesian philosophy applied to medicine that he encountered in the books of Regius. We find implicit evidence to support this thesis in the essay *De Aristotele et aristotelici*, published in the 1677 edition of the *Clavis*. Here, De Raey criticized Aristotle and his followers, because their theories potentially made the soul a corporeal being. He then proceeded to argue against those who wanted to correct these consequences by appealing to faith, because he thought that true philosophy taught the immateriality of the soul. But it was precisely Regius who held that the only proof we have of the immortality of the soul was in Scripture, prompting the indignation of both Descartes and Voetius.[40] De Raey, then, criticized Regius from very early on.

However, when we reach the late 1660s, the external enemy no longer seemed so threatening to De Raey. The internal enemy now came to the forefront. The methodological separation between common and philosophical knowledge now excluded the application of the rules of the *Discours* and of the Platonic dialectics not only in medicine, but also in law and theology. These disciplines could continue to use Aristotelian syllogistic logic, if they pleased. As is clear from the letter that closed the *Cogitata de interpretatione*, when Röell decided to develop a theology based on Cartesian principles, De Raey replied by repeating what he had written in 1668–1669, opposing most likely Lodewijk Meyer's *Philosophia Sacrae Scripturae interpres* without naming it.[41] However, when categorizing theology as common knowledge, De Raey distanced himself, not only from what has recently been called the "radical Enlightenment," but also from much less radical Cartesians including Wittich and Frans Burman, who were not entirely against using certain elements of the Cartesian philosophy to develop a theology free from scholastic mistakes.[42]

Returning now to our point of departure, the two letters published at the end of the *Cogitata de interpretatione*, we see that De Raey's intellectual autobiography distorted and adapted what he said in his youth to the future developments of his thought. He was considerably more accurate when describing the years 1668–1670. There is, however, a lexical and conceptual variation, which became important in the 1680 letter to Wittich, but which was already present, without being emphasized, in the 1677 edition of the *Clavis*.[43] In previous works, common knowledge is not intellectual, but sensible. In the *Cogitata*, however, it is referred to the understanding.[44] This change is problematic. It seems difficult to explain it by reference to any internal evolution in De Raey's thought, since it tended to weaken the distinction between art and science, the establishment of which was central to his project. Perhaps this was an effect of the *Tractatus theologico-politicus*? In this book, as is well known, Spinoza distinguished truth, which is the object of philosophy, from obedience, which is the purpose of religion. When opposing philosophical knowledge, which is theoretical and contemplative, to knowledge looking for the merely useful without seeking the truth, such as theology, medicine, and law, De Raey risked being assimilated to Spinoza. Referring common knowledge to the understanding was maybe an attempt to avoid downgrading theology to a discipline with no relationship to truth. Perhaps De Raey eventually understood that he was going from bad to worse: separating philosophy from other disciplines, making utilitarian knowledge out of them, did indeed protect them from attacks coming from philosophical rationalists who were becoming increasingly aggressive, such as Roëll. However, separating philosophy from other disciplines also amounted to depriving them of the capacity to reach the truth, like in Spinoza's *Tractatus theologico-politicus*. Conversely, establishing a kind of simultaneous presence of metaphysics and theology in the study of God and the mind, as proposed by Descartes in his *Notae in programma quoddam*, allowed assigning a truth content to theology and Scriptures, but ran the risk of subjecting them to reason.

What can we conclude from this overview of the texts of De Raey? First, we realize that the dynamics determining the reception of Descartes is complex. It includes both personal relationships and intellectual choices. The "Cartesian" philosophers often behaved as a community, consulting with each other on which strategies to follow, expressing their views on published books, and helping each other in their academic careers. As we can see in the case of De Raey, they did not break the united front. Sometimes they remained silent for a very long time about their internal disputes or they only evoked them implicitly, avoiding naming colleagues. Cartesianism is, however, not a uniform movement. Each Cartesian philosopher invented his own Descartes, choosing to emphasize this or that thesis or this or that part of his philosophy, placing these texts in a context which owed more or less to other philosophers, be they ancient or modern. These different kinds of Cartesianism evolved throughout the century, following the various controversies. From this, it follows that even when we find comments on or literal quotes from Cartesian works, they should always be interpreted taking into account the fact that those texts were addressed to a cultural environment often very far removed from the ideal audience that Descartes had in mind for his works. This also applied to the relationship between metaphysics and physics in De Raey's books. The relation between physics and metaphysics was of primary importance for Descartes. In the context of De Raey's work, however, we must examine the question in the broader—and for De Raey more important issue—of the relationship between philosophy, medicine, and theology. This was a problem that mobilized a large number of philosophers in the United Provinces during the second half of the seventeenth century, virtually unknown professors who were worried about the future of the discipline they taught, but also leading philosophers such as Spinoza.

Notes

1 On this controversy, see Roberto Bordoli, *Dio ragione verità: Le polemiche su Descartes e su Spinoza presso l'Università di Franeker* (Macerata: Quodlibet, 2009); on the relationship between philosophy and theology in reformed orthodoxy, see Aza Goudriaan, *Theology and Philosophy*, in *A Companion to Reformed Orthodoxy*, ed. Herman J. Selderhuis (Leiden and Boston: Brill, 2013), 27–63.

2 The use of the noun *Cartesiomania* was, however, an implicit reference to the title of two volumes published by Jacobus Revius in 1654 and 1655, criticizing Tobias Andreae (*Cartesiomania, hoc est furiosus nugamentum quod Tobias Andreae, sub titulo Assertionis Methodi Cartesianae, orbi literato obstrusit, succincte et solidum confutatum* [Lugduni Batavorum 1654]; *Cartesiomania pars altera, qua ad secundam partem rubiosae assertionis Tobias Andreae respondetur* [Lugduni Batavorum 1655]). This implicit reference clearly denoted a change of mind. In the late 40s, it was precisely Revius who had De Raey's public lectures suspended.

3 Johannes de Raey, *Cogitata de interpretatione. . .* (Amstelaedami: Apud Henricum Wetstenium, 1692), 665.

4 C. Louise Thijssen-Schoute, *Nederlands Cartesianisme*, ed. Th. Verbeek (Utrecht: HES, 1989), 134–42.

5 Heereboord himself tells us about a quarrel between De Raey and Stuart during a disputation. See Adriaan Heereboord, *Meletemata philosophica, maximam partem metaphysica* (Lugduni Batavorum: Francisci Moyardi, 1654), 18–19; Theo Verbeek, *Descartes and the Dutch. Early Reactions to Cartesian Philosophy,*

1637–1650 (Carbondale and Edwardsville: Southern Illinois University Press, 1992), 34–51 (on De Raey and his role, see especially 48–9).

6 On De Raey's teaching in Leiden, see Dirk van Miert, *Humanism in an Age of Science: The Amsterdam Atheneum in the Golden Age, 1632–1704* (Leiden: Brill, 2009), 97–100, 230–2, 242–5, 269–72.

7 Thijssen-Schoute, *Nederlands Cartesianisme*, 125–7; Wiep van Bunge et al., *The Dictionary of Seventeenth and Eighteenth-Century Dutch Philosophers* (Bristol: Thoemmes, 2003), t. II, *sub voce* by Paul Schuurman, 813–4.

8 On the *Cogitata de interpretatione* and this letter to Wittich, see Theo Verbeek, "Tradition and Novelty: Descartes and some Cartesians," in *The Rise of modern Philosophy*, ed. Tom Sorell (Oxford: Oxford University Press, 1993), 167–96; Theo Verbeek, "Les Cartésiens face à Spinoza: l'exemple de Johannes de Raey," in *L'hérésie spinoziste: la discussion sur le Tractatus theologico-politicus, 1670–1677, et la réception immédiate du spinozisme*, ed. Paolo Cristofolini (Amsterdam: APA Maaarsen, 1995), 77–88. Alexander Douglas ("Spinoza and the Dutch Cartesians on Philosophy and Theology", *Journal of the History of Philosophy*, 51 (2013) 567–88; *Spinoza and Dutch Cartesianism* (Oxford: Oxford University Press, 2015), 36–63) underestimates both the differences existing between Wittich and De Raey, and the changing in De Raey's mind.

9 De Raey, *Cogitata de interpretatione*, 657. On the evaluation and use of Aristotle's philosophy by De Raey, see Paul Dibon, *La philosophie néerlandaise au siècle d'or: I L'enseignement philosophique dans les universités à l'époque précartésienne (1575–1650)* (Paris, Amsterdam, Londres and New York: Elsevier Publishing Company, 1954); Paul Dibon, *Regards sur la Hollande du siècle d'or* (Napoli: Vivarium, 1990), 203–4; Verbeek, *Descartes and the Dutch*, 8–9; Theo Verbeek, "Descartes et les Pays-Bas: philosophie et théologie," in *Il vocabolario della République des Lettres: Terminologia filosofica e storia della filosofia. Problemi di metodo*, ed. by Marta Fattori (Firenze: Leo S. Olschki, 1997), 301–4; Andrea Strazzoni, "La fisica aristotelico-cartesiana di Johannes de Raey," *Giornale critico della filosofia italiana*, 80 (2011): 107–32.

10 De Raey, *Cogitata de interpretatione*, 659.

11 *Ibid.*, 660.

12 *Ibid.*, 661.

13 Antonella Del Prete, "Discussioni sul metodo nel cartesianismo olandese: Il caso di Johannes de Raey," in *La ragione e le sue vie: Saperi e procedure di prova in età moderna*, eds. Carlo Borghero and Claudio Buccolini (Firenze: Le Lettere, 2015), 146–67.

14 Johannes de Raey, *Clavis Philosophiae Naturalis, seu introductio ad naturae contemplationem, Aristotelico-Cartesiana* (Lugdunum Batavorum: Ex Officina Johannis et Danielis Elsevier, 1654), 7–8, 16–7, 21, 24, 36–7, 39–42.

15 See Nicholas Jardine, "Keeping Order in the School of Padua: Jacopo Zabarella and Francesco Piccolomini on the Offices of Philosophy", and Heikki Mikkeli, "The foundation of an autonomous natural philosophy: Zabarella on the classification of arts and sciences," in *Method and Order in Renaissance Philosophy of Nature*, eds. Daniel A. Di Liscia, Eckhardt Kessler and Charlotte Methuen (Aldershot: Ashgate, 1997), 183–209 and 211–28; Charles B. Schmitt, "Aristotle Among the Physicians," in *The Medical Renaissance of the Sixteenth-Century*, eds. Andrew Wear, Roger K. French and Iain M. Lonie, (Cambridge: Cambridge University Press, 1985), 1–15.

16 Johannes de Raey, *Disputatio Philosophica de Constitutione Logicae. . .* (Lugduni Batavorum: Apud Viduam et Haeredes Johannis Elsevirii, 1668), 1.

17 De Raey, *Cogitata de interpretatione*, 540.

18 *Ibid.*, 596–7.

19 De Raey, *Disputatio Philosophica de Constitutione Logicae*, 1–2. On Clauberg's *Logica vetus et nova*, see Massimiliano Savini, *Johannes Clauberg: Methodus cartesiana et ontologie* (Paris, Vrin, 2013), 197–268; Jacqueline Lagrée, "Qu'est-ce qu'être un logicien cartésien?" in *Qu'est-ce qu'être cartésien?* ed. Delphine Kolesnik-Antoine (Lyon: ENS Éditions, 2013), 143–58; Andrea Strazzoni, "A Logic to End Controversies: The Genesis of Clauberg's Logica Vetus et Nova," *Journal of Early Modern Studies* 2 (2013): 123–49.

20 De Raey, *Disputatio Philosophica de Constitutione Logicae*, 2–3.

21 *Ibid.*, 3.

22 *Ibid.*

23 *Ibid.*, 4–5.

24 *Ibid.*, 5–6.

25 *Ibid.*, 6. The second edition published in 1692 (*Cogitata de interpretatione*, 606), showing the author's final position, did not display any significant difference, apart from a stronger doubt about whether physics, thus understood, was really a science. On the classification of natural philosophy as a science, see *De sapientia veterum* (Amsterdam, 1669), in *Clavis Philosophiae Naturalis Aristotelico-Cartesiana. Editio secunda. . .* (Amstelodami: Daniel Elsevir, 1677), 727.

26 De Raey, *Disputatio Philosophica de Constitutione Logicae*, 6. In 1692, the same assessment is not written as a corollary.

27 De Raey, *Clavis Philosophiae Naturalis Aristotelico-Cartesiana: Editio secunda*, 412. For a similar comparison of logic, first philosophy, and the study of the principles of human knowledge, see the dissertation *De Aristotele et aristotelici*, 223.

28 René Descartes, *Principles of Philosophy*, Preface to the French Edition, in AT IXB 14, CSM I 186.

29 Franco Burgersdijk, *Institutionum Metaphysicarum libri duo. Opus Posthumum. Omni cura ac diligentia ex ipsius Authoris manuscripto collectum* (Leide: Jerôme de Vogel, 1640), 9; Franco Burgersdijk, *Collegium Physicum, Disputationibus XXII absolutum . . . Editio secunda* (Leyde: Elsevir, 1642), 9. This important partisan of Aristotelianism has been studied in *Franco Burgersdijk (1590–1635): Neo-Aristotelianism in Leiden*, eds. Egbert P. Bos and Henri A. Krop (Amsterdam and Atlanta: Rodopi, 1993). Burgersdijk's *Institutiones logicae* were very important during De Raey's stay in Amsterdam: Van Miert, *Humanism in an Age of Science*, 242–5.

30 Burgersdijk, *Institutionum Metaphysicarum libri duo*, 2, 4, 5. A history of the dissemination of Suarez's metaphysics in Reformed Universities is outlined by Charles H. Lohr, "Metaphysics," in *The Cambridge History of Renaissance Philosophy*, eds. Charles B. Schmitt and Quentin Skinner (Cambridge: Cambridge University Press, 1988), 537–638.

31 Verbeek, *Descartes and the Dutch*, 6–7; Aza Goudriaan, *Philosophische Gotteserkenntnis bei Suarez und Descartes: im Zusammenhang mit der niederlandischen reformierten Theologie und Philosophie des 17. Jahrhunderts* (Leiden and Boston: Brill, 1999); Aza Goudriaan, *Reformed Orthodoxy and Philosophy, 1625–1750: Gisbertus Voetius, Petrus van Mastricht, and Anthonius Driessen* (Leiden and Boston: Brill, 2006).

32 De Raey, *Clavis Philosophiae Naturalis Aristotelico-Cartesiana: Editio secunda*, 714; Burgersdijk, *Collegium Physicum*, 6; Franco Burgersdijk, *Idea philosophiae naturalis, sive Methodus definitionum et controversiarum Physicarum: Editio novissima* (Lugduni Batavorum: Ex Officina Elseviriorum, 1645), 11.

33 De Raey, *Clavis Philosophiae Naturalis Aristotelico-Cartesiana: Editio secunda*, 716.

34 *Ibid.*, 716 and 720.

35 *Ibid.*, 714.

36 If we compare this claim with the end of the disputation *De constitutione logicae*, we must conclude that De Raey oscillated between two positions: sometimes he attributed a capacity to attain intellectual knowledge of some truths *per se notae* to physics, sometimes he attributed this capacity to metaphysics only, whereas physics was able to develop some arguments taking its point of departure in these truths, understood as hypotheses.

37 De Raey, *Clavis Philosophiae Naturalis Aristotelico-Cartesiana: Editio secunda*, 717.

38 *Ibid.*, 719.

39 *Ibid.*, 720. The ideal of knowledge through the causes was obviously derived from the Aristotelian philosophy. Note, however, that Descartes' *Principia* proposed to find the reasons of the effects by their causes, and not the contrary. See AT VIII-1 81.

40 De Raey, *Clavis Philosophiae Naturalis Aristotelico-Cartesiana: Editio secunda*, 231–2.

41 On this point of detail, I cannot follow Theo Verbeek's otherwise excellent analysis of the *Cogitata de interpretatione* in "Les Cartésiens face à Spinoza: l'exemple de Johannes de Raey." It seems simpler to attribute the evolution of De Raey's thought to the influence of Meyer's book, which addressed these issues, than to suppose a reference to Spinoza's *Principia Philosophiae cartesianae*, the purpose of which was quite different. Verbeek also mentions Meyer, along with Hobbes, as a possible cause of De Raey's position in his *Dutch Cartesian Philosophy*, 176. On the reception of Meyer's work, see Roberto Bordoli, *Ragione e scrittura tra Descartes e Spinoza: Saggio sulla Philosophia S. Scripturae Interpres di Lodewijk Meyer e sulla sua ricezione* (Milano: Franco Angeli, 1997).

42 On the relationship that Wittich and Burman established between philosophy and theology, see Antonella Del Prete, "Oltre Descartes: filosofia e teologia nella Theologia pacifica di Christoph Wittich," in *Immagini filosofiche e interpretazioni storiografiche del cartesianismo*, eds. Carlo Borghero and Antonella Del Prete (Firenze: Le Lettere, 2011), 25–45.

43 In the dissertation *De Aristotele et aristotelicis*, this is a matter of a "*communis sensus et vulgaris intellectus*," which are given us for the necessities of life. At the beginning of *De sapientia veterum*, De Raey argued that all the arts and disciplines have their own truth and are scientific. Later in the dissertation, however, he continued to oppose, as he did before, the search for philosophical truth and the search for usefulness, typical of the other disciplines (De Raey, *Clavis Philosophiae Naturalis Aristotelico-Cartesiana: Editio secunda*, 228, 723–4).

44 De Raey, *Cogitata de interpretatione*, 657.

9 The Materialist Reception of the Cartesian Physics in Naples

Pierre Girard

The Naples Context

A number of concurring historical testimonies indicate that Descartes' works were physically present in Naples from around 1650 when Tommaso Cornelio, returning from his travels, brought a copy of Descartes' works with him.[1] Cornelio had visited Northern Italy where he frequented students of Galileo, some of whom, later, became members of the famous *Accademia del Cimento* in Florence which had decisive influence on the development of experimental science in Italy.[2] However, if we go beyond this official image of Cornelio returning to Naples with his new books—an official image that in many ways also conceals the historical reality—we must take into account the very specific context that shaped the introduction of modernity in Naples. Three decisive aspects of this context should be highlighted.[3]

The first was the exceptional geographical and natural location of Naples. The location fascinated travelers and scientists alike all over Europe. Naples was one of the privileged destinations for anybody undertaking a *grand tour* of Europe in the eighteenth century.[4] The volcanic activity of Vesuvius, very intense at the time,[5] was studied with great interest by scientists, but it also gave rise to popular superstition. This latter point must be underlined since it gave a very specific flavor to scientific modernity in Naples. Because of the superstitious beliefs related to such natural phenomena, practicing science for learned men in Naples was not just a question of grasping, theoretically, the laws of nature. It was also, and much more concretely, about avoiding natural disasters, opening the eyes of people, and opposing the political and religious forces that encouraged such superstition.

This immediately leads us the second characteristic aspect of the situation in Naples: Masaniello's so-called "revolution."[6] This urban revolt set off a wave of violence in Naples in 1647. The event is an important part of the context for the introduction of scientific modernity because of the emblematic status it acquired in the collective memory of the city. Behind Masaniello appeared the figure of the so-called *lazzari*, the great populace, always on the verge of revolt but devoid of any precise political project. The question of the populace was of particular political importance to the scientists in Naples. They worked against the political and theological authorities that according to them were controlling and maintaining the populace in a state of ignorance by encouraging their superstition. Consequently, scientific work necessarily had political resonances for the learned. Introducing scientific modernity was a way to acquire the means of building a new political class with a precise political objective. It is remarkable how often scientific research went hand in hand with political and social activism. From this perspective, Galileo[7] and Descartes[8] were never quietly received in the scientific academies but used as scientific instruments in the service of precise political and social projects.

The third and last characteristic aspect of the Naples situation concerns the horrible plague epidemic in 1656.[9] This epidemic ravaged most of Europe but was particularly virulent in Naples, where the population was very dense and people lived in deeply unhygienic conditions. At the time, Naples was one the most populated cities in Europe with approximately 400,000 to 450,000 inhabitants. But it lost more than half its population, around 250,000, in only a few months during the plague. On the worst days of the epidemic, there were two to five thousand deaths every day. This traumatic event strongly determined the scientific research undertaken, including the introduction of Cartesianism and scientific modernity. For those among the learned who survived the plague—given that some of them died, among others Marco Aurelio Severino, a medical doctor and anatomist who taught both Tommaso Cornelio and the famous Leonardo Di Capua, the introduction of scientific modernity was not something that could take place merely in the scientific academies. It was a matter of political urgency that acquired its significance in the context of a concrete historical situation.[10]

Methodological Materialism

Nonetheless, an important part of the context for the reception of Descartes in Naples concerned the different scientific academies. The best known among them was the *Accademia degli Investiganti*.[11] It was officially founded in 1663 by Tommaso Cornelio but held informal meetings as early as 1650, the year Cornelio returned to Naples. Detailed historical work regarding this academy cannot be undertaken here, but I shall outline some aspects of its history that pertain directly to the reception of Descartes.

The first aspect is the ubiquity of collective practices of experimentation. The members of the academy were clearly influenced by Galileo, in particular by the scientist's proclamation of a "return to things," and declared suspicion toward words. More generally, they were influenced by Galileo's caution when it came to any fiction or rhetoric liable to separate the scientist from the direct apprehension of reality. In practice, this "return to things" consisted in a return to *bodies* and to *matter*. The body was no longer a veil, nor a prison for the mind, but the direct expression of the laws of nature. The body was a laboratory for studying the laws of nature. To take a precise example, one would refer to the famous *Parere* by Leonardo Di Capua, one of the most well-known *investiganti*.[12] The text is a manifest for scientific modernity in medicine. It considers medical science to be the exclusive study of the material and mechanical functions of the body. Di Capua argued that only the consideration of the complexity of the body would allow us to escape the empty discourse of Scholasticism. This could also to some extent be achieved by studying Cartesian physiology, especially the *Traité de l'homme*, even though Di Capua deemed Descartes' work a little too simplistic. Hence, for Di Capua, knowing the nature of man was not a question of searching out essences. It was about grasping the material mechanisms that made the body function, such as the digestive apparatus or the circulation of the blood.[13] The particular attention that the *investiganti* paid to bodies can also be illustrated by a famous episode, where some fifty members of the academy solemnly went to the Lake of Agnano to study the allegedly bad effects of soaking linen in the lake. As was often the case, it was a question of putting an end, through (collective) experimentation, to the various prejudices and superstitions explaining the attribution of malefic properties to this particular lake.[14]

Although faithful to Descartes, the *investiganti* were mainly concerned with solving particular problems. The particular attention they paid to details prompted suspicion toward methods that were too general and in particular toward all forms of *mathesis universalis*. This pragmatic approach to science was only the more legitimate to the extent that the problems they addressed were often politically charged. However, because of their practical perspective, their reinterpretation of Descartes necessarily reoriented and reshaped the thought of the latter. At no point did they study Descartes' thought in its totality and unity. They only deemed the physics worthy of interest. Hence, apart from referring to Descartes' status as a symbol of the *libertas philosophandi*,[15] what the *investiganti* adopted from the French philosopher was above all the stress he had put on the *res extensa*. As for Descartes' metaphysics, it was simply considered useless nonsense, an archaic Scholastic leftover that placed the French philosopher far below Galileo in terms of radicality and boldness. Descartes' metaphysics was nothing but a foil, allowing him to advance undercover. It was a cunning tool for deceiving in an "honest" way and not shocking the Church. Once the metaphysics was put to one side, Descartes' thought was for them clearly connected to the resurgence of atomism and in particular to the thought of Lucretius, a figure who was widely discussed and read by the Naples intellectuals.[16] Thus, in summary, Descartes was appreciated for the scientific and philosophical modernity he symbolized. He did not represent a set of philosophical theses as much as a weapon to be used against tradition. Moreover, he was esteemed for the model he had elaborated for analyzing the *res extensa*, since it contributed to the reactualization of ancient materialism, and in particular Lucretius whose thought was otherwise mainly associated with the philosophy of Gassendi.[17]

If one pays attention to the materialism practiced in Naples, it will, however, be noted that it acquires a very special methodological status. Hence, the urgency of the situation and the necessity of developing a science capable of responding to natural disaster, the *novatores* were obliged to neutralize useless polemics from the outset and attempt to liberate science from the dead weight of tradition. In this respect, the materialism of the *novatores* was never initially meant to be provocative. It was grounded in a wish to liberate science from a stifling theoretical framework. This is particularly clear from a clandestine text by Francesco D'Andrea entitled *Apologia in difesa degli atomisti* that circulated extensively in Naples toward the end of the seventeenth century.[18] From the outset, D'Andrea underlined the specificity of the *novatores*'s position:

> This philosophy is the enemy of disputes, and to the extent that it focuses on the truth about things, it considers useless controversies about words with scorn. And since we cannot acquire knowledge about things otherwise than through the senses, this philosophy is entirely turned towards experiments and the sensible lessons that are founded upon them. Consequently, many call it free philosophy, since it is not bound by any of the Scholastic principles, while others call it elective philosophy, since it preserves what is good in every philosophical sect. And yet others call it modern philosophy, on account of its innovative experiments, unknown to the Ancients. But it can better and more commonly be called the philosophy of sense, since, guided by the senses, it only regards as true about sensible things propositions that are demonstrated by means of sensibility.[19]

The principle defended by this text consists in rejecting all explanations that are not grounded in the phenomenon that one is in the process of observing. The epistemological consequence of this requirement is the elaboration of an explicitly materialist position:

> Consequently, the matter from which all natural things derive has no being other than that of matter itself, the matter we see and touch. Its essence consists in nothing but a substance that is extended, that is to say, that has the three dimensions from which all bodies derive, in such a way that being matter and being body is one and the same thing. . . . For this reason, all the changes that occur in matter must be corporeal, that is to say, composition, transposition and movement of the parts of this matter. And the things we call forms, natures, or appearances of natural bodies, depend on this matter.[20]

This passage summarizes very neatly what "investigative" materialism consists in. We clearly recognize the basic principle of sensible knowledge already noted above. But when considered more closely, D'Andrea also stresses another and maybe more decisive point. The idea is the following: The matter from which all things in nature derive has no other being than the kind of being that we see and touch. We should, however, here note a nuance. In fact, D'Andrea does not simply state that matter *is* nothing but such things. More importantly, he affirms that they *need not be* anything else (*non ha da essere altra*).[21] Referring to matter and its principles, subsequently described in the text, was for him *sufficient* to account for the experience derived from the senses. D'Andrea's polemical target was clearly the Scholastic method, which consists in searching out explanatory principles in substantial forms, in entities that do not derive their being from matter itself. In contrast to this, "the atomist who considers things such as they truly are admits no other entity in nature (when it comes to corporeal things) than matter alone, divided into very small particles."[22] Hence, D'Andrea was brought to adopt a materialist—or, more precisely, atomistic—position by way of his critique of Scholastic fictions, their vain speculations, and the use they made of other things than matter to explain the appearance of shapes in matter. His position thus corresponded to something one could call "methodological materialism," to the extent that his appeal to atoms and matter was mainly grounded in his wish to elaborate explanatory principles that were immediately applicable and useful *hic et nunc*. It was not grounded in the development of an ontology of matter. For these purposes, materialism appeared as a good methodological choice, because it was efficient in practice and allowed explaining natural things without getting lost in occult and artificial qualities. This practical aspect of materialism came up repeatedly and explicitly among the *investiganti*. Appealing to materialism was never about provocation, about scandalizing people by adhering to a wicked doctrine. Granted, provocation is an aspect of the doctrine that we find in Pietro Giannone, but up until, and including, Giambattista Vico, materialism was endorsed mainly because it represented a particularly efficient method for understanding natural things without appealing to anything but matter, while avoiding getting lost in speculations about false qualities. This concern for practical efficacy must be understood in contrast to the multitude of explanations of natural things at the time that referred to occult qualities, secondary qualities, magical qualities, and so on.

Materialism thus mainly constituted an epistemological framework, a kind of basic requirement guiding scientific research still in the making. Methodological materialism has an advantage stemming exactly from fact that it is not grounded *a priori* in an explicit theory of matter, making it extremely flexible and malleable. Indeed, the idea that science could always progress, always be modified, was at the heart of the research undertaken by the *investiganti*. They were united by their shared effort to establish a free form of investigation, a kind of research opposed to any *a priori* theoretical principles liable to limit its practical efficacy. It is in this sense that the instrument of materialism, under the form of an atomist doctrine, seemed useful and necessary to them. The problem, however, was that the very same flexibility that was the force of the *investiganti*'s materialism was also, from another viewpoint, its weakness. The problem was very clearly explained by D'Andrea himself:

> As for myself, I think that the atomist philosophy—that is to say, the philosophy founded on the senses, the philosophy that is real and that applies itself to seek out the truth of things, as opposed to the philosophy that loses itself in useless disputes about words—will be corrupted when exposed to public controversy. For people will then no longer seek the true, or even plausible, things that this philosophy contains, but only seek the approbation of the vulgar by appealing to innovation, considering that they should never yield, and then begin employing terms and distinctions that are understood neither by those who reject them nor be those who defend them.[23]

The passage is surprising to the extent that the risk of disputes here seems to become double. D'Andrea envisages two distinct risks. On the one hand, he adopts the classic position of the *investiganti* wanting to overcome the vain disputes of the Scholastics, deeming that they lead nowhere and separate us from "the truth of things." But he then adds another element: the same risk also exists for the atomistic theory when exposed and placed on the stage of public debate. The risk for the atomistic doctrine was thus double and appeared both at the beginning and at the end of its constitution. At the beginning, it was becoming involved in vain discussions about outdated theories that lead nowhere. But at the end, the major risk appears: atomism should necessarily remain a dynamic method in search for the true or the plausible (*indagare il vero, o almeno il più verosimile*). It should never turn into a systematic doctrine closed upon itself. Even when it did not turn into such a doctrine, however, the risk was always there. For when the materialist method was put to use by committed scientists, in a context where urgency dominated the picture, it could always and easily be presented as something more than a simple method, especially in the context of a polemic and accusation. In other words—and I insist on this point because, to my mind, it is one of the most important ambiguities of the movement of *novatores* in Naples in the second half of the seventeenth century, D'Andrea was aware of the fact that the flexibility, the lack of definition that was the very force of the *investiganti*'s rehabilitation of materialism, could be forfeited in polemical contexts, when reacting to accusations. In the context of polemics, materialism risked becoming increasingly rigid and theoretical and thus losing exactly the character which at the outset constituted its very force and efficacy.

The Logic of Quarrels and the Birth of the Radical Enlightenment

Surprisingly, in order to refute Descartes, the opponents of the Cartesian philosophy took recourse to the same interpretive mechanisms as those that the proponents of Descartes considered to reveal the true nature of that philosophy, namely the doctrinal assimilation of Cartesianism to materialism. The reductive move was always the same, namely downplaying if not outright ignoring the metaphysical substratum of Descartes' philosophy. Several strategies for doing this were used. Sometimes the metaphysics was deemed feeble, confused, and contradictory, containing only a residual element of coherence, namely the doctrine of the *res extensa* and, behind that, materialism. Sometimes the metaphysics was regarded as a hypocritical veil, designed to dissimulate a materialist commitment, indifferently associated with Epicurus, Spinoza, and the atheists. Regardless, however, of whether these interpretations stressed Descartes' intellectual weakness or intentional malice, he was consistently reduced to a materialist philosopher. Such critical materialist reductions of Cartesianism were not simply a result of abstract speculation. They took shape in the context of concrete and sometimes very violent controversies and conflicts. The most important among them was the controversy opposing the *novatores* and the Jesuit Giovanni Battista De Benedictis, also known as L'Aletino. I will not relate the details of the exchange, but simply provide a sketch.

L'Aletino was a very influential Jesuit in Naples. He was close to the political and religious leaders. L'Aletino's attack targets from the outset the materialist aspects of Cartesian philosophy. Hence, from 1688 onward, beginning with his *Philosophia peripatetica*, L'Aletino attacked Descartes and Gassendi simultaneously, arguing that they were hidden materialists and the direct heirs of Epicurus and ancient atomism. This first polemical text was followed up in 1694 by his *Lettere apologetiche* in which L'Aletino proposed a severe and direct criticism of Di Capua's *Parere*.[24] Once again, in this second work presented as a collection of letters, his refutation of Descartes was part of a more general refutation of materialism. Hence, the third letter was dedicated to the refutation of Descartes, while the fourth undertook to refute his heirs, the modern atomists. These writings occasioned a lively polemic in the form of replies, counter replies, and anonymous pamphlets. These exchanges are important to us because of the status that Cartesian mechanism acquired in it. L'Aletino reduced mechanism to pure materialism. According to the Jesuit, when defending a conception of matter involving no divine intervention, Descartes was already moving toward Spinoza's *Deus sive natura*. This resulted in a classic criticism where L'Aletino turned Descartes into a hidden atheist and direct heir of Lucretius, Machiavelli, Vanini, Hobbes, and Spinoza. The tone of the letters was exceptionally violent. L'Aletino presented Descartes as an "impostor," a "rotten Epicurean" who had opened the "royal road to Epicureanism and atheism."[25]

This brief presentation of the controversy prompted by L'Aletino provides a fairly good idea of the conflicting and yet converging images of Descartes in Naples. Regardless of whether they were championing or opposing Cartesianism, they all agreed on one thing, namely the materialist status of his thought. For some, Descartes was, in a positive way, part of a materialist tradition that he renewed forcefully by his conception of the *res extensa*. For others, Descartes had to be strongly refuted exactly because of his materialism, a materialism that he had not avoided, either because of his metaphysics' weakness, or because, in his metaphysics, he had cunningly tried to dress his philosophy up as something other than materialism. In both cases, a paradoxical Descartes emerged, a Descartes reshaped by his reception in Naples and by the various polemics in which he was evoked either as an offensive weapon or, on the contrary, as an accusation. What resulted from this was an entirely new Descartes deployed within new systems of philosophy and from which new original forms of Cartesianism emerged.

I will give an example of this by briefly considering the philosophy of Pietro Giannone. Giannone was born in Puglia in 1676 and arrived in Naples in 1694 to study law. He attended the classes of Domenico Ausilio and took part in the *Accademia di Medinacœli*, which followed the orientation of the *Accademia degli Investiganti*. Here, he met several Cartesians who, as is most often the case, were in fact theoretically situated somewhere between Gassendi and Descartes.[26] As we can learn from his autobiography, Giannone's Cartesianism was part of the standard reception of mechanism in Naples. He endorsed the Cartesian physics but was deeply distrustful of his metaphysics. Giannone himself claimed to be an heir to both Descartes and Gassendi. The first—qualified as a "divine mind"—was synonymous with reason; the second—the reading of whom procured him "unspeakable pleasure"—was synonymous with experience.[27] Giannone's materialist reading of Descartes would gradually become more and more radical and take on a political dimension, to such an extent that Jonathan Israel has proclaimed him one of the proponents of the "Radical Enlightenment."[28]

The first stage in Giannone's progressive radicalization was the publication of his *Istoria civile del Regno di Napoli* in 1723. This work in forty books retraced the history of Naples since the Romans up to Giannone's own time. The explicit aim of the work was to denounce, by means of a historical and philological work, the power of the Church and the feudal structures in the kingdom of Naples. Giannone wanted to depict the mechanisms that historically had allowed the Church to evolve from a simple religious and spiritual authority into a moral authority and, consequently, a political force. The Church reacted violently to the book and condemned Giannone as a heretic. After his condemnation, Giannone was forced to leave the kingdom of Naples and sought refuge in Vienna. His stay there was crucial for his philosophical training and, in particular, for the interpretation of Descartes he later developed in his *Triregno*. In Vienna, Giannone had the opportunity to read texts that were unknown or banned in Naples, in particular the works of Toland and Spinoza. He began working on his best-known work, the *Triregno*, probably written in the period between 1731 and 1736. The text was widely circulated in Naples and in Jansenist milieus. It is difficult to summarize the work, but one can highlight certain points that illustrate its radicality, including the Spinozist and materialist influences. Thus, in the first part of the work, Giannone undertook a refutation similar to Spinoza's of the notion that Moses was the author of the Pentateuch. He also attempted to prove, now following a model taken from Lucretius, the mortality of the soul. According to Giannone, the Bible itself clearly demonstrated that God never promised man immortality but only a terrestrial kingdom. The development of these theses culminated in a violent rejection of the temporal powers of the Church, which, according to Giannone, had stifled civil society with age-old corruption.

Without going into the details of Giannone's thesis,[29] let us now see how a materialist interpretation of Descartes was part of the construction of this radical position. The interpretation is mainly put forward in chapter III of the first book of the *Triregno*, entitled *Del nuovo sistema di Cartesio intorno alla creazione del mondo, formazione dell'uomo e natura di questo spirito*. Giannone's conclusions were quite nuanced. He declared his admiration for Descartes, the "incomparable philosopher," who had managed to expulse final causes from physics and who had succeeded in formulating an explanatory model for physics that was both fruitful and efficient. Finally, and most importantly, Descartes had developed an admirable physiology.[30] But Giannone's admiration quickly turned into criticism. What is remarkable in this context is not so much the criticism itself as the direction it takes. According to Giannone, Descartes should not be reproached for the materialist undercurrents of his philosophy. On the contrary, he admired Descartes for having succeeded, taking his point of departure in a simple concept of matter, "in giving birth to everything that is admirable in this world."[31] Descartes should, however, be reproached for not having done this more explicitly and more radically. In a sense, for Giannone, only a materialist radicalization of Descartes' philosophy would make it truly coherent. Giannone affirmed, for example, that when reading the *Traité de l'homme* and the *Passions de l'âme*, it was immediately evident to him that, on Descartes' model, the totality of phenomena pertaining to human life could be explained through the sole consideration of simple matter. Consequently, he criticized Descartes for entertaining the futile project of looking for a second explanatory principle. By attempting to escape strict materialism, Descartes was so to speak unfaithful to himself. Moreover, Giannone denied the soul any metaphysical privilege. There was for him no difference whatsoever between the nature of man and the nature of animals. In this respect, he anticipated La Mettrie's thesis about the man-machine. Only a materialist principle explained the superiority of the first in relation to the second. Hence, according to Giannone, man was equipped with a "corporeal machine that is better organized than that of animals."[32] Man's mind was simply more active and was materially purer than that of an animal, but it did not manifest a "different nature."

Giannone's interpretation radicalized the Cartesian position by reducing it to what he took to be the only possible coherent interpretation, namely the materialist one. Hence, for Giannone, Descartes was a materialist in denial. Only his personal weakness, if not his ambition, prompted him to pose as an idealist. According to Giannone, Descartes was not sufficiently strong to draw out the full materialist force of his own propositions. And while Descartes claimed to fight against prejudices and eradicate them from our minds, he did not succeed in fighting the specific prejudice—compared by Giannone to a "raging river" uprooting everything in its course—that consisted in thinking that souls are immortal and separate from the body.

Let me finally make a few points in conclusion. The first is directly concerned with the specificity of the situation in Naples. The combination of exceptional natural surroundings and historical events such as the plague constituted an extremely determined framework for the reception of Descartes' philosophy in Naples. No philosophy can be received only abstractly under those conditions! From the outset, the Cartesian philosophy was used both scientifically and politically, as is clear from the numerous polemics and quarrels that marked the end of the *Seicento* in Naples. These various contextual factors largely contributed to transforming Descartes into a link in the chain of the materialist tradition, in both a negative and a positive sense. The Cartesian controversies in Naples aimed at removing archaisms from Cartesian thought and highlighting the aspects that were, for them, genuinely radical and innovative. I have clearly exemplified this kind of interpretive mechanism by the interpretation of Descartes developed by Pietro Giannone. For Giannone, Descartes, in order to be philosophically consistent and to live up to his own denunciation of prejudice, ought to have formulated the underlying materialism of his system. Only in this way could his system become coherent. This materialist necessity ought to have brought Descartes to adopt a more radical position. But only his most courageous successors, such as Giannone himself, dared to do so. From this interpretation arose a double and ambiguous critique of Descartes. Due to weakness or opportunism, Descartes had not acknowledged his own materialism and stopped short of thinking about politics. Hence, the numerous controversies that Descartes' thought gave rise to in Naples opened up a space for a materialist and political reading of his philosophy.

Notes

1 Francesco D'Andrea, *Avvertimenti ai nipoti*, ed. Imma Ascione (Napoli: Jovene Editore, 1990), 203: "Onde, venuto in Napoli l'anno 1649 il nostro signor Tomaso Cornelio a cui la nostra città deve tutto ciò oggi si sa di più verisimile nella filosofia e nella medicina, io fui il primo che abbracciassi quella maniera da lui propostaci di filosofare, con far venire in Napoli l'opere di Renato delle Carte, di cui sino a quel tempo n'era stato a noi incognito il nome." See also Costantino Grimaldi, *Istoria dei libri di Don Costantino Grimaldi scritta da lui medesimo*, in *Memorie di un anticurialista del settecento*, ed. V. I. Comparato (Firenze: Olschki 1964), 4–5: "Quando si era introdotto in Napoli il buon gusto nelle scienze, da Tommaso Cornelio, e Lionardo di Capua, uomini celebri per la loro rara letteratura, e per l'opere cacciate alla luce, ambedue medici di professione; [per] cui si erano sparse le dottrine nuove di Cartesio, e di Gassendi, ricevute avidamente da' Napoletani, amanti delle buone cose,

come lumi mai veduti." On the agreement between these various testimonies (we find a similar account in the *Istoria civile del Regno di Napoli* published by P. Giannone in 1723), see Giulia Belgioioso, *La variata immagine di Descartes: Gli itinerari della metafisica tra Parigi e Napoli (1690–1733)* (Lecce: Milella 1999), 42 and ff.

2 On this point, see Edgar William Knowles Middleton, *The Experimenters: A Study of the Accademia del Cimento* (Baltimore and London: Johns Hopkins Press, 1971); Paolo Galluzzi, "L'Accademia del Cimento: 'gusti' del principe, filosofia e ideologia dell'esperimento," *Quaderni storici* 16, no. 48 (1981): 788–844.

3 On this point, please refer to my article: Pierre Girard, "Matérialisme et politique: les enjeux de la réception du cartésianisme à Naples à l'Âge classique," *Corpus* 61 (2011): 113–32.

4 See Atanasio Mozzillo, *La frontiera del Grand Tour: Viaggi e viaggiatori nel Mezzogiorno borbonico* (Napoli: Liguori, 1992); Paolo Gasparini, *Un viaggio al Vesuvio: Il Vesuvio visto attraverso diari, lettere e resoconti di viaggiatori* (Napoli: Liguori, 1992).

5 See Atanasio Nazzaro, *Il rischio Vesuvio. Storia e geodiversità di un vulcano* (Napoli: Guida, 2009).

6 See Aurelio Musi, *La rivolta di Masaniello nella scena politica barocca* (Napoli: Guida, 2002); Silvana D'Alessio, *Masaniello. La sua v e il mito in Europa* (Roma: Salerno Editrice, 2007); Alain Hugon, *Naples insurgée. De l'événement à la mémoire, 1647–1648* (Rennes: Presses universitaires de Rennes, 2011).

7 See *Galileo e Napoli*, ed. Fabrizio Lomonaco and Maurizio Torrini (Napoli: Guida 1987).

8 On the reception of Descartes in Naples, see Belgioioso, *La variata immagine di Descartes*. See also Ettore Lojacono, *Immagini di René Descartes nella cultura napoletana (1644–1755)* (Lecce: Conte, 2003). See finally Pierre Girard, "Les usages de Malebranche dans la réception du cartésianisme à Naples," in *Les Malebranchismes des Lumières: Études sur les réceptions contrastées de la philosophie de Malebranche, fin XVIIe et XVIIIe siècles*, ed. D. Kolesnik-Antoine (Paris: Honoré Champion 2014), 225–46.

9 See Salvatore De Renzi, *Napoli nell'anno 1656, ovvero documenti della pestilenza che desolò Napoli nell'anno 1656, preceduti dalla storia di quella tremenda sventura narrata da Salvatore De Renzi* (Napoli: De Pascale, 1867).

10 On this phenomenon more generally, please see my forthcoming book: Pierre Girard, *Matérialisme et radicalité politique dans les premières Lumières à Naples (1647–1744)* (Paris: Champion, forthcoming).

11 See Max H. Fisch, "L'Accademia degli Investiganti," *De Homine* 6 (1968): 17–78; Maurizio Torrini, "L'Accademia degli Investiganti. Napoli 1663–1670," *Quaderni storici* 16, no. 48 (1981): 845–83.

12 Leonardo Di Capua, *Parere del signor Lionardo Di Capoa, divisato in otto ragionamenti, ne' quali partitamente narrandosi l'origine, e'l progresso della medicina, chiaramente l'incertezza della medicina si fa manifesta* (Napoli: Antonio Bulifon, 1681).

13 See Salvatore Serrapica, *Per una teoria dell'incertezza tra filosofia e medicina: Studio su Leonardo Di Capua 1617–1695* (Napoli: Liguori Editore, 2003).

14 On this polemic, see Maurizio Torrini, "Un episodio della polemica tra antichi e moderni: la disputa sulla macerazione dei lini nel lago di Agnano," *Bollettino del Centro di studi vichiani* 5 (1975): 56–70.

15 See Eugenio Garin, "Cartesio e l'Italia," *Giornale Critico della Filosofia Italiana* 4 (1950): 385–405.

16 We should note here in particular the scandal that A. Marchetti's Italian translation of Lucretius's *De rerum natura* gave rise to. On this figure, see Mario

Saccenti, *Lucrezio in Toscana: Studio su Alessandro Marchetti* (Firenze: Olschki 1966).

17 See Paolo Cristofolini, "Tommaso Cornelio et l'histoire du matérialisme," in *Gassendi et l'Europe (1592–1792)*, ed. Sylvia Murr (Paris: Vrin, 1997), 335–46.

18 Imma Ascione notes that even though *L'Apologia* initially wan known and circulated only in a small circle, it still remained "una sorta di manifesto politico e, forse, persino come un programma di governo" (*Il governo della prassi: L'esperienza ministeriale di Francesco D'Andrea* (Napoli: Jovene Editore, 1994), 222). On the importance of such clandestine circulation of manuscripts in Naples, see the remarks by Harold Samuel Stone in *Vico's Cultural History: The Production and Transmission of Ideas in Naples, 1685–1750* (Leiden: E. J. Brill, 1997), 52–4.

19 D'Andrea, *Lezioni*, cit. in Antonio Borrelli, *D'Andrea atomista: L'Apologia e altri inediti nella polemica filosofica della Napoli di fine Seicento* (Napoli: Liguori, 1995), 148: "Questa filosofia è nimica delle dispute, poiché attendendo alla verità delle cose, sdegna come inutili le contenzioni delle parole. E perché delle cose noi non potemo haverne altra cognizione, che per mezzo de' sensi, perciò tutta è dedita alle sperienze et al discorso sensato, che fondasi sopra le medesime. Quindi non giurando nelle parole di alcuno, da molti è chiamata filosofia libera, perché non sta ligata a' principii delle scuole; da altri filosofia elettiva, perché eligge da tutte le sètte quello che ciascheduna ha di buono; da altri filosofia moderna, per la novità delle sperienze, non conosciute da gli antichi; ma meglio di tutti è chiamata più comunemente filosofia sensata, perché, colla scorta de' sensi, non ammette per vero nelle cose sensibili se non quello che per essi sensi ne vien dimostrato."

20 Francesco D'Andrea, *Risposta a favore del sig. Lionardo De Capoa contro le lettere apologetiche del P. De Benedictis gesuita 1697*, quoted in Borrelli, D'Andrea atomista, 1: "La materia dunque della quale tutte le cose naturali costano, non ha da essere altra che quella materia stessa, che veggiamo, e che tocchiamo. La cui essenza non in altro consiste, che in essere una sostanza ch'habbia estensione, cioè, che habbia le tre dimensioni delle quali costano tutti i corpi, sì che l'esser materia, e l'esser corpo sia una medesima cosa . . . E per ciò tutte le mutationi che si fanno nella materia, han da esser similmente corporee, ciò è compositione, traspositione e movimento delle parti di essa materia. E da questa dipendono tutte le forme, o nature, o apparenze, che dir vogliamo de' corpi naturali."

21 The conception according to which the analysis of matter alone suffices for explaining the appearance of shapes is constantly reiterated. It is considered the only legitimate principle of explanation: "Onde non sarà bisogno introdurre altri principii nella natura, che la materia e il moto; intendendo che il moto non sia né sostanza, né accidente, ma un modo della medesima materia" (D'Andrea, *Lezioni*, quoted in Borelli, *D'Andrea atomista*, 151).

22 D'Andrae, *Risposta . . . alle* Lettere apolegetiche *di Benedetto Aletino in difesa della filosofia del Signor Leonardo Di Capua nell'anno 1720 in Napoli Quagliarelli*, quoted in Borrelli, *D'Andrea atomista*, 43: "Gli Atomisti non considerando le cose, se non per quello, che veramente sono, non ammettono in natura altre entità (parlanno delle cose corporee) che la sola materia in minutissime particelle divisa."

23 D'Andrea, cit. in Nino Cortese, *I ricordi di un avvocato napoletano del Seicento: Francesco D'Andrea* (Napoli: Lubrano, 1923), 39–40: "Io crederò che la filosofia atomistica, cioè la sensata, la reale, quella che mira a ricercar la verità delle cose, non di trattenersi nelle vane dispute delle parole, si corromperà quando comincerà ad esporsi alle pubbliche altercazioni. Poichè non si attenderà più ad indagare il vero, o almeno il più verisimile; ma solo ad acquistar l'aura del volgo colla novità, et ad estimarsi di non ceder mai et a provedersi di distinzioni e di vocaboli che non s'intendono nè da quei che l'oppugnano, nè da quei che la difendono."

24 Giovanni Battista De Benedictis [pseud. L'Aletino], *Lettere apologetiche in difesa della Teologia Scolastica, e della Filosofia Peripatetica di Benedetto Aletino dedicate all'Illustriss. & Eccelentiss. Signore, il signore D. Carlo Francesco Spinelli Principe di Tarsia, &c.* (Napoli: nella Stamperia di Giacomo Raillard, con licenza de' Superiori, 1694); *Difesa della terza lettera apologetica* (Roma: A. De Rossi, 1705). The *novatores*, and first of all Constantino Grimaldi, responded to these *Lettere* in a number of texts: *Risposta alla lettera apologetica in difesa della teologia scolastica di B. Aletino. Opera nella quale si dimostra esser quanto necessaria ed utile la teologia dogmatica e metodica; tanto inutile, e vana la volgar teologia scolastica* (Colonia: S. Hecht, 1699); *Risposta alla seconda lettera apolegetica di B. Aletino: Opera utilissima a' professori della filosofia: in cui fassi vedere quanto manchevole sia la peripatetica dottrina* (Colonia: S. Hecht, 1702); *Risposta alla terza lettera apologetica contra il Cartesio creduto da più d'Aristotele di Benedetto Aletino: Opera, in cui dimostrasi quanto salda e pia sia la filosofia di Renato delle Carte: e perché questo si debba stimare più d'Aristotele* (Colonia: S. Hecht, 1703). These different rejoinders to L'Aletino were gathered and reedited by C. Grimaldi in 1725 under a new title: *Discussioni istoriche, teologiche e filosofiche di Costantino Grimaldi fatte per occasione della risposta alle Lettere apologetiche di Benedetto Aletino* (Lucca: 1725). On this new edition, see Vittor Ivo Comparato "Ragione e fede nelle *Discussioni istoriche, teologiche e filosofiche* di Costantino Grimaldi," in *Saggi e ricerche sul Settecento*, ed. Ernest Sestan (Napoli: Istituto italiano per gli studi storici, 1968), 48–93. On this polemics, see Girolamo De Liguori, "Teologia, filosofia e fisica di Cartesio nella Difesa della terza lettera apologetica dell'Aletino (1705)," in *L'ateo smascherato. Immagini dell'ateismo e del materialismo nell'apologetica cattolica da Cartesio a Kant* (Firenze: Le Monnier Università, 2009), 63–94.

25 Giovanni Battista De Benedictis [pseud. L'Aletino], *Lettera terza apologetica contra il Cartesio creduto da più d'Aristotele*, addressed to "*Al signor Lionardo Di Capoa*," in *Lettere apologetiche in difesa della Teologia Scolastica, e della Filosofia Peripatetica di Benedetto Aletino dedicate all'Illustriss. & Eccelentiss. Signore, il signore D. Carlo Francesco Spinelli Principe di Tarsia, &c.* (Napoli: nella Stamperia di Giacomo Raillard, con licenza de' Superiori, 1694), 122: "*marcio epicuro*"; 124: "*un'impostore*"; 136: "*apre una strada reale al Caso di Epicuro, e all'Ateismo.*" On De Benedictis's attack, see the study by Ettore Lojacono: "Immagini di Descartes a Napoli: da Valletta a Costantino Grimaldi (parte I)," *Nouvelles de la République des Lettres* 2 (1999): 67 and ff.

26 On Pietro Giannone, see the classic study by Giuseppe Ricuperati, *L'esperienza civile e religiosa di Pietro Giannone* (Milano and Napoli: Riccardo Ricciardi, 1970). See also Giuseppe Ricuperati, *La città terrena di Pietro Giannone: Un itinerario tra "crisi della coscienza europea" e l'illuminismo radicale* (Firenze: Olschki 2001).

27 Pietro Giannone, "Vita di Pietro Giannone," in *Opere di Pietro Giannone*, eds. Sergio Bertelli and Giuseppe Ricuperati (Milano, Napoli: Riccardo Ricciardi, 1971), 49: "quel divino ingegno."

28 See Jonathan I. Israel, *Radical Enlightenment. Philosophy and the Making of Modernity 1650–1750* (Oxford: Oxford University Press, 2001), Part. 4, 35 (*The Radical Impact in Italy*).

29 Generally, on Giannone's reading of Descartes, see Maurizio Torrini, "Il Cartesio di Giannone," in *Pietro Giannone e il suo tempo*, ed. Raffaele Ajello, vol. 1 (Napoli: Jovene, 1980), 415–30.

30 Giannone, *Opere*, 624.

31 *Id.*, 628.

32 Pietro Giannone, *Il Triregno*, ed. Alfredo Parente (Bari: Laterza, 1940), 54: "una più perfetta e ben organizzata macchina corporea."

10 Physics in the Broad Sense

Boyle, Newton, and the Baconian Metaphysical Physics

Philippe Hamou

The purpose of this chapter is to consider and contextualize some early-modern British views on the relation between physics and metaphysics—namely in Bacon, Boyle, and Newton. I do not claim to offer a detailed description of their positions here. The aim is to sketch the big picture and try to find out whether these influential figures share a common ground. There are good textual reasons to think that Boyle's and Newton's views were at least partly framed as critical responses to Descartes' understanding of the relation. However, these responses should not be interpreted in the positivist guise that was long prevalent in the standard account of British science. The positions of Newton, Boyle, and a number of other "experimental philosophers" are distinctive not so much because they reject what Daniel Garber[1] has labeled "*Descartes' metaphysical physics*," but because they provide an alternative conception of the intimate relation between physics and metaphysics, which is partly grounded in the Baconian understanding of the architecture of natural knowledge. This allows physics, *in an enlarged sense of the word*, to include the metaphysical consideration of primary and final causes, as well as *forms* and essences.

In Descartes' view, metaphysical considerations are prior to physical ones, both in a constitutive sense and in an epistemic one. Metaphysics, or *philosophia prima*, contains all the foundations of physics,[2] and the knowledge of these foundations should be acquired before delving into physical considerations. *Philosophia prima* provides, in the first place, a standard of *certainty* (the *cogito*) and a *rule of truth* according to which things are just as our clear and distinct ideas represent them to be. Second, it furnishes us with a clear and distinct idea of matter as tridimensional extensional quantity, the parts of which (individual bodies) together with their modes (shapes and motions) are the primary object of physics. Third, it considers the prime source of motion, God's efficiency and His immutable will, from which the most general laws of nature—which describe the conservation and communication of motion—are safely and directly (that is *a priori*) deduced, together with a number of general effects concerning the general structure of the world and its elements.[3] Although Descartes concedes that a brief "history of the phenomena" is necessary to go further into physical investigations,[4] and that the discovery of the specific mechanisms underlying the more particular phenomena of nature requires that one proceeds through hypotheses and experimental confirmation, he adamantly rejects any *early* use of such an experimental method. Refined experiments, to one who is not already acquainted with first principles, cannot be properly interpreted and are very likely to lead astray.[5]

I want to point out two striking features in these well-known Cartesian conceptions. One concerns order: metaphysics is no longer the science "that comes after the physics" and the crowning of the whole enterprise of natural knowledge, as it was in the Aristotelian view; it is the science *to begin with*, however arduous and abstract it may seem, a prime philosophy, without which nothing can be properly secured in the edifice of knowledge. The second feature is organic continuity. In the tree of the sciences, roots and trunk are made of one and the same continuous substance. Metaphysical considerations are not only preliminaries; they are an integral part of the physical enterprise. Natural philosophy is at the start, or *in principle*, a *metaphysical physics*. And this is true, whether Descartes' metaphysical interests were wholly invested in the project of founding a new physics, or whether they were not—which is a controversial issue.[6] While these two features are tightly knitted in Descartes, we should bear in mind that they are in principle independent. It would be possible for metaphysics (the knowledge of the primary causes, and what counts as the basic constituents of nature) to be an integral part of physical enquiries and yet be considered as second in an epistemic sense if it succeeded empirical investigation and derived from it its epistemic value. As we shall see, this is exactly what holds true for some of the leading British natural philosophers.

It is however customary to present the British empiricist scientific tradition as directly opposed to the Cartesian view of the metaphysical foundations of science, not only at the methodological level but also at the substantive one. There are several ways of making that point, depending on which author one considers as primary. A Boyle-oriented perspective may want to express the opposition to Descartes in the contrast between "speculative natural philosophy" and "experimental natural philosophy": Descartes' *metaphysical physics* would be an example (among others—notably the scholastic physics) of an enquiry into nature that has been corrupted by an overconfidence in the powers of reason and a neglect of experiments. That would amount to rejecting a *speculative* physics and replacing it with an experimental one. Another way to make a strong case for the opposition is to stress the importance of Newton's *Principia* for a new definition and new understanding of physics as *mathematical physics*. Newton's carefully crafted title, "*Mathematical principles of natural philosophy*," is taken here as a direct reference to and rebuttal of Descartes' [metaphysical] *Principes de la Philosophie*. Metaphysics is no longer the science which gives us an intuitive, self-evident access to physical principles. The definition of matter and space, the laws of motions, and the characterization of forces are not derived from our innate notions of the nature of substance or God, nor derived from any other metaphysical considerations, but are cautiously introduced in the first book of the *Principia* as "mathematical hypotheses"—that is, an initial set of definitions, axioms, and postulates, in quite the same manner as Archimedes would have done in other branches of rational mechanics. From such mathematical principles, one can predict and quantify the behavior of imagined bodies in various circumstances. The agreement of these predictions with experience—for example, with the way actual planets behave in our solar system—is what makes these mathematical principles the true principles of natural philosophy. Nothing more is to be wished for. So, according to this view, at no point, neither at the start nor at the end, does metaphysics have a role to play—mathematical principles are the *only principles* of natural philosophy that one needs.[7]

Finally, one may also appeal to Locke and consider his own contribution to epistemology as being in effect a patient dismantlement of the Cartesian or Cartesian-like *metaphysical physics*. In the *Essay Concerning Human Understanding*, in a well-known passage of the Epistle to the Reader, Locke presents himself as the "under-laborer" employed, for the sake of master-builders such as Huygens, Boyle, or Newton, in "removing some of the rubbish that lies in the way to knowledge."[8] It is tempting to interpret this famous declaration as follows: while men such as Huygens, Boyle, or Newton make their positive contribution to the construction of a new physics purged from any metaphysical ingredients, Locke would do the dirty work, showing why, since our intellectual equipment is what it is, metaphysical physics—call it speculative physics, aprioristic physics, etc., is doomed to failure.[9]

These approaches to the opposition between Descartes and the British on the nature of physics may be called, with some caution, positivist approaches, at least if the term is construed as capturing the notion that physical science in its mature form should be cleared of any metaphysical ingredients. What may perhaps lend some support to the idea that such a positivist view prevailed amongst these authors is the fact that they very seldom use the term *metaphysics*. When they employ it, it is often in a pejorative sense, as a synonym for what is abstruse, verbose, conceited, and scholastic. Thus Boyle opposes the superfluity of "Logical and Metaphysical Notions and Niceties" to the solidity of "Physical Observations and Reasonings."[10] However, one should not draw conclusions too hastily. First, it does not seem that Descartes' practice of *philosophia prima* is the target of these pejorative uses. The adjective "metaphysical" is most often used to refer to scholastic verbosity and to either trivial or seemingly incomprehensible metaphysical definitions, such as the definition of motion as "the act of what is potential, in so far as it is potential." For the denunciation of such metaphysical niceties, it seems that the Englishmen could easily recruit Descartes on their side. One may even argue that it was Descartes who actually taught them to despise "the learned but frivolous use of uncouth, affected, or unintelligible terms"[11] that supposedly prevailed in scholastic metaphysics. Second, these semantic facts certainly show that metaphysics as an autonomous discipline had somehow fallen into disgrace in the second half of the century, and this perhaps may be revealed in certain shifts in the academic curriculum of English universities during the time. However, this is not to say that the metaphysical enterprise as such had fallen into oblivion. Metaphysical pursuits could still be undertaken in other less discredited disciplines, such as theology, or taken afresh in other parts of the philosophical curriculum—namely in logic, in ethics, and of course in *physics*. As a matter of fact, the very idea that physics, understood in a broad sense, should incorporate at least part of the traditional metaphysical program had been stated and vigorously defended in Bacon. As this source is more likely to have been a common ground for the later British natural philosophers than any Cartesian or anti-Cartesian commitment, it provides an appropriate starting point for discussing British accounts of the relation between physics and metaphysics.

Bacon's Metaphysical Physics

A description of the division and hierarchy of the sciences was to Bacon of central importance to his very project of "great reformation." Indeed, whoever wanted to invent and advance human learning had to become acquainted with the true state of our knowledge, make an inventory of what was known and what was still unknown, and have a sense of how the sciences are related to one another and ordered. Bacon undertook this description twice: first in English, in the second book *Of the Advancement and Proficiency of Learning* (1605), and then in the expanded Latin version *De Dignitate et Augmentis scientiarum* (1623), which became what was certainly his most widely read book.

On the place allotted to metaphysics, the Baconian account did not change substantially between 1605 and 1623. Bacon considered that the old concept of metaphysics had to be revised, and that its place and unity had to be reconsidered. He was aware that such reconsideration was so drastic that it might have led him to change the very name of the discipline. However, at least in these texts, he refrained from such a change, both because of his respect for antiquity,[12] and, I would argue, because he thought that the very name metaphysics has a truth about it, insofar that it implies a strong connection to physics. His redefinition of metaphysics aimed precisely at restoring this proximity to physics that the traditional discipline, as developed by the Aristotelians, did not reflect sufficiently. Bacon's central tenet was that metaphysics is a part of natural philosophy and deals with natural beings just as specifically as physics and natural history do. For that reason, the discipline should be stripped of foreign elements that the tradition has incorporated into it, namely *natural theology* and what Bacon called *philosophia prima* or *summary philosophy*.

Natural theology, the knowledge of God through His works, is no part of the knowledge of nature; it properly belongs to "divine philosophy", it aims at a (provisional) knowledge of God and helps us to shun atheism—but it is not concerned with the elucidation of natural beings.

Bacon's *philosophia prima* can be defined as the science of "common notions." It deals with the most abstract axioms that are in use in every science, such as the axioms concerning "quantity, similitude, diversity, and the rest of those extern characters of things."[13] Although his knowledge may be of some concern for physical and/or metaphysical enquiries, it appears no more specifically related to natural beings, than it is to *human* or *divine*. So here, it is because of its extreme generality and abstraction that *philosophia prima* is excluded from the realm of metaphysics.[14]

In *De Augmentis* (book III, chapter IV), Bacon explains somewhat more specifically in which sense many of the traditional objects of the old "metaphysics" should be forwarded either to *prima philosophia* or to *natural theology*: the common notions and the so-called "transcendentals" (listed as *Multum*, *Paucum*, *Idem*, *Diversum*, *Possibile*, *Impossibile*), which used to be subject matters for a general metaphysics (or what will be called later *ontologia*) belong only to *prima philosophia*, whereas the traditional objects of the so-called *metaphysica specialis*, such as *Deus*, *Unus*, *Bonus*, *Angelus*, and *Spiritus* are indeed the affair of natural theology. After such a severe trimming, metaphysics appears reduced to what should be its sole and unique purpose: the study of natural beings. Thus, physics and metaphysics belong to one and the same enterprise. They do not differ in subject matter, but simply in the mode of consideration: metaphysics looks at what is essential and permanent in natural beings, whereas physics deals with the more "transitory" and "accidental" aspects of matter and bodies. Bacon (using the old division of causes) also says that metaphysics is assigned to the study of formal and final causes, whereas physics is concern with material and efficient ones. Thus understood, physics and metaphysics are two connected parts of the "speculative" theory of nature. They both rest on natural history, and form with it the pyramid of natural knowledge:

> For knowledges are as PYRAMIDES, whereof HISTORY is the basis: So of NATURAL PHILOSOPHY, the BASIS is NATURAL HISTORY: the STAGE next the BASIS is phisicke; the STAGE next the vertical point is METAPHISICKE. As for the verticall point, *Opus quod operatur Deus a principio usque ad finem*,[15] the Summary law of nature, we knowe not whether Mans enquirie can attain unto it.
>
> (*Advancement of Learning*, 85)

Here, speculation is the upward (preferably slow) movement along the three stages of the pyramid which leads from material facts of natural history to the disclosure of physical (efficient) causes, from them to the apprehension of forms and essence, and finally reaches (if possible) the ultimate apex of metaphysical enquiry (the final cause of all things, or the "condensed law of nature" to which Bacon equates Salomon's words: *opus quod operatur Deus a principio usque ad finem (the work that God made from the beginning to the end)*.

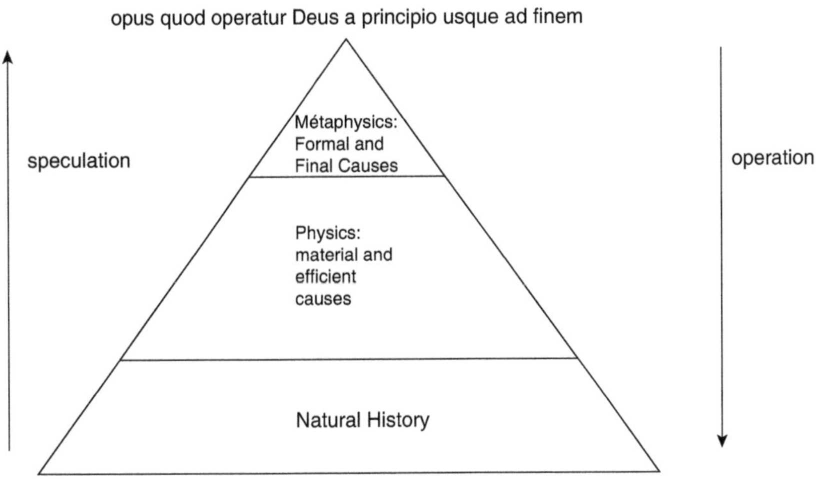

The contrary move, from metaphysics to physics and from physics to the discovery of new facts, deeds, and useful devices, is the "operation." Speculation aims at intelligibility. Operation aims at utility. *Pace* the usual Baconian vulgate, it is to be stressed that, to Bacon, the farther we go into speculation, the better we can choose our means and promote operation, and this is what make metaphysics an especially *useful* science:

> It [Metaphisicke] doth enfranchise the power of Man unto the greatest libertie and possibilitie of workes and effects. For Phisicke carrieth men in narrow and restrained waies, subject to many accidents and impediments, imitating the ordinaroe flexuous courses of nature. But *Latæ undique sunt sapientibus viæ*;[16] to sapience (which was anciently defined to be *Rerum divinarum, & humanarum scientia*) there is ever a choice of Meanes. For *Phisicall causes* give light to newe invention in *Simili materia*. But whosoever knoweth any forme knoweth the utmost possibilitie of *superinducing* that *Nature* upon *any varietie of matter*; and so is less restrained in operation, either to the *Basis* of the *Matter*, or the *condition* of the efficient.
>
> (emphasis in original, *Advancement of Learning*, 85–6)

In describing metaphysics, Bacon insists that it should be pursued with confidence, as something that is not, as skeptics often believe, out of reach of the human understanding:

> For METAPHISICKE, we have assigned unto it the enquirie of FORMAL and FINAL CAUSES; which assignation, as to the former of them may seeme to be Nugatorie and voide, because of the received and inveterate Opinion, that the inquisition of Man is not competent to finde out *essentiall formes* or *true differences*; of which opinion we will take this hold: That the invention of Formes is of al other Parts of

> Knowledge the worthiest to bee sought, if it bee Possible to be found. As for the possibilitie, they are ill discoverers, that thinke there is no land, when they can see nothing but Sea.
>
> (emphasis in original, *Advancement of Learning*, 83)

Bacon adds to this an important *proviso*: we do not really want to inquire into the form of each natural species, because those are potentially infinite, but we want to find out the real alphabet out of which each one of these forms is made:

> to enquire the form of a Lyon, of an Oake, of Gold; nay, of Water, of Aire, is a vaine pursuite; but to enquire the *formes* of Sence, of voluntary motion, of vegetation, of colours, of Gravitie and Levitie, of Densitie, of Tenuitie, of Heate, of Cold, & al other Natures and qualities, which, like an *Alphabet*, are not many, & of which the essences (upheld by Matter) of all creatures do consist.
>
> (emphasis in original, *Advancement of Learning*, 84)

So what we have here is a limited number of primitive natures and qualities whose diverse compositions make the whole of this natural world. The true end of Baconian metaphysics is to find out these elementary natures and to understand what natures there are and how they interact to compose all the phenomenal diversity.

I shall not enter here in the description of how Bacon tried to implement this metaphysical program. I do not think that his undertakings on this score were considered successful, even by his keenest supporters in the second half of the century. I do think, however, that the Baconian commitment to physics understood in a broader sense was still very much accepted and implicit in the way later experimental philosophers and British natural philosophers understood their own contribution to natural science. The new science of nature was called to take the place both of an actual physics that was still too narrow (following too closely the tortuous course of efficient causes) and of an actual metaphysics, which was ill founded and badly defined. This at least is what Bacon seemed to imply in his letter to Father Baranzan (June 1622):

> De metaphysica ne sis sollicitus. Nulla enim erit post veram physicam inventam.
>
> ("As for metaphysics, you should not be concerned: it will entirely disappear, when true physics is invented.")[17]

I take these words to mean not that the future would dispense altogether with metaphysical considerations, but rather that "true physics" will be the right location for them.

Some will find that this presentation of Bacon, as an advocate for a "speculative" and even "metaphysical" physics, flies in the face of the common view according to which Bacon was promoting experimental philosophy, precisely against a *speculative* approach to natural philosophy. As Peter Anstey has shown in an important paper,[18] the very word *speculative* tended to become, in the seventeenth century, an antonym for *experimental*, and these terms became the *actual* categories in which people of the time represented their aims and battles. This might have been true for subsequent authors, but Bacon did not oppose the *speculative* and the *experimental*. The duality that is singled out in the passages from the *Advancement of Learning* and *De Dignitate* is between speculative and operative; these terms, as we just saw, refer to two different legitimate aims, and two opposite directions, which are found in the practice of natural experimental science. Of course, Bacon's methodology, as stated for example in the *Novum Organum*, is full of strident warnings against what Bacon often calls the "premature flights" toward the more speculative parts of philosophy. This habit of going directly to axioms and conclusions, after considering superficially a small number of experiments, or even without any experimental grounding, simply on the basis of preconceived ideas or "idols"—is certainly one of the main obstacles on the road to true and certain induction. And this is also the *rashness in speculation* that will be eventually labeled the "speculative" way of natural philosophy, of which Descartes was supposedly the very paradigm, and against which the whole "Bacon-faced generation" of the early Royal Society period unanimously objected.

Nonetheless, neither in Bacon nor amongst the "experimental philosophers" of the Royal Society do we find the idea that physics should simply renounce its speculative aim and content itself with operation and usage. It is certainly the case that many Baconians and experimental philosophers thought that the time of speculation had not yet arrived, that the great business of the period was to reconstruct natural histories, and that the task of erecting higher superstructures was reserved for later generations of future philosophers. But these declarations, in addition to being sometimes rhetorical and hiding a not completely blank speculative agenda, did not mean that understanding nature "as it is" was not considered a proper aim for physical enquiries or that practical utility was their only purpose. This should appear all the more true when one considers that utility itself, or in Boyle's terms the "usefullness of natural philosophy,"[19] includes, as one of its most valuable parts, the pure benefit that true knowledge gives to the mind that possesses it.

Metaphysical Cosmogony and Final Causes in Boyle's "Corpuscularian hypothesis"

Robert Boyle is certainly the paradigmatic example of an experimental philosopher of the second half of the century. It could be said that, through his many writings and unremittent practice, he is the very one who framed the identity of the "experimental philosopher." In my view, this identity is somewhat mixed: it involves a Cartesian ingredient as well as a Baconian one. On the one hand, experimental philosophers were undoubtedly true Baconian disciples: they shared with Bacon the view that natural philosophy should be reformed, and that the only way this reform was going to succeed was to make it start where science first began, that is with sensible matters of facts, experiments, and natural history. On the other hand, experimental philosophers, no less undoubtedly, strongly felt the spell of Descartes' grand idea of mechanical philosophy, the idea that all that happens in the material realm can be explained as the result of the mechanical operations of insensible particles. They might not agree with Descartes about how one ought to justify this claim or how its details should be fleshed out—they may even have rejected it altogether, but in some sense they all recognized its attractiveness as far as intelligibility and heuristic values were concerned.

These two ingredients are no doubt present in Boyle, who advocated both the superiority and usefulness of the "experimental (natural) philosophy" and the "excellency of the mechanical hypothesis," and never found any contradiction in doing so.

Of course, Boyle was keen to distinguish his mechanical hypothesis from the Cartesian one both at a methodological level and in regard the very content of the hypothesis. Boyle has been rightly described as the "diffident naturalist,"[20] as having been extremely reluctant toward claims for evidence and absolute certainty in physical matters. As he writes in the *Excellency of Theology* (1674), hinting, no doubt, at Descartes: "the most even of the modern Virtuosi are wont to fancy more of clearness and certainty in their physical Theories than a Critical Examiner will find" (*The Works of Robert Boyle*, 8, 66). He then explains that the premises of most theories are only founded on moral certainty, and so are the inferences that are taken from them, "as geometrical as they could be" (ibid). In his methodological *Proemial Essay*, of his *Certain Physiological Essays* of 1669, Boyle defends the ideas that theoretical superstructures should be grounded on "a considerable number of experiments, in proportion to the comprehensiveness of the theory to be erected upon" (*The Works of Robert Boyle*, vol. 2, 14). Although this consideration applies here to local theories (particular mechanical explanations), there is no reason to think that it should not hold as well for the general one, that is, for the mechanical or "corpuscularian" hypothesis as such. And indeed it seems that Boyle's very program in natural philosophy was to provide some sort of massive inductive proof of the mechanical hypothesis, exhibiting a large range of experiments and observations of all sorts, showing again and again how far our understanding of them could be enhanced when set in the general framework of the mechanical hypothesis, taking only matter and motion as explanatory principles. So, to Boyle, the mechanical philosophy recommends itself inductively for its capacity to account for an increasingly wide range of phenomena. It is what Boyle calls its "comprehensiveness." Each step made in advancing mechanical explanations makes the general theory more plausible. But even though Boyle never seriously considered that another explanatory hypothesis could be a viable replacement candidate, the mechanical philosophy (and of course its metaphysical content) was deemed to remain a *hypothesis*.

Boyle usually presents his own "Corpuscularian hypothesis" in a careful way, always insisting on what makes it different from the Epicurean atomism on the one hand and the Cartesian mechanical hypothesis on the other. The three doctrines share the same basic "physical" idea, namely that whatever is produced in the material world is the direct effect of local motion on particles of matter, but they differ markedly in their underlying metaphysical premises. On the one hand, whereas Epicurean philosophers consider that the only reason why atoms go in such or such direction and make such or such composition is mere chance, Descartes and Boyle concur in assigning the first source of motion to God's will and efficiency. On the other hand, in Descartes, as Boyle reads him, it seems that no more is required to form the "system of the world" than the initial introduction into the world of an invariable quantity of motion, "the material parts being guided by their own unguided Motions to cast themselves into such a System." Boyle has another view of the matter, which he expresses (for example) in the *Excellency and Ground of the Mechanical Hypothesis*.

> But I plead onely for such a philosophy, as reaches but to things purely corporeal, and distinguishing between the first *original of things* and the subsequent *course of Nature*, teaches, concerning the *former* not only that God gave Motion to Matter but that in the beginning He so guided the various Motions of the parts of it, as to contrive them into the World he design'd they should compose, (furnish'd with the seminal principles and structures or Models of Living Creatures,) and establish'd those *Rules of Motion*, and that order amongst things Corporeal, which we are wont to call the *Laws of Nature*. And having told this as to the *former*, it may be allowed, as to the *latter* to teach, That the Universe being once fram'd by God, and the Laws of Motion being setled and all upheld by His incessant concourse and general Providence, the Phaenomena of the World thus constituted, are physically Produced by the Mechanical affections of the parts of Matter, and what they operate upon one another according to Mechanical Laws.
>
> (emphasis in original, *The Works of Robert Boyle*, vol. 8, 104)

The passage is consistent with many others similar pronouncements in Boyle's writings. It makes it clear that Boyle is not only committed to the "physical" thesis that whatever happens in the material realm is the result of corpuscular local motions. What he presents here in a condensed form is clearly a full-blown metaphysical physics, which offers the same kind of genetic—or, if I may say so, "cosmogonic"—intelligibility as the Cartesian principles of natural philosophy. As in Descartes, what is sought here is not only local mechanical explanations of the phenomena, but a general frame in which such explanations make sense and can be traced back to their first principles. Second, in this general frame, Boyle distinguishes the "first original of things" and the "subsequent course of nature." Whereas the description of the second phase (as the efficient course of motions ruled by (blind) mechanical laws and preserved by God's continuous concurrence) does not depart markedly from Descartes' views, the first phase is clearly different: in the first original of things, the first phase in the history of Creation, one has to consider not only God's efficiency in the production of motion, but also his "design," that is the ends He pursued in "directing the motions" of matter, arranging together the different parts of the universe, or following "Models" in the production of living creature of different species. Boyle mentions here the creation of "seminal principles," namely the (still unknown) grounds for the generation of living beings, principles that were put in our first parents and that are still acting now in the formation of each new living being. It seems clear that the hint here is that we need more to explain the living beings that simply matter and unregulated motion: some systems of matter must have been specifically organized in the beginning in order to allow for the generation of living Creatures and for the perpetuation of seminal principles. Admitting this however does not prevent those specifically designed seminal principles to act mechanically (as if they were some sorts of molds) in the processes of generation during the subsequent course of time.

This significant departure from Descartes amounts to a specific consideration of *final causes* in the account of the creation of natural beings, and specifically the account of the creation of *species* of living beings. As it turns out, Boyle's most cogent consideration of the role of final causes in physics, his *Disquisition about the final causes of natural things*, contains also his most explicit discussion of the disciplinary boundaries between physics and metaphysics and of how Boyle's conception of them relates to those of Descartes. In this text, Boyle takes Descartes to task for having excluded the consideration of final causes from physical enquiries. He considers in particular Descartes' answer to Gassendi in the Fifth replies, wherein, without denying that God had specific wills in creating the world, he says that those final causes are entirely hidden from us[21] and entirely foreign to the naturalist. Descartes reiterates the same argument in his Letter to Hyperaspistes of August 1641,[22] and then, most explicitly in *Principles* 1.28:

> When dealing with natural things we will then, never derive any explanations from the purposes which God or nature may have had in view when creating them and we shall entirely banish from our philosophy the search for final causes. We should not be so arrogant as to suppose that we can share in God's plans. We should instead consider him the efficient cause of all things; and starting from the divine attributes which by God's will we have some knowledge of, we shall see, with the aid of our God-given natural light, what conclusions should be drawn concerning those effects which are apparent to our senses.[23]

So, according to Descartes, we are not to consider the ends of God in creating such or such sensible thing, because our physical enquiries aim at explaining how those things are produced by discovering the chain of efficient, mechanical causes that concurred in producing the forms that we see. Any physical enquiry which would avail itself of teleological discourse would be altogether presumptuous, uncertain, anthropomorphic, and likely to revert to the old discarded Aristotelian view of nature, where things are deemed to be so and so because their production obeys some inherent form or *telos* that determines them to be so and so.

Boyle's answers to these Cartesian doctrines constitute the larger part of the first section of the *Disquisition*. In a first line of argument, he refutes and somehow inverts the accusation of presumption. To pretend to know (some of) God's ends "is not a presumption, but rather, to take notice of them is a Duty" (*Works of Robert Boyle*, 11, 89). He then explains:

> For there are some things in Nature so curiously contrived, and so exquisitely fitted for certain Operations and Uses that it seems little less than Blindness in Him, that acknowledges, with the *Cartesians*, a most wise Author of things, not to conclude that, tho' they *may* have been design'd for *other*, and perhaps higher, Uses; yet they *were* designed for *this* Use.
>
> (emphasis in original, *The Works of Robert Boyle*, 11, 89)

Boyle is clearly referring to living bodies and parts of living bodies, which are so exquisitely framed that it seems impossible (and somehow "blind") to conceive that chance lone could have produced them. We cannot understand the structure of the eye without considering its use in seeing. And seeing is such a manifest and obvious end that it would seem absurd to think that it is not for us to know it, or that it lays "hidden in the abyss of the divine wisdom."

A second line of argument depends on the diagnosis of why Descartes thought proper to exclude final causes in physics. To Boyle, such exclusion bears on a too-narrow conception of what may count as solid reasons in physics. Although physics strictly conceived deals specifically with mechanical reasons and efficient causes, its most fundamental principles are not, on Descartes' own account, strictly "physical," as they make use of God as an immaterial agent:

> I readily admit, that in Physicks we should indeed ground all things upon as solid reasons as may be had; But I see no necessity that those Reason should be always precisely Physical: especially if we be treating . . . of the *first* and general Causes of the world it self; from which Causes, I see not why the Final Causes, or Uses, that appear manifestly enough to have been design'd should be excluded. And to me it is not very material, whether or no, in Physicks or any other Discipline, a thing be prov'd by the peculiar Principles of that Science or Discipline;

> provided it be firmly proved by the common grounds of Reason. And on this occasion, let me observe, that the fundamental Tenents of Mr Des-Cartes's own philosophy are not by himself prov'd by Arguments strictly physical; but either by Metaphysical ones, or the more Catholick dictates of Reason, or the particular testimonies of Experience.
>
> (emphasis in original, *The Works of Robert Boyle*, 11, 91)

Boyle goes on to show that God, an immaterial being, is the efficient cause of motion in matter, since matter, whose essence does not include motion, must owe its motion to some being that is not material.[24] He also points out that when Descartes argues that God's immutability proves the conservation of the same quantity of motion, he is not using "a physical argument strictly so called, but rather a Metaphysical one." This of course it not an objection against the Cartesian proof (although Boyle, on other grounds, expresses some doubts about the absolute truth of the conservation thesis). It only shows that we should not exclude final causes from physical considerations in order to preserve some pretended "purity" of physics. Disciplinary boundaries do not apply here; a metaphysical argument can be used in physics, provided that it is founded on "the common grounds of reason." The point ironically has a Cartesian ring: all the sciences are one and proceed from the same natural light. Here is how Boyle puts it:

> And to me 'tis not very material, whether or no in Physicks or in any other Discipline; a thing be prov'd by the peculiar Principles of that Science or Discipline; provided it be firmly proved by the common grounds of Reason.
>
> (*The Works of Robert Boyle*, 11, 91)

The whole discussion calls for a few conclusive remarks. First, and quite strikingly, Descartes is not here taken to task for having made his physics too metaphysical, but rather the contrary. Descartes has unduly restricted the scope of natural enquiry to efficient causes. His metaphysical physics is all about efficiency and divine power, but provides no room for design and divine wisdom. To Boyle, the (metaphysical) consideration of final causes is necessary on two different grounds: on the one hand, it is necessary for the task of *physics* itself, insofar as it provides an explanatory resource for phenomena that are manifest everywhere in nature and in which there is evidence of design, fitness, and organization, especially in living beings, but also in the general frame and composure of the universe. In Boyle's view, "unregulated mechanism" (if I may so call the blind determinism that he tends to attribute to Descartes) is simply unfit for the task of accounting for such phenomena. On the other hand, final causes are also especially important to the physicist because they justify his calling on religious grounds. They make us aware of the superior wisdom, intelligence, and benevolence of the Creator, and the very fact that the study of nature offers a constant occasion for contemplating these divine attributes, and even for refining our understanding of them, is in itself a sufficient justification for the *study of the book of nature.*

So Descartes, according to Boyle, has missed the role that the physicist could play in promoting *natural religion* and refuting *atheism.* And this failure seems to be directly correlated with Descartes' dismissive attitude toward the idea that sound metaphysical conclusions may be drawn from the contemplation of nature. Metaphysics, for Descartes, begins in a meditative conversion, in which the mind, discarding the false testimony of the senses, looks into itself and finds there the very idea of the Infinite cause. In this perspective, there cannot be any empirical constraint on metaphysics: neither brute sensory experience nor experiments, however numerous and regulated, could yield a clear and distinct idea of the essence of matter, and even less could they provide a clear and distinct idea of the primary cause of all things. This is precisely why Descartes was extremely reluctant toward physico-theological arguments (and indeed to my knowledge never used them). He avoided them not because he thought nature was the reign of blind fate, but rather because these arguments, grounded as they are on sensory experience, were deemed to remain confused and their conclusions uncertain. Accordingly, they should be banned if one wants to keep physics on the safe tracks of science. Boyle, for his part, saw in the Cartesian aspiration to absolute certainty in physical matters a form of blindness—a misunderstanding both of the limits of human reason and of the fact that physics, or natural philosophy, serves nobler ends than the mere satisfaction of curiosity. It is deemed to express God's glory, to manifest His providence, and to contribute to disclosing what he calls "the great and universal system of God's contrivances." To achieve that objective one must engage in discussions that overlap with metaphysics and religion. Boyle expresses this most clearly in the *Excellency of Theology*:

> But neither the fundamental doctrine of Christianity, nor that of the effect of power and matter and motion seem to be more than an epicycle (if I may so call it) of the great and universal system of God's contrivances and makes but a part of a more general theory of things, knowable by the light of nature, improvable by the information of scriptures. So that both these doctrines, though very general, in respect to the subordinate parts of theology and philosophy, seem to be but members of the universal hypothesis, whose object I conceive to be the nature, counsels and works of God, as far as they are discoverable by us.
>
> (*The Works of Robert Boyle*, 8, 32–3)

Physics, or natural philosophy, gives only a partial view of the universal hypothesis; it is just like an epicycle, in the old astronomical systems, which offers only a partial representation of the movement of a planet. I take this comparison to mean that the same truth may be expressed in opposite, seemingly contradictory ways in different sciences, just as the retrograde motion of a planet may be represented as the composite effect of opposing but wholly compatible revolutions of several celestial circles. Similarly, apparent contradictions certainly may occur between physics and religion, between the knowledge of efficient causes and the knowledge of final causes, and between physics and metaphysics. But it would be foolish to think that we are forced to choose between these sciences and to reject one in favor of the other. At some level, all sciences are one and the contradictions must vanish, even though we might never be able in this life to understand how.

Newton's Empiricized Metaphysics

Newton's pronouncements on metaphysical matters are scant but quite significant. A striking one from a late manuscript, which was intended for a revision of the General Scholium, may provide a useful starting point. Here is I. B. Cohen's English translation of the passage:[25]

> What is taught in metaphysics, if it is derived from divine revelation, is religion; if it is derived from phaenomena through the five external senses, it pertains to physics; if it is derived from knowledge of the internal actions of our mind through the sense of reflection, it is only philosophy about the human mind and its ideas as internal phaenomena likewise pertain to physics. To dispute about the objects of ideas except insofar as they are phaenomena is dreaming. In all philosophy we must begin from phenomena and admit no principles of things, no causes, no explanations, except those which are established through phenomena. And although the whole of philosophy is not immediately evident, still it is better to add something to our knowledge day by day than to fill up men's minds in advance with the preconceptions of hypotheses.

This text lends itself to two opposite readings. On the one hand, it may appear as challenging the very idea of metaphysics: that discipline is entirely omitted and needs to be replaced by more legitimate enterprises. This reading makes sense when metaphysics is understood as a science whose main defining feature is an epistemic one. If one construes metaphysics as the science whose objects are addressed in a purely intellectual way (and somehow this is indeed the Cartesian construal), then metaphysics is an awakened dream, it is vain disputes about ideas or hypotheses "that fill up men's minds." The similarity between the two last sentences of the passage and other well-known methodological texts (especially the fourth *regula philosophandi*[26]), sheds some light on Newton's usual dismissive attitude toward *hypotheses*: they are not dismissed because they are probable conjectures or methodological tools of reasoning (since Newton in fact constantly used *hypotheses* in these senses). Rather, "hypotheses," in the pejorative sense, refers to empty reasoning about pure ideas (with no reference to empirical content), which is especially prevalent in metaphysics. It was particularly apparent in Cartesian metaphysics, in which one is supposed to access, in a purely intellectual way, some fundamental truths about the world.

The second reading is more positive. If metaphysics is defined not by an epistemic trait, but rather by its subject matter (eminent objects of knowledge, such as essences, forms, and primary causes), then it is not so much dismissed as redistributed into three distinct disciplines. Revealed religion, physics, and what Newton calls the "philosophy about the human mind and ideas" would each have a metaphysical part that is grounded on phenomena. The three sciences would nevertheless be distinct because the phenomenal realms on which they ground their conclusions are distinct: physics deals with the sensible external world, discovered to us by our five senses; philosophy of the mind considers the internal (but no less phenomenal world) discovered to us by reflection; and revealed religion also deals with something that is phenomenally given, and has to be read and interpreted, namely the Scriptures.

The passage is highly interesting. For one thing, it is a clear and rare expression of Newton's own view of the *distributio operis* that obtains in the intellectual world. In particular, one can recognize in Newton's description of the philosophy of the human mind a clear hint of Locke's enterprise in his *Essay concerning human understanding*. The exploration of an inner field of empirical phenomena, discovered through the inner sense of reflection, may be seen here as a sort of counterpart to Newton's own explorations of the external world. As Newton points out, following a suggestion of Locke himself,[27] the philosophy of mind belongs to physics understood in a broad sense and includes among its subject matters immaterial as well as material natural beings. Another striking point is the vindication of the methodological unity of these three great provinces of human science: they all derive their conclusions from the phenomena. Even theology finds in the words of the Scriptures its own sort of empirical constraints: there is no room for dogma here, no more than there is any in physics or in the philosophy of the mind. Finally, and it is the main point here, all three disciplines may be conceived as having a metaphysical part: they are not only descriptive disciplines, showing how the phenomena are connected, but they ought to offer some basis for speculation about true essences and primary causes.

Now, if we accept this reading, the question for us is how this "metaphysical part" could be developed in the specific case of physics (the science of bodies). I would suggest that we have two possible answers: whereas most of the published texts offer a rather "Boylian" or physico-theological answer to this question, unpublished manuscripts present a somewhat different and perhaps stronger version of the Newtonian metaphysical physics.

It would perhaps not be amiss to mention that neither the *Principia* nor the *Opticks* proved to be very hospitable places for metaphysical considerations. For example, Newton deliberately suppressed any references to

God that were still lingering in his first versions of the *scholium* on space. Newton's aim in the *Principia* was precisely circumscribed:

> For the basic problem [*lit.* the whole difficulty] of philosophy seems to be to discover the forces of nature from the phenomena of motions, and then to demonstrate the other phenomena from these forces.[28]

The purpose is not to find out "the physical causes and sites of forces" (*Principia*, 407)—a question that Newton considers out of his scope—but rather to establish from phenomena the very "fact" of forces, to exhibit their abstract (or mathematical) structure, and finally show how they could be applied to the explanation of other phenomena. This is all the *Principia* is concerned with, and this is certainly enough of a task for one book and for one man. The main text of the *Opticks*—whose object is not forces but rather *the abstract constitution* of light—is similarly devoid of any explicit metaphysical considerations. Whatever metaphysical content is present in both texts, its official expression arrives only at the very end and appears as second thoughts, queries, or appendices.

In the *Scholium Generale* (published in the second edition of the *Principia*, 1713), Newton writes of God: "we know him by his most wise and excellent contrivances of things and final causes." And after a couple of pages about how God ("pantokrator") exerts his dominion on the material world by being substantially present to it, he famously concludes: "This concludes the discussion of God, and to treat of God from phenomena is certainly a part of natural philosophy."[29] Similarly, in two final queries of the *Opticks*, after describing phenomena that include the attraction and the providential disposition of organs in living beings, Newton offers a few considerations (including the famous reference to space as God's sensorium) about how God might have formed matter at the beginning and how he is still present and providentially acting in the world.

Other interesting public pronouncements may be found in the 1692 correspondence with Richard Bentley, which took place just after Boyle's death, when Bentley was preparing the first Boyle Lectures.[30] Bentley asked Newton for suggestions about how his new natural philosophy might be put to use for promoting natural religion against the dangers of atheism and materialism; Newton was quite keen to detail for him a few interesting examples of such possible uses, and to show that the disposition of planets and the very measure of the laws of attraction were fit for providing the right sort of dwelling place for mankind, and thus could not be the result of mere natural causes.[31]

These texts quite undisputedly show that the late Newton had an interest in metaphysical questions, especially in the kind of physico-theological considerations that were so important to Boyle. These interests engaged him in the same sort of cosmogonic considerations that we found in Boyle: how did God arrange matter at the beginning? How did he dispose it in space, and to which ends? It may seem, however, that Newton's Boylianism (if it may be so named) is still a rather superficial feature of his physics. Although Newton said to Bentley, in his first letter, that he had always had an eye for the service that physics could do to religion,[32] Newton's explicit considerations on this specific topic are terse, and their public expression came quite late in his career. All in all, the defense of natural religion does not seem to have been such a strong driving force behind all of Newton's enterprises, as it was indeed in the works of Boyle, the "Christian virtuoso." Newton's interests in revealed theology were certainly much stronger, and it could be argued that they have helped to frame the actual contents of his view of God.[33]

A good case for the thesis that Newton actually had a metaphysical physics of a stronger sort (concerned not only with the design and arrangement of matter but also with its creation and very essence) may be made through an examination of the manuscript *De Gravitatione*.[34] Although the text is most probably[35] of an early date and was written before Newton came to the idea of universal gravitation, its contents are consistent with a number of other metaphysical hints posterior to the publication of the *Principia*. For that reason, it cannot be considered as some kind of youthful foray into metaphysics, which was soon to be entirely dismissed in favor of a sounder and soberer approach to physics.[36] The manuscript, for the most part, is a dialectical discussion of central concepts of Cartesian physics, showing that they are mutually inconsistent and, in fact, contradicted by Descartes himself. It argues that it is simply impossible to assign any determined speed or direction to motion, if motion is defined, as it is in Descartes, as the translation of the body from the vicinity of one contiguous body to the vicinity of another. If motion is something real, we need to refer it not simply to other bodies, but to an immobile being, pure extension, or absolute space. This in turn imposes metaphysical strictures on the conception of such immobile space, namely the idea that it exists "without subject" and thus cannot be a substance or an accident of a substance. As space is the very system organizing the relation of "places" of existent beings, it should be rather conceived as an emanative effect (that is a necessary concomitant) of the existence of beings. As Newton says, "when a being is posited, space is posited"[37] and so an infinite, absolute space has to be the emanative effect of the existence of a first, eternal, infinite being, the one that occupies all places, in all times. Even though the very idea of an omnipresent God might have a theological origin, the inferential process used in the argument has a very characteristic *regressive* or *analytical* dimension. It begins with a consideration of the phenomenon of motion, and it shows that in order to make coherent the various features of this phenomenon, absolute space is needed. That conclusion in turn presupposes a number of other metaphysical decisions concerning substantiality, existence, and God's omnipresence. So what we have here is indeed an upward speculative movement from the phenomena, ending in a consideration of the primary cause. It is not simply the logically spurious inference from effects to cause, as in the physico-theological argument.

Rather each new step expresses the condition of the possibility, or the essential presupposition, of the preceding one.

The *De Gravitatione* comes then to a striking discussion of how bodies could have been created. Again, considering that impenetrability is the phenomenal datum that makes bodies differ from space, Newton offers a sort of metaphysical fable, in order to account for how bodies (or something indistinguishable from bodies) could have been created. If God had chosen by an act of his will to render some regions of space impenetrable (to the real bodies) and had decided to transfer continuously this field of impenetrability from one part of space to another in accordance with the laws of the communication of motion (the ones that we know in effect to obtain among the phenomena), these seemingly moveable impenetrable regions of space would be indeed undistinguishable from "actual" bodies, at least for all their mechanical properties. This account of the creation of matter is assuredly adventurous, and it raises a number of questions that cannot be examined here.[38] Newton recognizes that, since God's will is contingent, God could have made bodies in another manner. The discussion is presented only as a conjecture, set in the frame of a cosmogonic fable and concerned only with *quasi bodies*.[39] But nevertheless, it could be argued that it is a discussion of how *physics* drives us naturally to conceive of the essence of bodies and bodily interactions, and how they could, or perhaps should, be constituted in nature to behave in such and such a manner. Here again, the style of reasoning is remarkable: Newton is not trying to deduce impenetrability from any innate idea of the essence of bodies, and even less the laws of collision from a preconceived idea of what God's immutability amounts to. The impenetrability and the laws of collision are taken for granted: they are matters of fact provided by empirical description and rational mechanics. Metaphysics enters the picture only when one asks the question: how could we genetically explain the fact that bodies have such or such qualities or obey such or such laws? What sort of essential constitution must have been given to them? As it appears, the answer in the *De Gravitatione* draws heavily on God's will and God's continuous action, and seems to dispense entirely with any substantial substratum that is understood as some sort of elusive *materia prima* in which the qualities of extension and impenetrability inhere. As a quasi-substantial subject, space suffices for sustaining the bodily qualities that God's will imparts to it.[40]

In his chapter on Newton's metaphysics, Howard Stein has suggested that we should take the metaphysics of the *De Gravitatione* as a kind of template for understanding the (still implicit) metaphysics of the *Principia*. As he writes:

> If all this is brought into relation to the metaphysical analysis in *De Gravitatione et aequipondio fluidorum* it implies that in creating a body, God, or in the constitution of a body, nature, must impose, not only the field of impenetrability and the laws of motion appropriate thereto, but other fields as well, with their laws characterizing forces of

> interaction of the kind that have been described—which fields, according to the preface to the *Principia*, it becomes the presumed task of natural philosophy to discover.[41]

Stein's descriptions are certainly quite suggestive. They make Newton a rather radical metaphysician, for whom the basic constituents of material reality are not substances or accidents, but fields of force (that is, dispatches of divine will) lawfully distributed in pure space. However, I am not sure that such a picture could always be easily reconciled with all of Newton's physical or metaphysical pronouncements. For example, in the last queries of the *Opticks*, Newton seems to hesitate between various causal interpretations of attraction. Attraction might be indeed the immediate action of God's will, which keeps the planets in their orbits and creates the field of force simply by being omnipresent to them. But it might also be operated through the mediation of some *active principle*, of a still unknown nature, which permeates space and interacts with bodies in a way that is also still unknown to us.

In any case, it cannot be doubted that there was a legitimate place, in Newton's eyes, for a metaphysical physics, provided that we understand its method in the Baconian sense—that is a physicalized metaphysics, an empirically constrained metaphysics of nature. As for its actual content, considering how scarce, terse, and sometimes cryptic the textual evidence is, the interpretive debate is still open and probably deemed to remain so.

Conclusion

Bacon's original views on the metaphysics of physics, together with Descartes' principles of philosophy, set the stage for British science in the second half of the seventeenth century. By no means were the British philosophers of nature hostile in principle to the idea that physics aims at discovering the true constitution of matter and the final cause of its arrangements. This speculative aim was still very much a driving force behind their devotion to physical investigations, together with service to natural religion and practical utility. They may have been more or less optimistic[42] about the prospect of such metaphysical speculations, but they never failed to recognize them as an integral part of their physical undertakings.

Descartes' role in framing the very idea of what a modern metaphysical physics should be cannot be underestimated. In order to present their metaphysical hypothesis, both Newton and Boyle borrowed from Descartes the device of a cosmogenetic fable. To make intelligible how things are constituted in nature, it shows how the course of efficient causes could have made them, if God, rather than creating the world and the creatures in their actual shape (as Genesis tells us), had chosen to create some other, simpler state of things, together with their laws—for example, brute matter and regulated motion, brute matter and unregulated motion, or infinite space and dispatches of God's will. There is no need to say that for Descartes, Boyle, and

Newton the cosmogenetic account is markedly different, but its form has an indisputable common stamp in all three authors.

This said, the Baconian strand of metaphysical physics is certainly the dominant feature in both Boyle's and Newton's accounts. Both considered that the metaphysical part of physical enquiries is dependent on instructions from phenomena. This shared conviction is the source of their common rejection of Descartes' ways in metaphysics, his belief that the basic truths about nature were to be found in a purely intellectual way, through an examination of only the content of our ideas. This intellectualist bias was to Boyle the very reason why Descartes missed the main metaphysical benefit of experimental philosophy, i.e., the disclosure of God's wisdom and design through the contemplation of his works. In this, Newton certainly concurred, and he said so in a number of public declarations. However, his main motive for rejecting Descartes' way of ideas in metaphysical physics seemed to have been somewhat more epistemological and somehow more internal to the physical project itself. Physical certainty, if any such is ever to be found, must be grounded on matters of facts and on the rules of induction, and certainly not on preconceived ideas of bodies, space, or even God. Physics, strictly conceived, aims at establishing the truth of the facts (such as the "fact" that matter everywhere attracts matter according to a certain law). In this sense, physics *stricto sensu* is nothing more than a refined (and mathematically instructed) way of extending the testimony of the senses to a larger and richer range of phenomena. But physics, in the broad sense, includes *metaphysical physics*, when one comes to the question of how, once the truth of the facts has been fully demonstrated, we ought to determine the meaning of our basic physical concepts (such as space, time, motion, essential qualities of matter, force) so that they may be used together consistently in our account of these (new) facts. Here, it seems that the metaphysical enquiry, however grounded on phenomena, is not so much, as it was still perhaps in Boyle, an inductive inference (a generalization from a large number of similar facts) or a retroductive move (such as the physico-theological argument concluding from the effects to the "presumed" cause). Rather, it assumes (at least in the example of the *De Gravitatione*) the form of *conceptual analysis*; it is a regression from conditioned to condition, where each step is (or ought to be) a necessary one. So it seems that, when it is properly conducted, this way of inferring from phenomena should yield not so much plausible conclusions, as most certain ones. This is the kind of achievement that Newton (who was in this regard certainly more Cartesian than any other of his British contemporaries) always sought.

Notes

1 Daniel Garber, *Descartes' Metaphysical Physics* (Chicago: University of Chicago Press, 1992).

2 See Descartes to Mersenne, 11 November 1640, AT III 297–8: "je vous dirai, entre nous, que ces six Méditations contiennent tous les fondements de ma physique."

3 Here is how Descartes describes this order: "*Mais l'ordre que j'ai tenu en ceci a été tel. Premièrement, j'ai taché de trouver en général les principes ou premières causes de tout ce qui est ou qui peut être dans le monde, sans rien considérer pour cet effet que Dieu seul qui l'a créé, ni les tirer d'ailleurs que de certaines semences de vérités qui sont naturellement en nos âmes. Après cela, j'ai examiné quels étoient les premiers et plus ordinaires effets qu'on pouvoit déduire de ces causes; et il me semble que par là j'ai trouvé des cieux, des astres, une terre, et même sur la terre de l'eau, de l'air, du feu, des minéraux, et quelques autres telles choses, qui sont les plus communes de toutes et les plus simples, et par conséquent les plus aisées à connoître*" (*Discours de la méthode*, AT VI 43).

4 "*breuem historiam praecipuorum naturae phaenomenon*," *Principia philosophiae*, AT III 4.

5 "*[J]e remarquois, touchant les expériences, qu'elles sont d'autant plus nécessaires qu'on est plus avancé en connoissance; car, pour le commencement, il vaut mieux ne se servir que de celles qui se présentent d'elles-mêmes à nos sens, et que nous ne saurions ignorer pourvu que nous y fassions tant soit peu de réflexion, que d'en chercher de plus rares et étudiées: dont la raison est que ces plus rares trompent souvent, lorsqu'on ne sait pas encore les causes des plus communes*" (*Discours de la méthode*, AT VI 43).

6 See the discussion in Denis Kambouchner, *Les Méditations Métaphysiques de Descartes I* (Paris: Presses universitaires de France, 2005).

7 These views are well illustrated in I. B. Cohen's classical studies of Newton's *Principia*, for example in Cohen (1980), and the introduction to I. B. Cohen and G. Smith, *The Cambridge Companion to Newton* (Cambridge: Cambridge University Press, 2002). See also the discussion of these views in Andrew Janiak, *Newton as Philosopher* (Cambridge: Cambridge University Press, 2008), chapter 2.

8 John Locke, *An Essay Concerning Human Understanding*, ed. Paul Nidditch (Oxford: Oxford University Press, 1975), The Epistle to the Reader, 10.

9 A less positivist and more proper assessment of Locke's views on physics, metaphysics, and their relation to the philosophy of the mind is beyond the scope of this chapter. Let us simply remark that Locke's own definition of physics, in the last chapter of the *Essay*, is broadened to the effect of including God and finite spirits among the objects of this science: "The knowledge of things, as they are in their own proper beings, their constitution, properties, and operations; whereby I mean not only matter and body, but spirits also, which have their proper natures, constitutions, and operations, as well as bodies. This, in a little more enlarged sense of the word, I call *Phusike*, or natural philosophy. The end of this is bare speculative truth" (*Essay*, IV, 21, 2).

10 "The Origine of Formes and Qualities," in *The Works of Robert Boyle*, eds. Michael Hunter and John Davies (Pickering and Chatto, London, 1999), vol. 5, 289.

11 Locke, *Essay*, 10.

12 See *The Advancement of Learning*, The Oxford Francis Bacon, IV, ed. M. Kiernan (Oxford: Clarendon Press, 2000), 80–1: " I use the word METAPHISICKE in a differing sense, from that, that is receyued: and in like manner I doubt not, but it will easilie appeare to men of judgment, that in this and other particulers, wheresoever my Conception & Notion may differ from the Auncient, yet I am studious to keep the Auncient Termes . . . I am otherwise zealous and affectionate to recede as little from Antiquitie, either in tearms or opinions, as may stand with truth and the proficience of knowledge."

13 Bacon defines it negatively, as "a Receptacle for All Such Profitable Observations and Axioms, as Fall Not Within the Compass of Any of the Special Parts of Philosophy, or Sciences, But Are More Common, and of a Higher Stage" (Bacon, *Advancement of Learning*, 77). He also says that it is the "inquirie touching the

operation of the Relative and adventive Characters of Essences, as Quantitie, Similitude, Diversitie, Possibilitie, and the rest." (82).

14 It should be clear that Bacon's *philosophia prima* is not primary in the fundational sense that Descartes' *prima philosophia* is. It owes its precedence only to the fact that its results are so general that they may be applied to every science. But the generality of *philosophia prima* concerns only the "Relative and adventice Characters" of things, not their true essences. It does not seem that there is any necessity to know thoroughly everything that could be known concerning these common accidents before coming to terms with the more particular sciences. This explains why the deficiencies in *philosophia prima*, noted in the *Advancement*, do not entail *ipso facto* deficiencies of the sciences that come second in the Baconian ordering.

15 "The work which God works from beginning to end." Eccles. 3:11.

16 "Broad are the ways on all sides to the wise." Cf. Prov. 4:11–12.

17 *The Works of Francis Bacon*, eds. J. Spedding and al., (London: Longman, 1857–1874), 14. 375.

18 Peter Anstey, "Experimental Versus Speculative Natural Philosophy," in *the Science of Nature in the Seventeenth Century*, eds. Peter Anstey and John Schuster (Dordrecht: Springer, 2005).

19 Cf. Boyle, *Some Considerations touching the Usefulness of Experimental Natural Philosophy*, in *The Works of Robert Boyle*, vol. 3, 189 ff.

20 As in Rose-Mary Sergeant's fine book, *The Diffident Naturalist, Robert Boyle and the Philosophy of Experiment* (Chicago: University of Chicago Press, 1995).

21 See *Reply to the Fifth Objections*, AT VII 375: "toutes les fins de Dieu sont toutes également cachées dans l'abîme impénétrable de sa sagesse."

22 See AT III 431–2. Boyle (*Works*, 11, 92), quoting the letter to Hyperaspistes (in Clerselier's French translation) admits that Descartes, in this letter, is only opposing "the reasonings of those who think that God has no other ends in creating the world but that of being praised by men." According to Descartes: "it would be childish and absurd for a metaphysician to assert than God as some vain glorious human being, had no other purpose in making the universe than to win men's praise." To Boyle however, Descartes overstates his case, and fails to consider our duty of praising God for his Creation.

23 See AT 8A 15, CSM I 202. For a balanced discussion of Descartes' views on final causes, see Margaret Osler, *Divine Will and the Mechanical Philosophy* (Cambridge: Cambridge University Press, 2005).

24 Of course, this "cosmological" proof of the existence of God, with which Boyle tends to credit Descartes, is nowhere to be found in Descartes' work. In the Third Meditation, the '*preuve par les effets*' was strictly restricted to the unique fact of which we could be absolutely certain (the fact of our own thinking). For all we know, motion, a sensible phenomenon, might only exist in our dreams.

25 Cf. I. B. Cohen's Introduction to *The Principia, Mathematical Principles of Natural Philosophy*, trans. I.B. Cohen and A. Whitman (Berkeley: University of California Press, 1999), 54. Here is the original Latin text, in *Newton's Mathematical Papers*, VIII 459: "*Quod in Metaphysica docetur, si a revelatione divina deducitur, religio est; si a Phaenomenis per sensus quincibus externos, ad Physica pertinent; si a cognitione actionum internarum mentis nostrae per sensum reflexionis, philosophia est de sola mente humana & ejus ideis tanquam Phaenomenis internis & ad Physicam item pertinet. De Idearum objectis nisi quatenus sunt phaenomena somnium est. In omni Philosophia incipere debemus a phaenomenis, & nulla admittere rerum prinicipia nullas causas nullas explicationes nisi quae per phaenomena stabiliuntur. Et quamvis tota philosophia non statim pateat, tamen satius est aliquid indies addiscere quam hypothesewn praejudicijs mentes hominum praeoccupare.*"

26 *Principia*, book III, 796.
27 See the text quoted supra note 9.
28 *Principia*, Preface, 382. In the last page of the *Opticks*, Newton similarly described the same two-stage enquiry: one part was resolutive from compound to ingredients, from motions to force, or from effects to their (immediate) causes, and one part was compositive, which goes in the other direction and extends the phenomenal realm of the explanation.
29 *Principia*, 943.
30 Under the terms of his will, Boyle endowed a series of lectures or sermons (originally eight each year) to be held at Saint Paul's Cathedral, with the explicit purpose of proving "the truth of the Christian religion against infidels, without descending to any controversies among Christians; and to answer new difficulties, scruples, etc." Richard Bentley's *Confutation of Atheism*, whose last part draws heavily on Newton's answers, was the first outcome of the lectures. Boyle's Lecture rapidly became the main stronghold, in the early eighteenth century, for the defense of a latitudinarian physico-theology, against the rise of freethinking. See Margaret Jacob, *The Newtonians and the English Revolution, 1689–1720* (Hassocks, Sussex: The Harvester Press, 1976).
31 The four Letters to Richard Bentley, together with the last part of Richard Bentley's *Confutation of Atheism from the Origin and Frame of the World* (London, 1693) are reprinted in *Isaac Newton's Papers and Letters on Natural Philosophy*, eds. I-B. Cohen and R. E. Schofield (Cambridge, MA: Harvard University Press, 1958), 279–352.
32 See the opening sentence of his first letter, "Sir, When I wrote my Treatise about our System, I had an Eye upon such Principles as might work with considering Men, for the Belief of a Deity; nothing can rejoice me more than to find it useful for that Purpose" (*Newton's Papers and Letters*, 280).
33 To some commentators, the tone of absolute certainty that Newton uses when he says how we should conceive of the divine attributes is more likely to have been derived from his own (and sometimes rather idiosyncratic) reading of the Scriptures than from any considerations of physical appearances, or phenomena. As Andrew Janiak, who defends this thesis, puts it, the conception of God found in Newton's writings is entirely "immune to revision regardless of any developments within physics" (*Newton as Philosopher*, 48).
34 *De Gravitatione et Aequipondio Fluidorum, in Isaac Newton, Unpublished scientific papers of Isaac Newton*, Alfred Rupert Hall and Marie Boas Hall ed. (Cambridge: Cambridge University Press, 1962). We are particularly indebted here to the inspiring chapter of Howard Stein, "Newton's Metaphysics," in *The Cambridge companion to Newton* (Cambridge: Cambridge University Press, 2002).
35 According to the first modern editors (Hall & Hall), the manuscript is from the mid-1660s. Although it has been argued since (notably by B.-J. Dobbs) that it comes from a later date, contemporary with the composition of the *Principia*, the early dating remains more plausible (see the discussions in Stein, "Newton's Metaphysics," and J-A Ruffner, "Newton's *De Gravitatione*: A Review and Reassessment," *Archive for History of Exact Sciences* 66, no. 3 (2012): 241–64).
36 The ontology of space as a divine affection appears in a later manuscript, posterior to the first edition of the *Principia*. The manuscript has been edited and translated in J-E. McGuire, in "Newton on Place, Time, and God: An Unpublished Source," *British Journal for the History of Science* 11 (1978): 114–29). According to the testimony of Pierre Coste, the translator of Locke's *Essay*, the account of the creation of matter that is found in *De Gravitatione* was still in Newton's mind at the time of his acquaintance with Locke, that is, after 1690. Locke cryptically refers to it in the second edition of his *Essay*. See *Essai sur l'entendement humain*, traduit par Pierre Coste, Amsterdam, 1729 (2nde éd.), Coste's note to paragraph 4.10.18.

37 Newton, *De Gravitatione*, 103: "*Et hinc sequitur quod spatium sit primario existentis effectus emanativus, quia posito quolibet ente ponitur spatium*" "And hence it follows that space is an emanative effect of what primarily exist, since when any being is posited, space is posited" (our translation).

38 See the rather severe discussion in Jonathan Bennett and Peter Remnant, "How Matter must first be made," *Canadian Journal of Philosophy* 4 (1978): 1–11; and, for a qualified defense of Newton's argument, Lisa Downing, "Locke's Metaphysics and Newtonian Metaphysics," in *Newton and Empiricism*, eds. Zvi Biener and Eric Schliesser (Oxford: Oxford University Press, 2014), 97–118.

39 The whole discussion, and especially this cosmogenetic fabulation (which evokes Descartes' genetic way of presenting his physics in *Le Monde, l'Homme*, and his *Discourse on Method*) shows how much Descartes' spirit and style was still present and influential in the *De Gravitatione*, even though the intellectualist bias of the Cartesian metaphysical physics was the main target of it.

40 See Newton, *De Gravitatione*, 106–7. "*Quod ad horum entium existentiam non opus est ut effingamus aliquam substantiam non intelligibilem dari cui tanquam subject forma substantialis, inheareat: sufficient extensio et actus divinae voluntatis. Extension vicem substantialis subjecti gerit in qua forma corporis per divinam voluntatem conservatur, et effectus iste divinae voluntatis est forma sive ratio formalis corporis denominans omnem spatii dimensionem in qua producitur esse corpus.*" Here is a revised version of the somewhat defective translation of this very significant passage by Hall and Hall (140): "For the existence of these beings, it is not necessary that we feign some unintelligible substance to exist, in which a substantial form should inhere as in a subject; extension and an act of divine will are enough. Extension takes the place of the substantial subjects in which the form of the body is conserved by the divine will. And the effect of this divine will is the form (or formal reason) of bodies, denominating every dimension of space where body is to be produced."

41 Stein, "Newton's Metaphysics," 288–9.

42 Even Locke, who was perhaps more pessimistic than his peers about how far we could make physics "scientific" (that is, raise it beyond the status of uncertain hypothesizing), still defines physics, not so much as a pragmatic enterprise, but as a "speculative" one, which aims at discovering the true constitutions of beings (see supra note and the reference to *Essay*, 4.21). Although he often declares that he does not want to "meddle" with physical considerations (*Essay* 1.1.2), the *Essay*, in many ways, makes room for them, at least at the level of probability, dealing both with the essence of the mind and that of matter. Since it shows how our metaphysical options are somehow constrained by the results of an empirical enquiry (the reflective enquiry on ideas), it would not be improper to say that Locke's *Essay* also belongs to the Baconian tradition of "metaphysical physics" that has been illustrated here.

11 Continuous Creation, Occasionalism, and Persistence

Leibniz on Bayle

Jean-Pascal Anfray

Introduction

According to the doctrine of conservation as continual creation (CCC henceforth), no creature can persist in time if it is not conserved by God's sustaining causal activity, and this causal activity is the same as God's initial creative action. God must continuously act in order to sustain the world in being. CCC was a common tenet within Christian theology.[1] It was eventually held among early modern philosophers, especially Descartes and the Cartesians, where it notoriously became a central item of metaphysics and physics. Indeed, God's role as a creator and sustainer of the world is a prime example of a metaphysical doctrine that has physical impact. In Descartes, it is well known that the doctrine plays a role not only for establishing God's existence, but also for understanding God's role as the ground of the laws of nature and his causal involvement in the motion of bodies.[2] But CCC is not a mere Scholastic inheritance. As Steven Nadler suggested, one key motivation that may explain the importance of CCC within Cartesianism and seventeenth-century philosophy more generally was its role in attempting to reconcile the world's radical dependence on God with divine transcendence.[3]

However, CCC is also a radical doctrine. It serves as the central premise in one of the main arguments for occasionalism, an argument that was first devised by Louis de La Forge and made popular through Malebranche's version in his *Dialogues on Metaphysics and Religion*. In short, the argument from CCC starts from the principle that what God continuously sustains are fully determinate bodies—including their distance relations with other bodies. From this and an assumption excluding causal overdetermination, the argument concludes that no particular body can move on its own. Generalizing this argument to minds would entail the denial of genuine causal activity to all created substances.[4]

Leibniz subscribes to the claim that God conserves all things in being and to the idea that conservation is a kind of continuous creation:

> [Actual things] depend on God for existence, since all things are freely created by God, and are also conserved by God; and it is not incorrectly

> taught that divine conservation is a continued creation (*continuatam creationem*), as the sun rays continuously (*continue*) flow from the sun, even if creatures do not emanate from God's essence nor necessarily.
> (*Causa Dei*, §9, GP VI, 440)[5]

This passage suggests at least a verbal adherence to CCC, illustrated by the sun metaphor, but it does not settle the exact way Leibniz understands it. Another passage from the *Theodicy* gives a less metaphorical description:

> What can be said for certain on the present subject is that the creature depends continually upon divine operation, and that it depends upon that no less after the time of its beginning than when it first begins. This dependence implies that it would not continue to exist if God did not continue to act; in short, that this action of God is free.
> (T 385)

Conservation is here understood minimally as meaning a constant ontological dependence of creatures on the free activity of the creator. Continued existence is not the "default option," so to speak, for creatures, for they would not persist if God ceased to will their continued existence. At any time of its duration, the dependence of creatures on God is the same as its dependence at the time of creation, thus conservation can be called a creation. This passage, however, remains neutral on some crucial points and so does not settle the issue of the nature of continuous creation. Some see that Leibniz and the Cartesians share the same, strong conception of CCC. According to this reading, Leibniz's main problem is how to reconcile this account with the thesis that substances are truly active and generate spontaneously their states. But even if Leibniz is able to solve this problem, there still remains the question of reconciliating a strong reading of CCC with the persistence of things.[6] This has led others to the claim that Leibniz is actually committed to a weaker, deflationary account of CCC, which does not generate the problem of persistence.[7] It is hard to adjudicate between these opposing interpretations. While difficulties generated by a literal reading tend to favor the latter, textual evidence seems to fit the former. I think that Leibniz subscribes to a strong view of CCC. I will try to defend this interpretation against the objections that such a view leads either to occasionalism or to the denial of the persistence of created things. To establish this, it is necessary to study Leibniz's criticisms of the Cartesians' understanding of CCC. The *locus classicus* on these questions for this are the extensive remarks in the *Theodicy* in relation to Bayle's occasionalist argument from CCC. These passages have been often commented upon, but I will show that we can get a better understanding of Leibniz's own position by studying some of his reading notes on Bayle for the *Theodicy*. I will start with some preliminary distinctions on the nature of CCC related to the Scholastic origins of the doctrine. Then I will discuss Bayle's own version of the argument from CCC, which differs significantly from Malebranche's version. The last three parts of the paper will be devoted to an examination of Leibniz's reply.

Admittedly, Leibniz's approach to CCC is highly abstract and seemingly detached from its immediate physical consequences concerning force and motion, since he is concerned with God's conservation of mind-like substances. Despite this, his own conception of CCC constitutes the metaphysical underpinning of his conception of enduring forces which manifest their own action in the various bodily motions.

Two Readings of CCC: Strong and Weak

Conservatio means the sustaining of a thing's being. A conservative cause is a *causa secundum esse*, which is distinct from a *causa secundum fieri*. The effects of a *causa secundum fieri* can persist even if the cause no longer acts, as a house persists once the builder has finished it. Again, conservative causes can be either immediate and *per se* or mediate and *per accidens*. A candle's flame can persist only if there is some oxygen, but the presence of oxygen acts only as an accidental conservative cause. Aquinas illustrates the same point with the case of an adult's preventing a child from falling in a precipice, whereby the adult acts as a preventive cause of destruction.[8] By contrast, a direct, *per se* conservative cause is such that the very being of its effect depends on it. God is the only *per se* conservative cause.[9] Moreover, things depend on God for their conservation in the sense that if God were to stop willing that something exist, then it would stop existing: no positive action is required on the part of God in order to annihilate a creature.[10] Finally, God is the only conservative cause of the substantial being of creatures.[11]

Conservation so defined does not differ from creation and is in fact a continued creation. Suárez presents the doctrine in the following way:

> Conservation is not an action that differs from creation with respect to any reality or real mode, but rather it differs from creation only by reason of a certain connotation or implied negation—that it is only conceptually distinct from creation. For "creation" designates the effecting of an entity while connoting that the entity has not previously existed, whereas "conservation" designates that same effecting while connoting that the entity has already existed beforehand.
>
> (*Disp. Met.* disp. XXII, sec. 2, n.2)

Thus, God creates x at t just in case God causes x to exist at t and there is no t' prior to t such that x exists at t'. On the other hand, God conserves x at t means that God causes x to exist at t and there is some t' prior to t such that x exists at t'. This characterization of CCC can be read in different ways.[12] Indeed, Suárez adds a qualification in the definition, namely that creation and conservation designate one and the same action. This defines the weak reading of CCC. According to it, God conserves things in existence by continuing the very same act by which He caused their existence in the first place. As Aquinas writes explicitly:

> The conservation of things by God does not occur through some new action; but through the continuation of the action by which he gives existence (*Conservatio rerum a Deo non est per aliquam novam actionem; sed per continuationem actionis qua dat esse*).
>
> (ST I, q.104, a.1, ad 4)

Suárez illustrates this by analogy with the identity of the act by which the sun produces light with light's propagation (ST I, q.104, n.3). The claim that God's conservative action is token-identical with the creative action is characteristic of a weak reading of CCC.

The strong reading of CCC (or "strong CCC") by contrast affirms that God literally *re*-creates each creature at every instant. The terms of these mini-creative acts form a series of distinct, successive, momentary states, or temporal parts, whose summation correspond to the duration of everyday creatures. Conservation is a *continual* creation insofar as it is an uninterrupted repetition of discrete mini-creations. Within Islamic thought, the strong reading was defended by the Ash'arîte theologians. They derived it from the atomic structure of time: since time is composed out of time atoms or instants, God's conserving activity is bound to single instants. From this, some concluded that bodies do not persist. It seems that they argued as follows: a thing's persisting is its continuing to exist; but existence is an accident and no accident is permanent (i.e., accidents are essentially momentary entities); therefore, bodies can persist only by being constantly re-created. Accordingly, bodies themselves do not endure beyond an instant. Richard Sorabji (following Maimonides's report) summarizes the Ash'arîtes' central claims:

> Every time-atom, God creates an entirely new set of accidental properties, although they may be accidents of the same kind as before. If he omits to create new accidents, the substance which bore them will cease to exist. This shows that the blackness which we think is introduced in the cloth by the dye must in fact be created and re-created every time-atom by God.[13]

So they derived strong CCC from time atomism and saw occasionalism and the denial of enduring substances as its necessary corollaries.

Nevertheless, despite the fact that strong CCC appears a natural consequence of temporal atomism, as in Ash'arîte thought, some partisans of strong CCC also subscribed to a continuist picture of time. Bonaventure, for instance, argued that immutable substances like angels, whose duration is called *aevum*, are nevertheless temporally extended, insofar as their duration is successive, just as a ray is constantly emitted from the sun.[14] Conservation is then accounted for as the successive but continuous process whereby a thing's existence is continued. It is a process whereby one and the same creature is *continuously re*-created. Moreover, the combination of strong CCC with time atomism does not automatically yield the denial of persistence. Thus, among medieval Scholastics, Peter John Olivi seems to have defended a position of this kind.[15] Moreover, both Bonaventure and Olivi claim that what is conserved is a permanent entity, not an *ens successivum*. One way of characterizing the difference between a permanent and a successive being was that a permanent entity exists all at once (*totum simul*), while a successive being does not.[16] Created substances are wholly present at each successive moment of their conservation. This conception of persistence corresponds fairly well to the notion of a continuant in an endurantist account of persistence.[17]

Nonetheless, these were minority views. Suárez represents the majority view when he refuses strong CCC. He argues that it is superfluous to assume a plurality of conservation acts given the sameness of the effect, namely the persistence of the substantial being of the creature.[18] However, it is interesting to remark that though he regards it as superfluous, he still accepts as possible God's successively creating and annihilating the selfsame creature. Thus, Suárez would concede to Bonaventure (and Olivi, if he could have had access to his work) that strong CCC does not entail *per se* the denial of persistence or even of an endurantist account of persistence.[19]

Descartes on CCC

As is well known, Descartes bases his second proof for the existence of God on CCC and derives it from the assumption of the independence of the parts of time. From this, it follows indeed that that my present existence does not entail my existence at future times, "unless there is some cause which as it were creates me afresh at this moment—that is, which conserves me (*me quasi rursus creet ad hoc momentum, hoc est me conservet*)."[20] To this, Gassendi objected, first, that the parts of time form an "inviolable series and connection"; and second, that the parts of time are external to things and cannot affect their existence.[21] Descartes' reply makes clear that what is separable are the parts of a thing's duration or "the duration of the enduring thing" (*duratio rei durantis*): "you will not deny that the individual moments of this time could be separated from those next to them, that is, that the enduring thing could at any single moment cease to exist" (AT VII, 369–70, CSM II 254–5). The fact that Descartes refers to the innumerable parts of time as "moments" has led many commentators to ascribe him temporal atomism, and from this to conclude that he endorsed strong CCC.[22] Besides, many commentators judged that this entailed the negation of genuine persistence and causation among creatures. Norman Kemp Smith thus saw occasionalism as the logical consequence of CCC: "Bodies cannot be capable of causing changes in one another: not having sufficient reality to persist, they cannot have sufficient force to act."[23] More recently, Jonathan Bennett holds that "the continual preservation of things through time . . . is really the continual creation of successors to them."[24] His argument for this conclusion rests on the use of the separability criterion of real distinction: a cut at any point in one's lifespan distinguishes an earlier and a later part. Given that each part can exist without the other, they are really distinct and make up two substances.[25] Bennett then concludes:

> Descartes ought to hold that what we call a single enduring mind or body is really a series of ontologically distinct things which compose a kind of pseudo-substance because of their qualitative and spatiotemporal continuity. I accept this, and affirm that Descartes ought to have accepted it.[26]

Therefore, CCC excludes an endurantist account of persistence and leads to occasionalism. To sum up, according to the "traditional" interpretation, Descartes either accepts or is logically committed to the following four claims:

A. Time is composed out of independent temporal atoms (time atomism).[27]
B. God conserves some particular creature x over a time interval $[t, t']$ by successively re-creating x at each moment comprised between t and t' (strong CCC).[28]
C. God's re-creating x over a time interval amounts to his creating distinct particulars $x_t, \ldots, x_{t'}$ at each moment $t, \ldots, t'$ (negation of endurance).
D. No particular body in motion at t can cause some motion in another body at any time comprised in the interval $[t, t']$ (occasionalism).

While many have thought that (B) depends on (A), this is not shared by all "traditional" interpreters, and the two issues are distinct in principle. Thus, Bennett does not think that CCC commits one to time atoms.[29] Moreover, while Kemp Smith derives occasionalism from the lack of persistence of bodies, the historical occasionalist argument from CCC as it appears in La Forge, and more clearly in Malebranche, rests neither on time atomism nor on strong CCC. Indeed, Malebranche affirms time's continuity and claims that the "instant of creation" never passes away, so that God's conservation is token-identical with his creation act.[30]

Leibniz on the other hand reads Descartes along the traditional interpretation. Thus in the *Theodicy*, he claims that CCC *seems* to entail the denial of persistence, linking the Cartesian doctrine to the Platonic thesis that corporeal beings are in a constant flow (T 382). Then he diagnoses Descartes' argument for CCC, which is based on the principle of the independence of the parts of time (T 383). Finally, he identifies these independent parts of time with time atoms, so that CCC leads to the labyrinth of the continuum

(T 384). This argumentative sequence is punctuated by Leibniz's own positive but ambiguous characterization of conservation as a kind of ontological dependence of creatures on their creator. Thus in order to evaluate his conception of CCC, it is necessary to examine his critical comments on Bayle.

Bayle's Argument

Pierre Bayle discusses the argument from CCC in various entries of his *Dictionnaire*, but his most extensive discussion is contained in chapter 141 of his book *Réponse aux questions d'un Provincial*, where it is used as part of an argument showing the incompatibility of CCC with human freedom of indifference, against Isaac Jacquelot.[31] Jacquelot combined CCC with genuine activity in the mind and defended a kind of concurrentism, claiming that God sustains the existence of the finite substances and their causal powers he has first created, and then concurs with them in order to bring about their effects. Bayle refutes this account by stressing the perfect symmetry between creation and conservation and the creature's passivity at the time of its creation (OD III, 788a). The basic structure of the refutation is as follows:

1. Conservation does not differ from creation (CCC);
2. At the time of creation, God causes the being of creatures and their modifications;

Therefore,

3. At any time, God is the total cause of any effect within the world and no creature is even a partial cause of any effect.

Let us examine the premises in turn. By contrast with Malebranche, Bayle understands (1) as strong CCC. He thinks of continuous creation as a series of similar but distinct creations. To establish this, he distinguishes conservation *qua* God's immanent volition, from conservation *qua* terminus of God's creative action. Considered from the first point of view, conservation is a simple and immutable action, whereas considered from the creature's point of view, conservation consists in a discrete series of successive states.[32]

Not only does Bayle subscribe to strong CCC, he also seems to defend temporal atomism, directly based on the analysis of the nature of time, as the following extract from the *Dictionnaire* shows:

> Two parts of time cannot coexist together. . . . There is no part of time that can coexist with any other; each one must exist alone; each must begin to exist when the former ceases to exist. Whence it follows that time is not divisible to infinity and that the duration of things is composed of times properly said, each of which is simple and indivisible,

perfectly distinct from past and future times, and contains only the present time.

(*Dictionnaire historique et critique*, art. "Zénon" rem. F, IV, 538b-539a: Popkin, 353–4)[33]

The argument proceeds from the successiveness of the parts of time to the conclusion that time is atomic. This might look like an argument from the independence of the parts of time. It could be that Bayle is simply assuming a move from independence to atomicity. However, it is time's successiveness which is important for him. And successiveness entails something stronger than ontological independence, namely exclusion: the existence of one part of time precludes the existence of any other. Now, it might look unpromising, for a process might be successive though some of its parts overlap. So the inference from successiveness to exclusion looks invalid, until one realizes that Bayle does not use a diachronic, but a synchronic, criterion of successiveness, as was common in medieval philosophy.[34] According to the diachronic criterion, an entity is permanent if it endures numerically the same from one time to another, whereas it is successive if different temporal parts of it exist at different times. Bayle implicitly uses the synchronic criterion instead: an entity is permanent if it exists all at once (*totum simul*), while it is successive if *no* two parts of it could exist all at once. Thus, time being successive, the existence of one part of time excludes any other time and there cannot be overlapping parts of time. But how does it entail temporal atomism? Bayle's reasoning can be reconstructed as follows. If time is infinitely divisible, then the existence of the prior part of a given duration excludes that of its posterior part ("each must begin to exist when the former ceases to exist"). And if the process is reiterated to infinity, then no part of it ever exists. Now, if no part of time ever exists, then time does not exist. But this is false, since at least the present time exists. Therefore, the present is an indivisible part of time and time is composed out of these indivisible parts. The independence of the parts of time is thus neither merely logical nor based on the tacit assumption of a principle of causal simultaneity, but specifically ontological.

A further consequence of Bayle's understanding of CCC is that the duration of creatures is also composed of independent, atomic temporal stages. And these stages, which are the *termini* of God's creative action, seem to be the basic particulars within Bayle's ontology. One might conclude nothing truly persists. Bayle discusses precisely such a skeptical hypothesis in his *Dictionary*:

> Your soul has been created. God must therefore renew its existence every moment, for the preservation of creatures is a continual creation. How do you know that this very morning God did not allow the soul, which he continually created from the very first moments of your life

> until now, to fall back into nothingness? How do you know he has not created another soul with modifications like the ones yours had?
>
> (*Dictionnaire*, "Pyrrho" rem. B, III, 733b, in Pierre Bayle, *Selections*, 204)

The scenario is the following: given CCC, there is no discernible difference between a soul persisting for a given span of time and a series of distinct but similar transitory souls (one can generalize to any substance). What lesson can be drawn from this skeptical hypothesis? A strong reading would conclude that, as a matter of fact, nothing truly persists. This seems too strong, however, for Bayle does not affirm the truth of the skeptical hypothesis. It looks thus preferable to adopt a weaker reading, according to which the skeptical hypothesis would support the claim that substances do not persist by enduring, that is by existing wholly at each moment of their career. Individual souls would be said to persist because at each time there exists a definite temporal stage, which exists all at once. And the similarity between these temporal stages goes proxy for persistence.[35]

Along with CCC, premise (2) is the crucial premise in the argument. As Todd Ryan has shown, Bayle actually produces two different lines of justification of (2).[36] The first appears exclusively in the *Réponse*. It relies on the Scholastic principle that in order for anything to act its existence is presupposed:

> What conserves itself acts; now what acts, exists and nothing can act before it has its complete existence; thus if one creature were able to conserve itself, it would act before it exists. This reasoning is not grounded on probabilities, but on the first principles of metaphysics, *non entis nulla sunt accidentia*; *operari sequitur esse*, which are as clear as day.
>
> (*Réponse*, OD III, 788a)[37]

The first step of the argument for (2) states that no substance can even partially sustain itself, but requires God as a total cause.[38] But then, it is impossible for a substance either to produce an effect or to concur with God in the production of such an effect. For, Bayle argues, at the initial moment of the creation of a soul, what God produces is a complete entity, that is a fully determinate substance or "thick particular," e.g., a soul with such and such thought or a body in such or such location. By symmetry, it must also be the case for any other moment of time, and thus God is at any time the total cause of any effect occurring at that time.

Still, there remains a gap within Bayle's argument, for it does not exclude that a creature is a partial cause at t of an effect occurring at a later time t', for instance my soul having a determinate thought now might have the power to produce another thought later. This sort of situation does not violate the *operari sequitur esse* principle. What entitles Bayle to exclude it is the tacit reliance on a requirement of temporal simultaneity of cause and effect. This thesis is considered as a principle by Descartes and lies at the heart of his argument for the independence of the parts of time.[39] Though it does not appear directly in the *Réponse*, there are other places where Bayle is very close to explicitly endorsing it.[40]

Now this additional principle of causal simultaneity is necessary only on the version of the argument based on the *operari sequitur esse* principle. But Bayle develops another line of argument, where the conclusion (3) follows directly from the premises. This second version of the argument rests on the thesis that modes are not really distinct from substances and that faculties or powers are strictly identical to their substances.

> One could not maintain that there is a real distinction between the substance of minds [*esprits*], their faculties, and the acts of those faculties without admitting inconceivable absurdities. Spiritual substance, its faculty of thinking, [and] the thought that it has at each moment are only one thing. . . . Thus, the creative act that brings about the substance of minds and their capacity for thought necessarily brings about their actual thought. It would be a contradiction that of three really identical things, God creates two and does not create the third.
>
> (*OD* III, 789a; translated by T. Ryan *Pierre Bayle*, 85)

Bayle has a very strong reading of the no real distinction thesis, which amounts to a pure identity. Ontologically speaking, modes are nothing over and above the modified substance.[41] The latter is in turn identical to the substance itself. Therefore, by transitivity of identity, the creation of a substance is identical to the creation of the modes of this substance.

Bayle's reductionism with respect to modes is very strong. He sees it as a natural consequence of his nominalism with respect to properties in general.[42] Modes stand on a par with relational properties. In his lectures on metaphysics, he envisages the particular mode of a body's local situation, which he treats as an extrinsic denomination, deriving from the distance relation between the situated body and a group of other bodies (OD IV, 502). Now, extrinsic denominations can change without a change in the denominated subjects (OD IV, 498). To use Bayle's example, a man in Paris does not acquire a new property compared to the same man in Roma (OD IV, 505). The crucial step is that such external denominations are treated like internal relations as resemblance.[43] Resemblance supervenes on the resembling items, without need of any additional entity. Similarly, distance relations supervene on the created things. Bayle then applies the same reductive analysis to the other bodily modes of shape and motion: shape depends on the various locations of the parts of the shaped body; motion is the successive recreation of a body in different locations (OD IV, 503). Therefore, all bodily modes are extrinsic properties, which represent no ontological addition for the body.[44] Change of shape is the paradigm case: my finger can be first straight and then bent. One can speak of modes insofar as there is a *fact of change*, but this change does not suppose the inherence of two distinct entities, straightness and bentness, in a subject.[45] The argument is then generalized to mental modes (OD IV, 505; OD III, 789a). In addition to the conceptual problems raised by this strong reductionism,[46] there are also problems of consistency with respect to other elements of Bayle's thought. One particularly pressing concerns the asymmetry between a substance and its modes. How can a substance differ from its modes if they are really identical with it? In other words, can Bayle avoid the charge of rendering all modes essential to the identity of a substance? Here is one of his attempts at an answer:

> It must be replied that such is pain with respect to the soul, that when it exists, it is not distinct from the soul, but that it does not always exist as long as the soul exists. One can therefore know the soul without knowing pain, because the state of pain is not part of the soul's essence, i.e., it does not essentially accompany it.
>
> (OD IV, 506)

As I see it, the difference between a mode or an accidental property and the essence of a substance would collapse if one considered a single moment of time, but it reappears from a diachronic point of view. Bayle's answer to this objection rests on the traditional assumption that substances persist, and that they do so by enduring. However, Bayle's skeptical hypothesis questions precisely this assumption, as we have seen. Thus, the answer to the objection fails and no substance can survive the slightest change. We can see that Bayle accepts claims (A) through (D). This threatens the very possibility that things persist, for Bayle needs a relation linking the various temporal stages in a single persisting object. Usually a causal relation is viewed as a necessary condition: namely, an object persists through time if earlier stages (at least partially) cause later stages. But Bayle's occasionalist argument denies this. Therefore, it could just as well be said that transitory objects are successively replaced by God's continuous creation: nothing can settle whether a sequence of temporal parts constitutes the same man or a successive replacement of distinct but similar men.[47]

Leibniz on Strong CCC and Occasionalism

It is now time to turn to Leibniz. His basic strategy against Bayle is to concede him his strong reading of CCC, while denying that it leads to occasionalism. This has led some commentators to claim that Leibniz solves the problem of occasionalism within the framework of strong CCC.[48] This interpretation is based on the following passage:

> Suppose that the creature is produced anew at each instant; grant also that the instant, being indivisible, excludes all priority of time. But let us notice

> that it does not exclude priority of nature, or what is called anteriority *in signo rationis*, and that this is enough. The production, or action whereby God produces, is anterior by nature to the existence of the creature that is produced; the creature taken in itself, with its nature and its necessary properties, is prior to its accidental affections and its actions; and yet all these things occur (*se trouvent*) in the same moment. God produces the creature in conformity with the demand (*exigence*) of the preceding instants, following the laws of his wisdom; and the creature operates in conformity with that nature that he returns to it in creating it always.
>
> (T 388)

Leibniz distinguishes here two dimensions of causation, synchronic and diachronic. God and the creature play a distinct role at each of these dimensions. Leibniz's reply grants strong CCC and time atomism. Whether this represents a purely dialectical concession or his own position cannot be adjudicated based on this single passage. But working under these assumptions, Leibniz blocks the occasionalist conclusion by introducing a conceptual or quasi-logical order within instants, according to which substance and its powers is prior to its modifications. Thus, God's conserving the whole being of a creature (including its modifications) at any time does not preclude the causal contribution from the creature. Adams glosses this text thus:

> what God (directly) produces . . . is not just the creature's nature or substantial form or *capacity* to produce, and not just the creature's nature *and* its affections and actions, but the creature's nature "operating" and thus *producing* its affections and actions.[49]

Although no creature can be the cause of the existence of a substance, this does not preclude creatures from being the true causes of their own modifications.

It is difficult, however, to determine exactly the causal role of created substances with respect to the two dimensions of causation, diachronic and synchronic. At first sight, it seems that, within a single moment, one cannot isolate the created substance's causal contribution from God's causing the existence of its states. This is where diachronic causation comes into play. What determines the production of a given mode at a time t is the state of the substance at a prior time t'. Leibniz talks of an exigency (*exigence*) of the prior states on the later states. This terminology is borrowed from Jesuits' discussions of divine concurrence.[50] The connection with this context is acknowledged by Leibniz, who writes in 1706 to Des Bosses:

> I maintain that in an active power there is a certain exigency (as your schools say) for action and hence for divine concurrence for action, albeit an exigency that can be resisted, which is grounded in the laws of nature established through divine wisdom.
>
> (LDB 11)

The idea of exigency can be understood in terms of ideal constraints on God's concurrence (and conservation), namely that God produces the series of modifications of a substance in view of its preceding states. To put it differently, the state of substance *s* at *t* determines the state of *s* at *t'*, but God actually produces both states by creating *s* at *t* and *t'*. In some passages from the 1680s, Leibniz already defended such a view of concurrence, where the creature's causal role is limited to this ideal-diachronic causation:

> In the great question, whether a secondary cause determines the first [cause], or whether the first [determines] the secondary, it must be replied that the first is determined by the secondary taken ideally, or that the idea of the secondary [cause] apprehended in the divine intellect determines the will of the first [cause]. But that the secondary [cause] is actually determined by the first [cause], or that it receives all its entity from the first [cause].
>
> (A VI, 4B, 1458)

Leibniz claims that creatures are only ideal causes, whereas actual causation, which we can understand as efficient causation, is restricted to God. But an ideal cause is dangerously close to an occasional cause, and created substances seem to be deprived of real causal efficacy. Some commentators ascribe to Leibniz a view very close to the one expressed by the preceding passage.[51] On this reading, Leibniz's position differs from Malebranche's occasionalism insofar as the latter introduces the idea of necessary connection as a requirement of true causal relations, whereas the former admits weaker connections among causal relata.

However, Leibniz never seems to have entertained seriously this quasi-occasionalism. For instance, in a passage from the later 1680s, he writes that each creature is "a true and real cause (*causa vera et realis*) of its immanent actions" (A VI 4B 1640). The two-dimensional analysis of causation in paragraph 388 of the *Theodicy* is precisely devised in order to maintain the causal contribution of creatures. A similar model appears elsewhere, for instance in remarks on Bayle's dictionary, written in 1702:

> For it is God who conserves and continually creates their forces, that is to say, a source of modifications that is in the creature, or indeed (*ou bien*) a state by which it can be judged that there will be a change of modifications.
>
> (GP IV, 568; L 583)

This passage is best read as proposing two alternative ways of considering the creature's causal contribution, which is either a synchronic productive causation or a diachronic ideal determination. The true productive activity and causal efficacy of created substances is confined to single moments of time. By contrast, the prior states put a constraint on God's contribution, which, given his wisdom, morally necessitates him.

The distinction between diachronic and synchronic causation reappears with the problem of persistence. Against the Cartesian thesis of the independence of the parts of a thing's duration, Leibniz objects that the connection between earlier and later states of a substance is morally necessary:

> One may answer that in fact it does not follow *of necessity* that, because I am, I shall be; but this follows *naturally*, nevertheless, that is, of itself, *per se*, if nothing prevents it. It is the distinction that can be drawn between the essential and the natural. For the same movement endures naturally unless some new cause prevents it or changes it, because the reason which makes it cease at this instant, if it is no new reason, would have already made it cease sooner.
>
> (emphasis in original, T 383)[52]

Between any two states of a substance, there is a *per se* connection. Leibniz had elaborated the notion of a *per se* inference many years earlier. In the *De affectibus*, he says that what follows *per se* from something else is what follows if there are no extrinsic impediments and gives as an example the descent of a heavy body (A VI 4B 1436f.). At first sight, then, Leibniz would be saying that if x exists at t, it will *per se* continue to exist at t' unless some cause prevents it from continuing to exist. This, however, seems close to Gassendi's extreme conservationism where God conserves a thing not by actually producing it continually but by removing impediments to its persistence. But God's conservation is a necessary condition for the thing's persistence. Thus, I think that a better reading of this passage is that the sequence of states is fixed within the thing's nature. *If* nothing prevents it, x will be F at t. But this does not mean that x will exist by itself. This requires in addition God's conservation as a necessary condition.[53] This interpretation fits nicely with what Leibniz says in §388 of the role of God's wisdom.

Elsewhere, Leibniz explains this natural connection between the successive states of a substance by the sameness of law generating the sequence of its states. This law, which he eventually came to understand as a principle of individuation (see GP IV, 518), accounts for the persistence of a substance, as he tells De Volder:

> For me nothing is permanent in those things, except the very law that involves the continued succession, which in individual things corresponds to that law that is in the whole universe.
>
> (*Letter to De Volder*, 01–24–1704, LDV 289)[54]

A substance persists insofar as its later states are the effects of its earlier states given its own law of the series, which is thus a necessary condition of persistence. Since the law specifies a given sequence of states, i.e., the history of the substance, it follows that no substance could be in a given state F unless F is a consequence of its law of the series. There are various ways of understanding which states are included in this law, in particular whether miraculous states pertaining to created substances can be deduced from this law of the series.[55] But one thing that is clear is that, as miracles "of the first rank" excluding any causal contribution by creatures, creation and annihilation are *not* included within the substance's law. Thus, a substance could not unfold a different history, unless it were annihilated by God. If x is F at t, the only way God could have made it so that x is G instead at t is by annihilating x and re-creating a substance y instead. Since it is impossible to deduce G from x's law, x is distinct from y.

But the law by itself is not a sufficient condition of persistence. Indeed, one may suppose that x and y, though distinct substances existing at different moments, nevertheless share the same law of the series, if we assume further that the sequence of states unfolds exactly as if this law pertained to a single substance. Leibniz envisages precisely such a possibility in the same letter to De Volder:

> But if anyone claims other substances that succeed prior ones are always produced by God and that they do not remain the same, he would be quarreling about a word, for there will be no further principle in things for deciding the question. The substance that succeeds is taken to be the same so long as the same law of the series, i.e., of the continual simple transition, persists that gives rise to our belief in the same subject of change, i.e., the monad.
>
> (LDV 291)

The fact that earlier and later states belong to the same enduring substance or to a succession of "transitory individuals," as R. Sleigh calls them,[56] seems a conventional matter at first sight, just as in Bayle the skeptical hypothesis entails the indiscernibility of the persistence of an individual soul and a succession of momentary souls. It could even be said that these transitory individuals are more basic particulars in Leibniz's ontology if Leibniz subscribes to strong CCC. Indeed, we have seen that God cannot change the course of events except by annihilation and recreation of a distinct particular. This condition is satisfied in our hypothetical case, since the same law links successively created and annihilated substance. If God's conservation is constrained only by exigencies deriving from laws of the series, then there is no reason to favor the claim that one and the same substance persists as a temporally extended object over the claim that each particular conserved by God is a distinct, momentary substance.[57]

From the preceding, we may conclude that the answer to the occasionalist argument based on the model of ideal causes is doubly incomplete. On the one hand, ideal causes tend to collapse into occasional causes. On the other hand, the moral necessity imposed on God's conservative action by the prior states of a substance is an insufficient ground for the true persistence of a substance. Causation as diachronic ideal determination, just like according to occasionalism, could be satisfied in a system of transitory individuals. Therefore, synchronic real production, to which I now turn, is an essential requirement of persistence.

Modes and Substances: The Ontology of Persistence

The role of synchronic real efficacy appears clearly in Leibniz's *reductio* argument of occasionalism to Spinozistic monism. This argument is frequently used by Leibniz and appears also in the *Theodicy*:

> It is well to beware, moreover, lest in confusing substances with accidents, in depriving created substances of action, one fall into Spinozism, which is an exaggerated Cartesianism. That which does not act does not merit the name of substance. If the accidents are not distinct from the substances; if the created substance is a successive being, like movement; if it does not endure beyond a moment, and does not remain the same (during some stated portion of time) any more than its accidents; if it does not operate any more than a mathematical figure or a number: why shall one not say, with Spinoza, that God is the only substance, and that creatures are only accidents or modifications? Hitherto it has been supposed that the substance remains, and that the accidents change; and I think one ought still to abide by this ancient doctrine.
>
> (T 393)

This line of argument is well known: substances are truly active beings; by depriving creatures from activity, occasionalism entails the denial that creatures are substances; from thence it follows that "God is the only substance" (see also *De ipsa natura* GP IV 509: AG 160). But there is a twist to the argument here. Leibniz equates the claim that creatures do not act with the thesis that substances are identical with their accidents. What justifies this equivalence is precisely Bayle's second version of the argument from CCC.[58] The distinction thesis is formulated in diachronic terms: accidents are transitory entities, whereas substances persist. Persistence is here incompatible with perdurance for, according to the preceding passage, a successive being (i.e., a being which has different temporal parts at different times) is equated with what does not "endure beyond a moment." A successive being like motion exists at various times, but it is never complete at any time, it is not wholly present at a given time. By contrast, a substance is a continuant, for it is wholly present when it exists. Thus, Bayle's occasionalism leads to the denial of the existence of substances.

Earlier in the *Theodicy*, Leibniz refers to the Duke of Buckingham as another proponent of the denial of the real distinction between substance and accident. It is useful here to compare Leibniz's criticism of Bayle with the argument in *De realitate accidentium* (A VI 4A, 994–6), where he mentions Buckingham too.[59] In that piece, Leibniz argues against real accidents, by showing that they lead to metaphysical difficulties with respect to the problem of change. In the final analysis, he subscribes to a kind of nominalism, where abstract terms do not refer to accidents, but are mere *compendia loquendi* (A VI 4A 996). Nominalism with respect to real accidents then appeared to Leibniz as a way of preserving enduring substances. However, Leibniz remains agnostic on the ontological ground of the facts of change. By contrast, Bayle's nominalism *identifies* the accident or mode with the substance. Leibniz shows that if we have an identity claim, there would not be substances but only transitory beings.[60] If the relation between the substance and its modes is that of a whole to its parts, the same conclusion still follows, because any change in a part will entail a change in the whole (A VI 4A 995).

Whereas by the end of the 1680s Leibniz judged that he had no adequate theory of the ontology of change, by 1710 he thought that modes or modifications provided him with the ontological framework required to solve the problem of change (LDB 234–5; 256–7). A modification represents no ontological addition, but is just a limitation of some permanent thing. Dynamics provided the model for this ontology and modifications correspond to derivative forces:

> And the qualities or derivative forces, or what are called accidental forms, I take to be modifications of the primitive Entelechy, even as shapes are modifications of matter. That is why these modifications are perpetually changing, while the simple substance remains.
>
> (T 396; see LDV 306–7; GM VI, 102–3: AG 254)

The relation of substance to its modifications is thought of as the relation of primitive to derivative force. Primitive active force is a persistent being. It is contrasted with derivative active force, which is the momentary state of a substance insofar as it strives toward new states. Primitive force is sometimes identified to the law of the series, the values of which are the states of the substance:

> Derivative force is the present state itself insofar as it tends toward a following state, i.e., preinvolves a following state, just as everything in the present is pregnant with the future. But the persisting thing itself, insofar as it involves all cases, has primitive force, so that primitive force is like the law of the series, while derivative force is like a determination that designates some term in the series.
>
> (LDV 287)

A little later in the same letter, Leibniz adds that the law of the series is the only permanent thing, as we saw above (LDV 288–9). However, we have seen that the law of the series considered in abstract is insufficient to ground an enduring substance. Thus, the primitive force is best understood as a concrete permanent power or disposition, which gives reality to the abstract law of the series (GP IV, 512: AG 162–3 where it is identified with *lex insita*). This could explain why Leibniz says that primitive force is "as it were" the law of the series, instead of simply identifying them. This is an important proviso, because while the law of the series is an atemporal abstract entity, primitive forces are concrete and ever present. Some support for this interpretation can be gained from an argument against occasionalism in *De ipsa natura* (GP IV, 507: AG 158). Even if one assumes that some present created effect is the product of a divine command, such a command is now past and cannot bring about an effect now. Therefore, for this command to be effective, it is necessary that some power acts now, which is nothing else than the inherent law or primitive force. One may generalize the argument to an atemporal entity like a law of the series considered *in abstracto*.[61]

Derivative force is the temporary disposition or, more precisely, a momentary state of the substance insofar as it strives toward new states. This temporary disposition flows from the primitive force, it is an aspect of it (see NE IV.iii.6; A VI 6, 379).[62] Thus when Leibniz claims that derivative forces are "what is momentary in action" (*quod in actione momentaneum est*, LDV 304), we should not think that this entails that the various derivative forces form a series of discrete, independent entities. What corresponds to such a series is the action, which is the product of the momentary forces through time.[63] Action is a successive being since it never exists all at once. By contrast, all momentary derivative forces flow from the primitive force. At each time, a distinct derivative force is the particular determination of the primitive force, which exists at any time. A passage in the *Theodicy* gives a crucial precision of the status of derivative forces with respect to the successive/permanent dichotomy:

> The permanent or lasting act is nothing but the Substantial or Accidental Form: the substantial form (as for example the soul) is altogether permanent, at least according to my judgment, and the accidental is only so for a time. But the altogether momentary act, whose nature is transitory, consists in action itself.
>
> (T 87)

The momentary act is a successive being: it is momentary not in the sense that it exists only for one moment, but in the sense that no two parts of it exist simultaneously. On the other hand, Leibniz implicitly uses the two Scholastic criteria for a permanent being: the accidental form or derivative force is permanent too insofar as it exists all at once, in a single moment. By contrast, the substantial form or primitive form is permanent insofar as it is wholly present at different moments of time and is thereby an enduring entity.

Continuous Creation and the Labyrinth of the Continuum

So Leibniz thinks that occasionalism can be resisted *ad hominem* even by the lights of strong CCC. This, however, is insufficient to ground the true activity of created substances. The latter presupposes persisting substances, on pain of falling back to the denial of the plurality of substances and from there, to Spinozism. What the preceding section has shown is that primitive force is an enduring entity, in the sense that it exists all at once at any time in the career of a given substance. However, strong CCC seems to entail that no creature truly endures beyond an instant. This has led some commentators to the conclusion that Leibniz must therefore reject strong CCC.[64]

One could ask "where does the incompatibility between enduring substances and strong CCC lie?" According to John Whipple, this is where the labyrinth of the continuum comes into play, as Leibniz himself admits in the *Theodicy* (T 384). Whipple formulates his objection as follows:

> if the duration of a substance were composed out of a series of instantaneous perceptual states (or a series of perceptual states of some duration or other), then the duration of a substance would be discontinuous. Durational atomism thus leads to the conclusion that substances do not endure, strictly speaking, beyond an instant (or a state of some duration or other).[65]

Hence, strong CCC entails that the duration of things is discontinuous, whereas something endures (or is a continuant) only if it is continuous. Despite its apparent cogency, this inference is not warranted. It is historically undermined, as my initial survey of Scholastic theories has shown. In particular, even an opponent of strong CCC like Suárez saw no internal inconsistency in the idea of an enduring object having an atomistic duration.

Historical antecedents provide neither sufficient vindication nor a decisive reason to ascribe a similar view to Leibniz. In the *Theodicy*, Leibniz takes for granted that strong CCC presupposes temporal atomism. To this, he opposes a conception of moments of time and spatial points as mere extremities or limits of the continuum. Unfortunately, he sidesteps this problem, and cautiously claims, "this is not the place for entering into that labyrinth" (T 384).

Although he does not take an explicit stance, Leibniz does not hide his preference. It is well known that he never accepted that a continuum could be composed out of indivisible parts. Time being continuous is not composed of instants. These are mere limits or extremities generated by the cuts we make within this continuous quantity. The parts that are so generated are arbitrary, and a continuous whole has indeterminate, possible parts (LDV 302–3; 320–1; 326–7, 332–3). What is actual is fully determinate: "In real things . . . unities are prior to the multitude, and multitudes do not exist except through unities" (LDV 327). Continuous quantities are therefore merely ideal beings. But time must be contrasted with duration as space is with extension. Whereas the former in each pair is ideal and continuous, the latter is actual and discontinuous. In the letters to De Volder, it is not settled whether substances or monads are related to extension, but they are explicitly linked to duration.[66] Therefore, even if Leibniz rejects temporal atomism, he seems committed to a kind of "durational atomism."[67] A number of passages support such a view, especially in relation with CCC:

> Time is also resolved through actual changes in unities of duration, or (*seu*) in as many creations infinite in number. For of time, only instants exist.
>
> (*Letter to De Volder*, 10–11–1705, LDV 327)

> Although the duration of creatures, like actual motion, consists in a mass of momentaneous states, nevertheless we must say that . . . time [is not at all composed] of instants.
>
> (*Letter to Sophie*, 10–31–1705, GP VII, 562)

> The duration of things, or the multitude of momentaneous states, is the accumulation [*l'amas*] of an infinity of bursts [*éclats*] of the divinity, of which each one at each instant is a creation or reproduction of all things, there not at all being a continuous passage [*passage continuel*], to speak properly, from one state to the next.
>
> (*Letter to Sophie*, 10–31–1705, GP VII, 564)

A natural reading of these passages is that the duration of substances is discrete and composed out of an infinity of moments. These moments are not temporally extended nor are they identical to instants conceived as extremities of time intervals. But this does not prevent the substance's being an enduring entity: it is temporally identical insofar as its primitive force is wholly present at each of these moments of duration.

Whereas Leibniz refuses to enter the labyrinth of the continuum in the *Theodicy*, the study of his reading notes on Bayle's *Réponses aux questions d'un provincial* written around 1707 confirms the position worked out in the preceding passages. Two reflections are especially interesting. In the first, Leibniz is struggling with the status of instants:

> It could be said that fundamentally instants are just conceptions and that nothing exists in a pure instant (*dans un instant tout pur*). They are only ceasings (*cessations*). There are only beings having appearances, and the appearances of such beings. In these appearances, there are other appearances, but there are never instantaneous appearances,

> *minima*. There is nevertheless this argument in favor of instants: since no two parts of time can both exist, no appearance can be said to exist save others have existed from which they result. Therefore, no appearance will exist save those that are instantaneous. And these instantaneous appearances will be continual productions in so far as they exist by the Divinity who makes all appearances harmonious. Substance can be said to persist (*on peut dire la substance subsistance*) in whichever instant, with this element of appearance. . . . You are allowed to call this action of the sovereign substance "creation."
>
> (LH I, 1, 4, f.93v)[68]

In the first part of the text, Leibniz assumes that at the phenomenal level, there is no final decomposition into minimal or "snapshot" appearances: any temporally extended perception can be further divided into a plurality of less extended sub-perceptions. There is an analogous claim with respect to extended phenomena in a 1704 letter to De Volder:

> Matter is not composed of constitutive unities; rather it results from them. . . . And there is no reality in anything except the reality of unities, and so phenomena can always be divided into lesser phenomena. . . . By contrast, substantial unities are not really parts, but the foundations of phenomena.
>
> (LDV 303)

Just as matter results from constitutive unities or monads, which are not its atomic parts, similarly, the passage on Bayle refutes the existence of phenomenal instants. By symmetry, one may assume that there are durational unities from which these temporally extended phenomena result. And this is just what Leibniz establishes in the second part of this passage. The argument here is very close to Bayle's argument for time atoms that was analyzed earlier in this chapter. It is based on the idea of successiveness: duration is a successive quantity (GP IV, 394; IV, 568); but no two parts of time exist together; therefore, any temporally extended interval contains incompatible parts. Therefore, no temporally extended part of time exists. And thus either duration does not exist or only indivisible parts of duration exist.[69]

A correct interpretation requires a distinction between two kinds of instants in Leibniz: the instants as limits or extremities of continuous stretches of time and the actual instants of duration whose discrete succession grounds the duration.[70] Admittedly, Leibniz sometimes seems to argue that what has a merely instantaneous existence is nothing, for instance in an argument establishing that there are no precise shapes in bodies: "Now I believe that what is only in a moment has no existence, since it begins and ends at the same time" (A VI 4B, 1613). From this, one may infer that the states of substances cannot be momentaneous.[71] This passage from the 1680s seems to run against my interpretation of the argument contained in the notes on Bayle. Maybe these passages are incompatible, in which case I would tend to consider that Leibniz abandons the assumption grounding his argument against precise shapes in bodies. However, I think that they can be harmonized provided one realizes that, strictly speaking, Leibniz does not discount instants of duration as such, but reduces to nothing what lasts only for a single instant. Finally, the end of the passage supports strong CCC.

These discrete moments are the *termini* of God's conservative action. And yet, this is compatible with the persistence of substances, as Leibniz shows in another note:

> Assuming this continual creation (*création continuelle*), how some substance is the same. This occurs when finding in an indefinitely proximate time (*dans un temps indéfiniment prochain*) a thing infinitely proximate to a given thing and continuing so in any approaching point, one falls in another, I say that these two things are the same.
> (LH I, 1, 4 f.93r, in Pelletier "Substance, corps et phénomène," 136)

This sketches a possible solution to the problem of persistence through recreation and to the problem of change, based on the law of continuity. This appears at first sight to be incompatible with durational atomism. However, one may regard the times as referring either to divisible intervals or to indivisible instants. But, given the argument in the preceding passage and the last clause, which refers to these times by speaking of "points," Leibniz is best understood as referring to actual instants, which thus form an infinite collection. Consider the case where something remains unchanged for some time: in this case, there is an infinite sequence of actual instants where this thing is re-created in the same state. Now, such a case never occurs, either in bodies or in monads: bodies are constantly in motion and monads constantly change by striving toward new perceptions (GP IV, 518). Consider a given change in a monad. For instance, I perceive the dark sky in the early morning and then I perceive a clear sky. According to our passage, these states belong to the same subject because there is a continuous path from one to the other. This means that, in conformity with the law of continuity, if one considers closer and closer instants, then the delivered states of the monad will be closer and closer. However, there is no state of change according to Leibniz, and change always involves a pair of opposite states (e.g., a sky with degree of darkness n and a dark sky having with degree $n' \neq n$).[72] This entails that the metaphysical ground of change must be constituted by a discrete collection of states, which assign the actual instants of duration. Although Leibniz does not explicitly say how this relates to the problem of endurance given strong CCC (compare Leibniz's talk of *création continuelle* with the constant reproduction of substance in the letter to Sophie), his complete reasoning would be the following: (i) continuity (and especially continuous change) is a criterion of persistence; (ii) but a collection of infinitely many states can approach sufficiently this condition of continuity; therefore (iii) strong CCC is compatible with the persistence of enduring substances. Premise (ii) coincides with Leibniz's claim that even if true continuity pertains only to ideal things, the law of continuity is nevertheless applicable to the actual world and its phenomena (GP IV, 568). In conformity with this, the law of continuity functions in premise (i) merely as an epistemic criterion of persistence. Continuity can thus be a necessary condition for our judging that something persists while it changes, even if the actual duration of the thing is not continuous.

Conclusion

I have thus defended the compatibility of Leibniz's commitment to strong CCC with the rejection of occasionalism and an ontology of enduring substances. Let us envisage again the four claims (A)–(D) which Leibniz ascribes to the Cartesians: I have tried to show that Leibniz adheres to (A) (temporal atomism) and (B) (strong CCC), and that he can consistently reject (D) (occasionalism) *and* (C) (denial of endurance), because he accepts that primitive force is wholly present at each temporal stage. To put it slightly differently, CCC can be harmonized with an endurantist account of persistence provided durational multilocation goes proxy for continuity. Much work remains to be done, however. In particular, I have constantly assumed that substances truly endure, in other words that there is a genuine monadic duration. This assumption would need to be more fully defended. Some texts favor the ascription of a monadic time (or quasi-time) to Leibniz.[73] On the other hand, there is support for the view that substances are no more in time than they are in space.[74] Moreover, the reductive view of time proposed by Leibniz, which is sometimes interpreted as a causal theory of time, seems to leave little prospect for the idea of monadic duration. But, first, there is a distinction between ascribing time and duration to monads. Time supposes a complete ordering of states with respect to the simultaneity relation. There can be no such simultaneity relations between the states of different monads. But this does not exclude a real succession of states internal to the monad. Second, the fact that Leibniz reduces temporal relations to more basic relations does not entail by itself that he completely eliminates succession. Finally, one of the main obstacles to attributing genuine duration to monads stems from its incompatibility with monadic simplicity.[75] What precedes has established that Leibniz can allow for an endurantist view of persistence, which is compatible with such simplicity.

Acknowledgment

I wish to thank Steve Nadler for his comments on an earlier version of this paper.

Notes

1 Thomas Aquinas, *Summa Theologiae* Ia, q.104, art.1, resp. "The being of each creature depends on God in such a way that, unless creatures are conserved in being by the operation of the divine power, they could not subsist for a moment but would be reduced to nothing." The doctrine can be traced back to Augustine; see, e.g., *De Genesi ad litteram* 4.12.22 quoted in Richard Sorabji, *Time, Creation, and the Continuum* (London: Duckworth, 1983), 303–4.

2 See Daniel Garber, "How God Causes Motion: Descartes, Divine Sustenance and Occasionalism," *Journal of Philosophy* 84 (1987): 567–80; Daniel Garber, *Descartes' Metaphysical Physics* (Chicago: University of Chicago Press, 1992), 302–5.

3 Steven Nadler, "Louis de La Forge and the Development of Occasionalism: Continuous Creation and the Activity of the Soul," reprinted in Steven Nadler, *Occasionalism. Causation among the Cartesians* (Oxford: Oxford University Press, 2011), 126–7.

4 Louis de La Forge, *Traité de l'esprit de l'homme (1665) in Œuvres philosophiques*, ed. Pierre Clair (Paris: Presses Universitaires de France, 1974), 240. Malebranche's argument figures most prominently in his Dialogues on Metaphysics VII, 7–10, OCM XII, 157–60: JS 113–6. For commentary, see Garber, *Descartes' Metaphysical Physics*, 300–1; Nadler, "Louis de La Forge"; Andrew Pessin, "Does Continuous Creation Entail Occasionalism? Malebranche (and Descartes)", *Canadian Journal of Philosophy* 30 (2000): 413–39.

5 See A VI, 4B, 1550; 1596; Gr 307; 330; 381; T 27; GP III, 566; IV, 588; M 47.

6 Robert M. Adams, *Leibniz: Determinist, Theist, Idealist* (New York: Oxford University Press), 94–9; André Robinet, *Archtectonique disjonctive et automates systémiques* (Paris: J. Vrin, 1986), 418–42; Sukjae Lee, "Leibniz on Divine Concurrence", *The Philosophical Review* 113 (2004): 81–118. Note that he treats conservation only incidentally, since concurrence is his main problem. I have defended a similar interpretation in Jean-Pascal Anfray, "Le labyrinthe temporel. Simplicité, persistance et création continuée chez Leibniz," *Archives de philosophie* 77, no. 1 (2014): 43–62.

7 Most recently, Jeffrey McDonough, "Leibniz: Creation and Conservation and Concurrence," *The Leibniz Review* 17 (2007): 31–60; John Whipple, "Continual Creation and Finite Substance in Leibniz's Metaphysics," *Journal of Philosophical Research* 36 (2011): 1–30. See also Catherine Wilson, *Leibniz's Metaphysics: A Historical and Comparative Study* (Manchester: Manchester University Press, 1989), 169–73 and Jacques Jalabert, *La theorie leibnizienne de la substance* (Paris: Presses universitaires de France, 1947), 167–91.

8 Thomas Aquinas, *Summa theologiae*, Ia, q.104, a.1–2.

9 See also Suárez, *Disp. Met.*, disp. XXI, sec.1, n.6–12.

10 *Id.*, n.14.

11 *Id.*, sec.3, n.4. However, Suárez grants that substances can be the conservative causes of their accidents.

12 See McDonough, "Leibniz: Creation," 47. McDonough formulates the same distinction between a strong and a weak reading of CCC.

13 Sorabji, *Time, Creation and the Continuum*, 297. See also Dominik Perler and Ulrich Rudolph, *Occasionalismus: Theorien der Kausalität im arabisch-islamischen und im europäischen Denken* (Göttingen: Vandenhoeck & Ruprecht, 2000), 48–54. Before Ah'arî, Al-Nazzam thought that bodies persist only by being endowed with the atomic accident of duration, without being an occasionalist. Evidence concerning the Mutakallimun comes mainly from Maimonides critical discussion in the *Guide for the Perplexed* I, ch.73, third premise, transl.

Shlomo Pines, 2 vols. (Chicago: University of Chicago Press, 1963). Leibniz read and annotated this chapter; see A VI, 4B, 2487–2488.

14 Bonaventure, *Commentaria in Quatuor libros Sententiarum*, II, 1, 1, 1, 3, *Opera omnia* 10 vols (Quaracchi: Collegium S. Bonaventurae, 1882–1902) II, 62b-63a. Suárez gives Bonaventure's argument from conservation in *Disp. met.*, disp. L, sec.5, n.5.

15 Peter John Olivi, *Quaestiones in Secundum Librum Sententiarum* q. 9, quoted in Richard Cross, "Eternity, creation, and conservation," *Religious Studies* 42, no. 4 (2006): 403–6 at 411.

16 See Suárez, *Disp. met.* disp. L, sec. 5, n.1 and sec.9, n.20. See also Robert Pasnau, *Metaphysical Themes. 1274d sec* (Oxford: Clarendon Press, 2011), chapter 18.

17 David Lewis popularized the terminological distinction endurance vs perdurance. See *On the Plurality of Worlds* (Oxford: Blackwell Publishers, 1986), 202f. According to endurantism, things persist by being wholly present at each moment of their existence. An enduring object is sometimes called a "Continuant." Endurance accounts are opposed to perdurantism, where an object persists by having distinct temporal parts spread out in time, just as it extends through space by having spatial parts. See also Katherine Hawley, *How Things Persist* (Oxford: Clarendon Press, 2001).

18 Suárez, *Disp. met.* disp. XXI, sec. 2, n.3.

19 *Id.*, n.6. Recently, Philipp L. Quinn has defended a similar idea, claiming that God's continuing re-creation of an individual moment by moment is consistent with its strict identity over time. See Quinn, "Divine Conservation, Continuous Creation, and Human Action," in *The Existence and Nature of God*, ed. Alfred J. Freddoso (Notre Dame: University of Notre Dame Press, 1983), 76.

20 Descartes *Meditation* III, AT VII, 49, CSM II, 33; *Principles* I, §21, AT VIII-a 13, CSM I, 200.

21 *Fifth objections*, AT VII, 301, CSM II, 209.

22 See in particular Martial Gueroult, *Descartes selon l'ordre des raisons* (Paris: Aubier, 1953), I, 272–85. Among other proponents of temporal atomism and/or strong CCC, see Norman K. Smith, *Studies in the Cartesian Philosophy* (New York: Palgrave Macmillan, 1902), 131–2; E. Gilson, *René Descartes: Discours de la méthode, texte et commentaire*, 4th ed. (Paris: J. Vrin, 1967), 340–2.

23 Smith, *Studies in the Cartesian Philosophy*, 73–4.

24 John Bennett, *Learning from Six Philosophers* (Oxford: Clarendon Press, 2001), 1–98.

25 According to Bennett, the independence of the parts of time is not the consequence of Descartes' acceptance of a principle of causal simultaneity, *pace* Jorge Secada, "Descartes on Time and Causality," *Philosophical Review* 99 (1990): 45–72. For the separability criterion of real distinction, see *Principles* I, art. 60 AT VIIIa-28–9, CSM I, 213.

26 Bennett, *Learning from Six Philosophers*.

27 Passages where Descartes talks of "next moment" or "next time" might seem to favor the atomist reading: see AT VII, 165, 370, CSM II, 116, 255; *Principles* I, 21, AT VIII-a, 13, CSM I, 185–6. Similarly in *Principles* II, 39, Descartes derives the second law of nature (i.e., that all motion is rectilinear by itself) from the fact that God conserves a body at any moment and that no motion occurs at an instant, suggesting that the moment is durationless (AT VIII-a, 64, CSM I, 242). But these passages are consistent with a continuist reading too; see Garber, *Descartes' Metaphysical Physics*, 271.

28 Some passages favor weak CCC instead. See especially *Principles* II, §42, AT VIII-a 66, CSM I, 243. Tad Schmaltz attributes a weak reading of CCC to Descartes. See his *Descartes on Causation* (New York: Oxford University Press,

2008), 82 and 125f. There is no need for a series of distinct creative acts. Note that this question is in principle distinct from the question of the discrete or continuous nature of time and duration.

29 For an overview of the debated question concerning the continuity or discreteness of Cartesian time, see Garber, *Descartes' Metaphysical Physics*, 266–73. This interpretive debate has been connected to the problem of persistence. For instance, Richard Arthur objects that if time were discrete and CCC meant the constant recreation of creatures, then nothing would truly persist. To illustrate his point, he considers the cloning of mammoths from their frozen cells, which would not be counted as a case of conservation (see Richard Arthur, "Continuous Creation, Continuous Time: A Refutation of the Alleged Discontinuity of Cartesian Time," *Journal of the History of Philosophy* 26 (1988): 349–75, at 357). Ken Levy objects that the mammoth case is not a case of conservation only because there is a real, measurable gap between the extinction of mammoths and their cloned descendants, which is not the case when one considers two adjacent atoms of duration (see Ken Levy, "Is Descartes a Temporal Atomist?" *British Journal of the History of Philosophy* 13 (2005): 627–74).

30 This point is emphasized by Nadler ("Louis de La Forge and the Development of Occasionalism," 126). Concerning the continuity of time in Malebranche, see *Search after Truth* I, 8, 2, OC I, 104: LO 38–9: "duration has no instants, as bodies have no atoms"). The token-identity of God's act of creation and conservation is affirmed in the *Dialogues on Metaphysics*, VII, 7: "The instant of creation past! . . . Beware of what you say . . . In reality, creation does not pass away because, in God, conservation, and creation are one and the same volition" (OCM XII, 157; JS 113–4).

31 OD III, 785a and DHC "Pauliciens," rem. F, 628a, Popkin 180. This focus marks an important twist compared with Malebranche's version of the argument because it directly concludes to the causal impotency of creatures in general and spiritual substances in particular. Todd Ryan rightly emphasizes this point; see his *Pierre Bayle's Cartesian Metaphysics* (London: Routledge, 2009), 81. However, given the scope of my paper, I will focus on this aspect of the argument and neglect what Bayle says about free will.

32 DHC, art. "Rodon," note D; "Zabarella," note F.

33 This is extracted from a discussion of Zeno's Arrow argument. It might be objected that Bayle is merely paraphrasing Zeno's reasoning. However, Bayle endorses the conclusion that time cannot be infinitely indivisible. Alternatively, we could give a hypothetical version of his own position: if time is real, it must be atomic.

34 See Pasnau, *Metaphysical Themes*, 377ff and chapter 18 passim.

35 Such an account of persistence can be compared to an "exdurance" view of persistence, according to which temporal stages persist by being related to their counterparts at other times. No particular is wholly present at different times (*pace* endurantism) and a persisting particular is not a collection of temporal stages (*pace* perdurantism). See Sally Haslanger, "Persistence Through Time," in *The Oxford Handbook of Metaphysics*, eds. Michael J. Loux and Dean W. Zimmerman (Oxford: Oxford University Press, 2003), 313–54.

36 Ryan, *Pierre Bayle*, 84–6.

37 Bayle quotes here the Jesuit Rodrigo Arriaga, *Cursus philosophicus*, Phys. Disp.9, sec.6, n.3. The same reasoning features in his lectures on metaphysics, OD IV, 483.

38 Bayle therefore excludes self-causation, at least for finite being. See also OD IV, 482.

39 Descartes, *First Replies*, AT VII 108; CSM II 78.

40 DHC, "Zabarella," rem. H; OD IV 478. See especially Ryan, *Pierre Bayle*, 83–4. He distinguishes very clearly the two stages in Bayle's argument: (i) no creature

can be a partial cause of its existence or of any accident at a given moment; (ii) no creature can be a partial cause of its existence or of any accident at any later time, since every effect must be simultaneous with its cause.

41 OD IV, 506: "Existence (esse), in so far as it modified in such a way, is not something beyond existence (*ens, prout taliter modificatur, non esse aliquid praeter ens*)."

42 I agree with Ryan's claim (*Pierre Bayle*, 86) that Bayle reasons from an apparently Cartesian, but actually strongly nominalist, ontology of substance to supply the missing premises to the argument from CCC.

43 This assimilation of external relations, like distance and location, to an internal relation, like resemblance, is well analyzed in Ryan, *Pierre Bayle*, 87.

44 Bayle tries to have his cake and eat it too, saying that all properties are extrinsic denominations (derived from external relations) and reducing all relations to internal relations.

45 OD IV, 505.

46 Thus, Bayle would hold there are true modal (i.e., accidental) predications, and there are "modal" facts, without ontological commitment to modes and properties generally.

47 In this respect, the implications of Bayle's view are similar to Jonathan Edward's theory of temporal persistence as based on a divine decree. See Paul Helm, "Jonathan Edwards and the Doctrine of Temporal Parts," *Archiv für die Geschichte der Philosophie* 61, no. 1 (1979): 37–61. It might be a very loose kind of perdurance theory, where causal continuity is replaced by mere similarity as a criterion of persistence.

48 See above, note 6.

49 Adams, *Leibniz*, 97. Other passages lend support to this interpretation. See GP IV, 588: "God continually produces or conserves in us that energy or activity which according to me constitutes the nature of substance and the source of its modifications." Thus, God conserves both the substantial form (or the nature of the substance) and its modes, but the creature causally contributes to the production of its modes. Concurrence is at play at this second level of causation.

50 In short, the Jesuits held that God concurred with the creature in the production of her free actions in conformity with what God knew the creature would do under hypothetical circumstances. The content of God's so-called middle knowledge (*scientia media*) somehow necessitates the nature of God's concurrence. See Suárez, *Disp. Met.* disp. XIX, sec.3, n.19 and 20: "from some general law and definite volition, God concurs from now on to these actions according to the exigency (*juxta exigentiam*) of the nature or secondary cause." Leibniz refers to exigency in relation to discussions on middle knowledge in his notes on Louis Béreur de Dôle from the late 1680s; see A VI 4B 1792. See also Francesco Piro, "L'action des créatures et le concours de Dieu chez Leibniz. Entre transcréationnistes et durandiens," in *Lectures et interprétations des Essais de Théodicée de G.W. Leibniz*, ed. Paul Rateau (Stuttgart: Franz-Steiner Verlag, 2011), 79–96.

51 See Lee, "Leibniz on Divine Concurrence," 230. Malebranche defines an occasional cause as "what determines the Author of nature to act in such and such a manner in such and such a situation" (OCM II, 312; Search, 448). A true cause, by contrast, "is one such that the mind perceives a necessary connection between it and its effect" (OCM II, 316; Search, 450). According to Lee, Leibnizian reasons differ from occasional causes for the following reason: occasional causes are neutral with respect to the outcome of God's volition, because they are divorced from the laws, which are identical to God's will. In contrast, the laws of nature are encoded within Leibnizian reasons (Lee, "Leibniz on Divine Concurrence," 235f.).

52 See also GP IV, 360; GP III, 588: "La connexion de deux états est une consécution naturelle, mais non pas nécessaire, comme il est naturel à l'arbre de porter des fruits, quoiqu'il puisse arriver par certaines saisons qu'il n'en porte point."

53 See Leibniz to Bourguet, 22 May 1714, GP III, 566. Jan Cover and John O'Leary Hawthorne suggest a similar interpretation in *Substance and Individuation in Leibniz* (Cambridge: Cambridge University Press, 1999), 238.

54 The idea goes back to the 1670s, see A VI, 3, 326. It was worked out during the 1680s when the device of the complete concept was at the center stage of Leibniz's metaphysics; see A VI, 4, 556 and GP II, 43. But it came to the forefront during the 1690s, when the complete concept theory was replaced by the dynamical approach to the substance. See GP II, 136; 171: L 517 and GP IV, 518. For commentary, see Robert C. Sleigh, *Leibniz and Arnauld: A Commentary on Their Correspondence* (New Haven: Yale University Press, 1990), 127–36; Donald Rutherford, *Leibniz and the Rational Order of Nature* (Cambridge: Cambridge University Press, 1995), 151–4; Cover and O'Leary Hawthorne, *Substance and Individuation*, chapter 6.

55 Adams thinks that, except for miracles of the first rank, creation, conservation, and incarnation, all miraculous events are included in the substance's defining law (*Leibniz*, 85–94). Cover and Hawthorne provide models supporting both interpretations (*Substance and Individuation*, 237–42).

56 Sleigh, *Leibniz and Arnauld*, 132.

57 Thus, I disagree with McDonough. According to him, "even by the lights of (merely) strong essentialism we cannot assume that, having actually existed on Monday and Tuesday, I—as opposed to someone else otherwise quite like me—could have existed for merely a day, an hour, or a second. In maintaining that each and every substance enjoys a large and robust core of essential properties, Leibniz thus affords himself the resources for resisting what might otherwise seem like an irresistible argument for continuous recreation" (McDonough, "Leibniz: Creation," 52–3). Strong essentialism admits that no individual could have had different intrinsic properties from the ones it actually has. McDonough's claim entails that no substance could have had a distinct duration because this would entail a change in its intrinsic properties, and to its essence. But this entails that the existence of a substance at *t* entails its existence at any other time at which it actually exists, which is incompatible with the fact that even if a substance has existed until *t*, God could annihilate it (compare with Gr 331). The claim I have defended is an extension of this: God could annihilate this substance and create another substance in its place, and by symmetry, this is possible with earlier times.

58 T 381: "M. Bayle . . . goes so far as to deny action to creatures; he does not even acknowledge any real distinction between accident and substance."

59 Piro "L'action des creatures," 87. See also Massimo Mugnai, "Leibniz on Substance and Changing Properties," *Dialectica* 59 (2005): 503–16.

60 "I conceive that nothing can be properly said to endure, any longer that it remains just the same; for in the instant any part of it is changed, that thing as it was before, is no more being" (George Villiers Duke of Buckingham, *A Short Discourse on the Reasonableness of Men's Having a Religion or Worship of God* (London, 1685), quoted in Piro "L'action des creatures," 87).

61 This is necessary if the argument is to be effective against an occasionalist like Malebranche. For the latter denies that creation is truly past (*Entretiens* VII, 7; OCM XII, 157).

62 As Rutherford aptly says, it is "nothing more than the primitive force of substance, conceived as determined in some particular way" (Rutherford, "Leibniz on Infinitesimals and the Reality of Force," 277).

63 See GP IV, 509: AG 160; GM VI, 102: AG 253: "derivative force differs from action only as the instantenous differs from the successive. For there is already force in the first instant, while action requires the passage of time, so that action is the product of force and time, considered in every part of a body"; see also GM IV, 379 and 389. The latter passage is a letter to Hermann in which Leibniz says that "power is extended through time, since, in my sense, in and of itself it does not involve time (*tempus non involvit*), but is something momentary, which is replicated at any moment or is prolonged in time" (quoted and translated in Rutherford "Leibniz on Infinitesimals," 276. According to Rutherford, this means that properly force does not last for any length of time, in other word that it does not exist *for* a moment but *at* a moment. If this means that force (derivative or primitive) is outside of time, then I disagree with Rutherford. If, however, he means that force has duration, though no temporal extension, then I agree with this interpretation.

64 McDonough, "Leibniz: Creation," 52–3. I agree with him insofar as he considers that exigency is too weak to ground genuine causation (*pace* Lee, "Leibniz on Divine Concurrence"). See also Whipple "Continual Creation," 8–9.

65 "Continual Creation," 8, see also 18. One reason why the inference from strong CCC to the denial of endurance seems compelling is that it tacitly assumes that there are temporal gaps, as if God re-created things at distant times. Even if it is tempting to associate durational atomism with temporal gaps, the "strong temporal discontinuity" should not be confused with strong CCC. Levy, "Is Descartes a Temporal Atomist?" believes that Descartes is indeed committed to strong temporal discontinuity. This is refuted by Geoffrey Gorham, "Cartesian Temporal Atomism: A New Defence, A New Refutation," *British Journal for the History of Philosophy* 16, no. 3 (2008): 625–37. In the *Pacidius*, Leibniz claims that motion (and time) takes place through leaps without gaps or "indistant leaps" from one time to another; A VI, 3, 541, 559, and 567 and Levey, "The Interval of Motion."

66 LDV 289: "But all individual things are successive or are subject to succession." See also, for the parallel with respect to space, LDV 266–9. There remains a difficulty here. For (i) it seems that only phenomena, i.e., bodies and motions, are spatially located, while monads are only derivatively so. If (ii) Leibniz takes a similar stance on the relation of substances to time, then (iii) one should conclude that monads are only indirectly temporal (see LDV 324–5; GP III, 357; LDB, 254–5; C 14). There are two options here: either to deny (ii), by recognizing that the analogy of space and time breaks at this point; or more controversially perhaps, to reject (i) outright, by claiming that monads have an intrinsic spatiality. The conclusion (iii) is endorsed by many commentators: see Jan Cover and Glenn A. Hartz, "Space and Time in the Leibnizian Metaphysic," *Noûs* 22, no. 4 (1988): 493–519 at 507–8; Adams, *Leibniz*, 250, 255 and Rutherford, *Leibniz and the Rational Order*, 192 and 208, n.37. Recently, the thesis of spatiality of monads has been recently defended by J. McDonough; see "Leibniz and the Foundations of Physics: The Later Years," *The Philosophical Review* 125, no. 1 (2016): 22–6.

67 John Whipple coins the expression but criticizes this view ("Continual Creation," 8).

68 This passage has been edited by Arnaud Pelletier, "Substance, corps et phénomène dans la *Théodicée*," *Lectures et interprétations des Essais de Théodicée*, ed. Paul Rateau (Stuttgart: Franz-Steiner Verlag, 2011), 117–38.

69 Leibniz develops a similar argument about time in his correspondence with Clarke, but there the emphasis is more on divisibility than on successiveness; see GP VII, 402: AG 340. See also GP VII, 564; GP III, 457 and GM VI, 235:

AG 119: "For, strictly speaking, motion (and likewise time) never really exists, since the whole never exists, inasmuch as it lacks coexistent parts." For commentary, see Michael J. Futch, *Leibniz's Metaphysics of Time and Space* (New York: Springer, 2008), 139ff. Futch correctly stresses that this is not necessarily an argument for presentism or an A-theory of time. A B-theorist can make sense of the idea that the parts of a successive being are actual at themselves and not at other times. Another worry comes from the comparison of the conclusion of the arguments in the notes on Bayle and in the letter to Clarke. In the latter, Leibniz concludes that time is purely ideal, because any part is divisible and instants are not parts of time, but extremities: "For how can a thing exist of which no part does ever exist? Nothing of time does ever exist but instants, and an instant is not even itself a part of time" (AG 340; see also GP III, 588: "l'instant n'est pas une partie du temps"). He is concerned there with the instant related to a continuous whole. But in the notes on Bayle, Leibniz concludes to the existence of instants from which duration results. Time is not real because no part of it ever exists. Duration is real, on the other hand, because it results from these moments.

70 The sequence of such instants is dense, yet it is not continuous, because two instants are not joined by a common limit (see Aristotle, *Physics* V, 3, 227a10–13). Two successive instants are not temporally distant from each other, so that there are no gaps between them. As a consequence, objects existing at more than one of these instants are not discontinuous in the sense of being episodic or scattered. For a comparison with Leibniz conception of the continuum in 1676, see Samuel Levey, "The Interval of Motion in Leibniz's *Pacidius Philalethi*," *Noûs* 37, no. 3 (2003): 371–416.

71 Whipple, "Continual Creation," 8, and Whipple, "The Structure of Leibnizian Simple Substance," *British Journal for the History of Philosophy* 18 (2010): 379–410. Adams discusses the same text (Leibniz, 231). I share his interpretation, namely that Leibniz is not concerned here with monadic states.

72 A VI, 3, 541; A VI, 4, 307: "Change is an aggregate of two opposed states in one stretch of time, with no existing moment of change, as I have demonstrated in a certain dialog"; see also A VI, 4, 556; 869 and C 9: "Moreover, the change, or variation, itself . . . is nothing but a complex of two states which are immediate and opposed to one another, together with a force or ground for the change, which is itself a quality." These passages are analyzed by Levey, "The Interval of Motion," and Rutherford, "Leibniz on Infinitesimals," 278–9.

73 Letter to De Volder, 07–6–1699, LDV 107: "time is neither more or less a being of reason than space, but coexistence, and existing before and after are something real." See also the passage quoted in n.65.

74 The atemporal interpretation of monads is proposed by Jalabert, *La théorie leibnizienne de la substance*; Futch, *Leibniz's Metaphysics*, 167–70; Nicholas Jolley, *Leibniz* (London: Routledge, 2005), 87; and *Causality and Mind: Essays on Early Modern Philosophy* (Oxford: Oxford University Press, 2013), 169–82. According to Whipple, "a substance is distinguishable into successive states only at the phenomenal level, not at the metaphysical ground floor . . . each series of phenomenal states resolves into a monad that is simple and undivided at the deepest level of analysis" ("The Structure of Leibnizian Simple Substances," 407).

75 James E. McGuire thought that Leibniz was facing inescapable difficulties here ("*Labyrinthus Continui*: Leibniz on Substance, Activity, and Matter," in *Motion and Time, Space and Matter*, eds. Peter K. Machamer and Robert G. Turnbull (Columbus: Ohio State University Press, 1976), 290–326 at 313–318).

12 Sticking to the Middle Course

Intellectual Ethics and Scientific Practice in Leibniz's Metaphysical Physics

Mogens Lærke

Introduction

Leibniz's intellectual motto was *theoria cum praxi*.[1] Most often, and with good reason, that motto has been taken to express a conviction that scientific theory should have a practical application or, as Leibniz himself often put it, that science should be useful, promote the public good, and advance the glory of God.[2] In this chapter, however, I want to explore another aspect of the intricate relations between theory and practice in Leibniz. More precisely, I focus on the way that practical concerns in important respects determined Leibniz's method of scientific discovery and thus shaped the development of his theoretical system. Moreover, by practical concerns, I do not here refer only to public usefulness but mostly to the ethical aspects of Leibniz's conception of how to conduct scientific work. Hence, I will make the perhaps surprising claim that Leibniz took certain scientific theories to be true on the grounds that the motivations behind their formulation conformed to the rules of a practical code of intellectual conduct. By proposing this, I do not intend to make the strong claim that scientific theories were true for Leibniz *because* they expressed such a practical code of conduct, or that the ethical nature of some doctrine *as such* made that doctrine true. I rather make the claim that, for Leibniz, identifying certain practical attitudes among intellectuals, certain practical features of their philosophical or scientific arguments, was a means of gauging the measure of truth contained in those arguments. Hence, in brief, for Leibniz, good scientific practice provided a *guide* to recognizing true science.

The practical attitudes (of philosophers) and features (of arguments) in question relate to what I call Leibniz's "intellectual ethics." By this expression, I understand a set of rules and values that, according to Leibniz, ought to govern the Republic of Letters by regulating the exchanges and relations between intellectuals. As I have shown in some detail elsewhere,[3] in Leibniz, this intellectual ethics took the form of a classification of attitudes and corresponding behaviors. Leibniz himself called such attitudes and behaviors, respectively, "spirits" (fr. *esprits*) and "conducts" (fr. *conduites*). I here focus on a central value in that intellectual ethics, namely "the spirit of moderation." First, I explore the nature and origin of this notion of moderation. I show in particular how it contains an active component. Moderation was for Leibniz not just a passive "holding back," i.e., merely tolerating by not blaming or attacking, but an active effort to "moderate," or litigate, between opposing parties. In other words, the conduct corresponding to the attitude of moderation was for him a conciliatory scientific practice. Next, I show how Leibniz's ethics of moderation grounded a specific rule of conduct or practical rule of thumb for evaluating doctrines that I term the "positive motivations test." Finally, I show how Leibniz applied that rule when evaluating the arguments advanced in a particular controversy. This last part of the paper is dedicated specifically to the relations between physics and metaphysics. Retracing the history of Leibniz's comments on the Cambridge Platonist Henry More's anti-mechanism, I thus show how Leibniz dismissed a central element in More's doctrine, namely the so-called hylarchic principle, on the grounds that More's argumentation in favor of that principle failed to pass the positive motivations test. Moreover, by considering these same comments, I also argue that Leibniz's own metaphysical physics was conceived as a conciliatory construction aiming at actively moderating between formalists and mechanists, between Aristotelians and Cartesians, in order to find a "middle ground."

The Ethics of Moderation

In a draft letter to Bartholomew Des Bosses from September 1706, Leibniz recounts the story of the Jansenist Pasquier Quesnel, who had escaped to the Netherlands after the archbishop of Malines had put him to prison because of his theological opinions, concluding his account of this dramatic episode by exclaiming: "If only the learned could be persuaded to exercise moderation in their battles with each other!"[4] Similarly, when conversing with Johann Friedrich about the merits of his method for reuniting the Christian churches, Leibniz affirms "that there is nothing that makes a dispute more commendable than the moderation of the disputants; well, I claim that this moderation will be manifest here in a quite special and indisputable way."[5] Such "moderation" is sometimes described by Leibniz as a seemingly passive attitude, for example when advocating "this moderation which consists in not condemning"[6] in a letter to Madame de Brinon. In religious matters, however, moderation is also closely linked to a form of pluralism. For Leibniz, it consists in recognizing, and even embracing, the value of more than one interpretation of religion, thus allowing for the coexistence of different and even diverging confessions. For example, in his annotations to Saint Augustine's *Confessions*, Leibniz notes that "moderation" in relation to the interpretation of the Christian religion consists in recognizing that Holy Scripture "contains several meanings at the same time" and that, for that reason, one should "not reject others temerariously."[7] I have analyzed this religious pluralism elsewhere under the name of Leibniz's *credo maximum*.[8]

Leibniz's persistent focus on the virtue of moderation in debates and exchanges doubtless originated from his involvement in religious controversy.[9] But the intellectual ideal of "moderation" extended far beyond this domain into science and philosophy. One finds confirmation of this scattered everywhere in Leibniz's texts. For example, when writing to Gabriel Wagner about the status of *logic*, describing how he places himself between the Aristotelians who defend and the modern Cartesian philosophers who reject syllogistic logic as philosophically useful, Leibniz concludes by saying, "I hope these remarks, which have grown more extended than I had planned, will suffice to show you my thoughts and that they may serve to achieve a compromise or moderation [*Vergleich oder* Temperament], since both sides do presuppose the art of reason."[10] Or again, when discussing the disputes opposing mechanistic and scholastic philosophers on the question of causation, Leibniz proposes to reconcile them by "explain[ing] a little more distinctly how a middle way [*medium*] can be found, in my opinion, between the scholastic and the mechanistic basis for philosophy; or, better, in what sense there is truth on both sides [*quomodo ab utraque parte sit veritas*]."[11] This last passage in particular reflects an epistemological pluralism which is the scientific equivalent of the hermeneutic pluralism in relation to religion already briefly mentioned above, i.e., the conviction that the Bible "contains several meanings at the same time."

Leibniz's general commitment to the ideal of moderation goes back to his early youth, to the deep interest he took in the Aristotelian ethics developed by his teacher, Jacob Thomasius, mainly in the *Philosophia practica*, a book that Leibniz read and annotated around 1663–1664 at the age of 17 or 18.[12] It was a reading that stayed with him for the rest of his life and he still refers to it as late as the 1710 *Essais de théodicée*.[13] Moderation is of course inseparable from the Aristotelian *mediocritas*, this "middle way" which Leibniz also associates with the notion of justice.[14] By the same token—since for Leibniz justice is defined as "the charity of the wise"—moderation is also intimately linked to charity and with the golden rule prescribing to always put oneself in the other's place. In short, moderation consists for Leibniz in searching out a wise balance between the charitable consideration of the other's view and the prudent consideration of one's own in a controversy.[15] For Leibniz, then, "moderation" consists primarily in a certain intellectual behavior, or what he sometimes referred to as a "conduct" toward others in the context of an intellectual exchange.[16]

Leibniz summarizes one concrete aspect of such moderate conduct in a passage from the correspondence with De Volder, in which he is trying to reassure his correspondent about some disagreement that the latter had expressed in a previous letter:

> There is no reason for you to apologize for your dissent, for that is not a voluntary thing. The desire for truth, care in investigation, and candor

> combined with moderation in speech should be enough for us both, and they cannot be anything but useful and pleasing to men of good will.[17]

Here, "moderation in speech" represents a certain rhetorical or dialectical conduct. It involves, among other things, avoiding precipitating judgment, *ad hominem* argumentation, and keeping focus clearly on the matter under discussion. The aim of moderation is to assure that intellectual exchanges proceed in an orderly manner and to avoid such "confused disputes" where, as Leibniz writes in a *Promemoria* on the reunion of the churches,

> the disputing parties rant to the winds, busy themselves in punctilious discussions, swerve from the issue in digressions, change the order, answer only to that which they find convenient, mask the adversary's objections or solutions, try to escape them through derision or invectives, employ repetitions, do not distinguish the job of the respondent from that of the opponent nor that of the one who must prove from that of the one who must not.[18]

There is, however, also a deeper and more proactive conception of moderation in Leibniz according to which it consists in actively "moderating" an exchange. On that point, we can consult a short piece from around 1695, entitled *De arte dialogistica*, where he writes the following:

> It is usual to write dialogues in such a way that the author favors one side. The truly philosophical dialogistic art would be to write so that both sides dispute with equal art, and that those things that a ferocious adversary could say be actually said. Thus, ultimately, the triumph of the dialogue would be the triumph of the cause. Indeed, it would then be like a colloquium and a judiciary conference of the litigating parties—the dialogue's author acting, as it were, as a judge or, if you prefer, as president and moderator. *[praeside atque moderatore]*[19]

Leibniz is writing about fictive dialogues, a genre he frequently used himself. When writing such dialogues, he argues, in order to be equitable to all the positions involved, it is important not to assign the role of the judge to any of the interlocutors. The judge must be the author and situated above the dialogue, so to speak. Moreover, at no point is such a judge to take sides or decide who is right. The judge is only to assume the role of a "moderator" ascertaining that all parties present their position in the best possible way and that they follow the rules prescribed by the form of controversy.

The practice of moderating action, then, is tightly linked to an ideal of impartiality. Hence, in the unfinished *Dialogus de religione rustici* from 1673, where Leibniz recounts how he had himself "moderated" a dispute between a peasant, a servant, a skipper, and a Jew, he insists on the fact that, in order to take on this role, he "should be so alien to the parties commitments that he could not appear himself as a witness, even in those clear and adequate cases that define [unequivocally] the issue and permit its resolution."[20] Indeed, elsewhere, Leibniz contends that if "no one will be able to guess which party he belongs to," this "can be taken as a palpable sign of moderation and equity."[21]

The Spirit of Moderation

Leibniz sometimes assigned the role of the moderator to an actual person in a given controversy. This is, for example, the role he asked Ernst von Hessen-Rheinfels to take on in the correspondence with Arnauld, exactly because he was someone, as Leibniz writes, "whose moderation I have admired."[22] On 12 April 1686, he thus wrote to the Duke:

> If your highness were not a Prince whose enlightenment is as grand as is his moderation, I would not write you regarding these matters . . . and now, whom can we better report to than you, and since you have had the goodness to establish this commerce, how could we possible be so imprudent as to go chose a different arbiter?[23]

In the subsequent exchange with Arnauld, Leibniz frequently asked the landgrave to take on this role actively and "moderate" Arnauld whose "bad temper" and "sourness" threatened to derail the exchange.[24] For example, on 12 April 1686, Leibniz wrote, "Your highness might find an occasion to advise him that acting in this way only serves to put off people unnecessarily, so that he may in the future proceed with a little more moderation."[25]

This said, a moderator was for Leibniz primarily a kind of *persona* to be internalized by the participants in an intellectual exchange. Moderation must become a constant habit and constitutive aspect of their general intellectual attitude. In other words, moderate conduct must to root itself in a veritable "spirit of moderation."[26] Leibniz thus wrote about his own efforts to act as a moderator in the disputes between Aristotelians and Cartesians:

> Here, what usually happens to those who are moderate happens to me. The Aristotelians take me to be a Cartesian, and the Cartesians are astounded that I do not submit to all their alleged insights. For when I speak with people who stubbornly stick to the School and who treat Descartes with spite I emphasize his qualities, but when I am dealing with a Cartesian who is too zealous, I find myself obliged to change tone in order to bring down a notch the overblown opinion they have of their master.[27]

Everyone should become the moderators of themselves and of the others in dialogical exchanges by actively seeking to ascertain that their own discourse as well as that of the other is always directed toward the search for truth. This, however, does not imply that "moderation" is simply a state of mind, or that Leibniz's appeal to this virtue in the context of intellectual exchanges can be reduced to a vague effort to influence the conscience or the temperament of one's interlocutors. What is at stake is agreeing upon and respecting practical rules that are precise and constraining. For when an exchange is governed by such practical rules, it is transformed into a quasi-automatic process that in many ways represents Leibniz's alternative to the scholastic *disputatio*:

> the combatants would have their arms so bound that they could not move except in an orderly fashion and in due measure, and that they would be pulled by machines which would do all the work, as in a naval battle, where the movement of the vessel and the force of the gun rule over the combatants. Furthermore, anger would be out of question if one could not discern clearly between friend and foe.[28]

Leibniz thus proposes procedural rules for controversies that are designed to assure that "the nature of the dispute forces people to speak moderately in spite of themselves," as he puts it with an extraordinary formula in text from 1680.[29]

Thus speaking moderately in spite of oneself involves constantly and systematically shifting between positions, thus calibrating the exchange in order to assure an equitable presentation of all perspectives.[30] As Leibniz puts it, one should consider all sides before judging:

> this is what gives rise to the diversity of opinions, everyone considering each thing from a particular side. There are only a few persons capable of considering all sides [*faire le tour de la chose*] until they put themselves in the place of their adversary, that is to say, persons who undertake to examine the *pro* and the *contra* with equal application and without bias, in order to see to which side the balance should incline. . . . Normally, we have a certain spirit of contradiction, and . . . we make a point of opposing every judgment and desire of ordinary men.[31]

We can compare these practical considerations about the nature of moderation with the following statement that Leibniz made about his own system of pre-established harmony in a rejoinder to Pierre Bayle:

> Considering this system makes us see that when one gets to bottom of things, one finds more reason in most of the philosophical sects than one would think. [They] find themselves reunited like in the center of a perspective from here the object (out of focus when considered from

> elsewhere) shows the regularity and the conformity of its parts: most of it has been overlooked as a result of a sectarian spirit limiting itself to rejecting others.[32]

The opposite of the "spirit of moderation" is not only the "spirit of contradiction," but also more generally the "sectarian spirit" that Leibniz denounces constantly. It is, for example, something of which he frequently criticizes the more orthodox Cartesians.[33] Hence, the system of pre-established harmony is not just conceived as a set of theoretical propositions forming a "system" aiming a providing a truthful representation of the world.[34] It is also a practical and dialogical tool for moderating between conflicting positions in the sense of the *De arte dialogistica*, i.e., a tool for countering the "sectarian spirit" and calibrating the opposing positions so that the "truth on both sides" is honored, to use with the expression from the *Praefatio ad libellum elementorum physicae* quoted above.

But how exactly does one determine what on each side is truthful and what is not? How does one proceed concretely in order to pick out "the best of all sides"[35] of a doctrine? Consider the following passages drawn from the *Nouveaux essais sur l'entendement humain* (1704) and from a 1714 letter to Remond:

> I admit that it often happens to people to be wrong even when they engage in serious disputation and when they speak according to their opinions; nonetheless I have also noticed that in their speculative disputes, on matters which relate to their mind, *they are right on both sides except in the objections that they make to each other*, where they take the opinion of the other party in the wrong way [. . .] [36]

> *I have found that most of the sects are right in a good part of what they propose, but not so much in what they deny*. The formalists, Platonists and Aristotelians, for example, are right in seeking the source of things in final and formal causes. But they are wrong in neglecting efficient and material causes and in inferring from this, as did Henry More in England and certain other Platonists, that there are phenomena which cannot be explained mechanically. The materialists, on the other hand, or those who accept only a mechanical philosophy, are wrong in rejecting metaphysical considerations and in trying to explain everything in terms of sense experience.[37]

From these two texts, we can derive a practical rule of thumb that Leibniz employed when approaching and evaluating other doctrines—a rule we can call the "positive motivations test." This test involves distinguishing between the parts of a doctrine that are motivated by the points that the doctrine seeks to affirm, and those parts that are motivated by the attempt to deny the affirmations of some other, competing doctrine.

Leibniz then contends that, as a general rule, the former parts are true while the latter are false. Consequently—and again: as a general rule—one should pick out from each doctrine that which has a positive motivation while discarding that which has a negative one. By doing that, one will then generally have identified what Leibniz calls the "good side" of the doctrine.

The positive motivations test is primarily a concrete, practical rule designed to counter scientific sectarianism, for those who are driven by the sectarian spirit only "want to make others contemptible, they try to make their faults stand out; they suppress the good things they say and try to attribute them to themselves in some disguised form."[38] The test is designed to counter such tendencies by separating the features of a doctrine that are amenable to doctrinal conciliation from those that are not. What is most remarkable about that rule is, however, that it has not only a practical and ethical application but also a theoretical and epistemological one: Leibniz in fact suggests that the distinction between what is true and what is false can be established by reference to such practical criteria. He proposes to pick out the true knowledge contained in some doctrine based on a consideration of the intellectual behavior that the various parts of the doctrine incarnate: positive and negative doctrinal motivations provide a *guide* to truth and falsity in the situation of a controversy where one has to evaluate the affirmations of an interlocutor.

Leibniz, More, and the Mechanist-Formalist Controversy

So how does this work? For the remainder of this chapter, I shall consider an example. It concerns the dispute regarding metaphysical physics between formalists and mechanists already alluded to in the letter to Remond quoted above. More precisely, it concerns the kind of anti-mechanist discourse that can be found in later Henry More.

According to More, "there is no purely-Mechanicall Phaenomenon in the whole Universe."[39] In order to account for the ubiquitous non-mechanistic elements in nature, he introduced the notion of a "Spirit of Nature," defined as follows in *The Immortality of the Soul*:

> A substance incorporeal, but without Sense and Animadversion, pervading the whole Matter of the Universe, and exercising a Plastick powere therein according to the sundry predispositions and occasions in the parts it works upon, raising such *Phaenomena* in the World, by directing the parts of the Matter and their Motion, as cannot be revolved into mere Mechanical powers.[40]

Much for the same reason, in response to criticism by the Boylian experimental philosopher Johann Chistoph Sturm,[41] he also introduced the so-called "hylarchic principle" in the *Enchridion metaphysicum*.[42]

Now, whenever the mature Leibniz wrote about More, it was most often to address the hylarchic principle. For example, in 1698, Leibniz published *De ipsa natura*, a commentary on the natural and experimental philosophy of Sturm.[43] In that text, he disagreed with Sturm "that it is in agreement with reason to deny all created, active force inherent in things," but did agree, however, "that there is no such thing as the soul of the world." Leibniz then proceded to reject doctrines which, according to him, embrace such a soul of the world, noting that "the omniscient heat of Hippocrates, and Avicenna's Cholcodean giver of souls, the exceedingly wise plastic virtue of Scaliger and others, and the hylarchic principle of More are in part impossible, and in part unnecessary." [44] We should linger a bit on this double determination: in part impossible; in part unnecessary.

It is easy enough to grasp why Leibniz would say that More's hylarchic principle, if indeed comparable to a world soul, is impossible. Leibniz argues in many texts that, since the world is an aggregate of infinitely many bodies, it cannot have a single substantial form, since the notion of an infinite whole involves a contradiction.[45] To the extent, however, that a soul is nothing other than a substantial form, no such thing as a world soul can exist. It is not so clear, however, why Leibniz declares More's hylarchic principle "unnecessary." In his mature philosophy generally, Leibniz argues that one could adopt the mechanist thesis in the full sense according to which all physical phenomena should, in principle, be explained mechanically. In accounting for the metaphysical basis of these phenomena, however, one must appeal to incorporeal and formalist principles and consider final causes and the goodness and wisdom of the creator. Leibniz writes, for example, in the 1696 *Tentamen anagogicum*:

> The true middle term for satisfying both truth and piety is this: all natural phenomena could be explained mechanically if we understood them well enough, but the principles of mechanics themselves cannot be explained geometrically, since they depend on more sublime principles which show the wisdom of the Author in the order and perfection of his work.[46]

Specifically with regard to the mechanist-formalist controversy between More and Sturm, Leibniz nicely summarizes the intermediate position he occupies in a letter sent to Sturm via Otto Menche: "Although I think with Henry More that there is something in bodies other than geometrical notions, yet I consider his [i.e., Sturm's] refutation to be correct, because phenomena can be mechanically explained."[47] This is, of course, a very well-known position of Leibniz and I shall not enter the details of his argument. I am interested only in the *motivations* behind it and in *the rules that guided him* in the process toward formulating his position in relation to More in particular.

So let us work our way back in time, retracing the various texts wherein Leibniz addresses More and the hylarchic principle. Let me quote three passages in inverse chronological order, written in 1697, 1695, and 1690, respectively:

> My philosophical opinions are a bit closer to those of the late Madam the Countess of Conway, and *stays in the middle* between Plato and Democritus, because I think that everything happens mechanically as Democritus and Descartes want it, against the opinion of Mister More and other similar thinkers; and yet everything happens vitally following final causes, since everything is full of life and perceptions, against the opinion of the Democritians (emphasis added).[48]

> . . . even though I admit an active and, so to speak, vital principle superior to material notion everywhere in bodies, I do not agree with Henry More and other gentlemen distinguished in piety and ability, who use an *Archeaus* (unintelligible to me) or hylarchic principle even for dealing with the phenomena, as if not everything in nature can be explained mechanically and as if those who try to explain everything mechanically are thought to eliminate incorporeal things, not without suspicion of impiety, or as if it were necessary, with Aristotle, to attach intelligence to the rotating spheres, or as if one ought to say that the elements rise or fall by the virtue of form, a concise, but useless doctrine. With these views, I say, I do not agree. . . . In my opinion, *the middle way* in which one satisfies both piety and knowledge is the best (emphasis added).[49]

> And I am not of the opinion of the Reverend Father Fabry, of Mister More, and other able people who have held that one cannot explain the particular phenomena of nature by means of mechanical principles, that which have induced them to take recourse to a middle power, to a hylarchic principle, to a soul of the world, to a proportional middle light in between substance and attribute, and other similar things. As for me, *I stick to the middle course* and I am persuaded that all of the particular physics can be mechanically explained, that is to say, by magnitude, shape and movement, if only we know the hidden mechanisms behind it (emphasis added).[50]

These three texts testify to a stable position. According to Leibniz, More failed to take the "middle course" and this was one of the principal reasons why he was in error. Leibniz, on the contrary, suggests that we must stick to this middle course by granting that everything happens *both* mechanically *and* according to principles of divine wisdom, according to *both* efficient *and* final causes. We can easily hear Leibniz's spirit of moderation express itself in these determinations: it is clearly about affirming the "truth on both sides." The question is then, from the point of view of actual discovery, *what comes first*: the practical obligation to search out the middle course or the theoretical conviction that the truth conforms to such a middle course?

To get deeper into this, we have to go another ten years back, to around 1679–1680, shortly after Gottfried Shultze had sent Leibniz a copy of More's *Opera omnia*.[51] At that time, Leibniz wrote what, to my knowledge, is the earliest text in which he addresses More's hylarchic principle, namely the *Praefatio ad libellum elementorum physicae*:

> I know too that there are excellent and most learned men who cannot abide having all bodily phenomena explained mechanically. For they think that this injures religion, and they believe that if it was accepted, the world mechanism would need neither God nor any other incorporeal substance. This they rightly regard as absurd and dangerous. Hence some of them make use of an immediate intervention of God everywhere, while others introduce intelligences and angels as moving forces here and there. Some set up a kind of a world soul or a hylarchic principle. . . . If those who oppose mechanical laws had known that these laws themselves are finally resolved into metaphysical reasons and that these metaphysical reasons arise from the divine will or wisdom, they would not have so strongly opposed mechanistic explanations.[52]

Leibniz stresses how anti-mechanists such as More claim that mechanism "injures religion," how they consider it "absurd and dangerous." As should be clear, such determinations are practical, not theoretical. They refer to More's moral evaluations of mechanism, not his epistemological ones. Indeed, More introduced the hylarchic principle exactly *because* he "[could not] abide having all bodily phenomena explained mechanically," and thus ended up "strongly opposing" mechanism. Moreover, for Leibniz, his own alternative to More's position is better exactly *because* it does not involve such "strong opposition." Similar conclusions follow from a 1680 letter from the correspondence with Christian Philipp. While most of this correspondence is dedicated to denouncing the Spinozistic consequences of Descartes' philosophy, including Cartesian mechanism, Leibniz still insists, explicitly against More, that the mechanist principles are sound if correctly understood, since "wanting to explain everything mechanically is neither criminal nor impious, since God has done everything according to the laws of mathematics, that is to say, according to the eternal truths which are the objects of wisdom."[53] Here, More's position is not exactly depicted as wrong or untruthful, but rather as unfair, excessive, much too motivated by the rejection of his adversaries as criminal and impious. These non-conciliatory features of More's doctrine first led Leibniz to reject it and favor other options.

As it appears, when considering the actual genesis of Leibniz's rejection of More's hylarchic principle, what counts is primarily something like the rule of thumb later stated explicitly in the *Nouveaux essais* and the letter to Remond: the problem with More's theory is that the hylarchic principle fails to pass the positive motivations test. More did not, according to Leibniz, introduce the principle because he was motivated to do so by some positive reason internal to his own doctrine. He introduced it *only* to reject another doctrine, namely mechanism. Hence, in 1696, Leibniz writes in the *Tentamen anagogicum*:

> We know that while there have been, on the one hand, able philosophers who recognized nothing except what is material in the universe, there are, on the other hand, learned and zealous theologians who, shocked at the corpuscular philosophy and not content with checking its misuse, have felt obliged to maintain that there are phenomena in nature which cannot be explained by mechanical principles.[54]

For Leibniz, More clearly belonged among those "shocked" theologians who had introduced principles and entities exclusively designed to counter mechanism. And this, I think, is the sense in which Leibniz, in *De ipsa natura*, declared More's hylarchic principle not only impossible but also "unnecessary." It is unnecessary in the sense of being redundant. Making no positive contribution to science, it served only the purpose of providing an anti-mechanistic bulwark. As such, the hylarchic principle expressed a sectarian spirit, a spirit of contradiction, and was motivated by More's blame and rejection of mechanism. Consequently, it was contrary to the spirit of moderation, the characteristic feature of which was exactly to search out "the truth on both sides."

Conclusion

While resting at an inn during a voyage, Leibniz was asked to arbitrate in a religious discussion among common people of different confessional orientations. He began by reassuring his interlocutors: "*Ille vero: est aliquod inter utrumque temperamentum* [to be sure, there is some moderate combination of them all]."[55] By giving this reply, Leibniz certainly did not say too much. More precisely, he said just enough to keep everyone content. Maybe he responded in this elusive way so that he could get back as quickly as possible to writing one of those papers he liked to jot down while resting in inns. Nonetheless, it was emblematic of the intellectual ethics that runs like a red thread through his texts. Indeed, his response was not only emblematic of his attitude toward religious differences, but of his attitude toward intellectual differences generally, including when it came to questions of physics and metaphysics. In his own thinking, Leibniz himself attempted to comply with this ethics of moderation by always searching out the "middle way." This is clear from the way he navigated the Republic of Letters, how he evaluated his interlocutors, and how he elaborated his own position in relation to those of others. In all these aspects, Leibniz was constantly deploying an active effort to make the *medium* prevail, and thus to search out the "truth on both sides." And he evaluated his interlocutors according to their compliance with this ethics, by submitting their doctrines to the positive motivations test. Hence, it seems that, from the point of view of the actual genesis and from the point of view of the heuristic principles that guided him toward it, Leibniz's metaphysical physics was as much motivated by his desire that it should be a practical construction capable of accommodating both formalists and materialists as it was motivated by his conviction that it was scientifically correct. Leibniz's own solution was designed to deflect the sectarian spirit and promote the spirit of moderation. In other words, in its actual conception and development, pre-established harmony was as much a hybrid conception mounted to facilitate a conciliatory scientific practice as it was a theoretical system conceived to adequately represent the order of the world.

Notes

1 I use the following abbreviation: AC = *The Art of Controversies*, ed. and trans. Marcelo Dascal, Quintín Racionero and Adelino Cardoso (Dordrecht: Springer, 2006). Translations are my own unless otherwise indicated.

2 See, for example, this passage in one of Leibniz's outlines for the Berlin Academy: "*Wäre demnach der Zweck* theoriam cum praxi *vereinigen, und nicht allein die Künste une die Wissenschaften, sondern auch Land und Leute, Feldbau, Manufacturen und Commercien, und, mit einem Wort, die Nahrungsmittel zu verbessern, überdieß auc solche Entdeckungen zu thun, dadurch die überschwengliche Ehre Gottes mehr ausgebreitet, und dessen Wunder besser als bißher erkannt, mithin die christliche Religion, auch gute Policey, Ordnung und Sitte theils bey heidnischen, theils noch rohen, auch wohl gar barbarischen Völkern gepflanzet oder mehr ausgebreitet würden*" (Leibniz, *Denkschrift in Bezug auf die Einrichtung einer* Societas Scientiarum et Artium *in Berlin*, 24/26 March 1700, in *Dokumente zu Geschichte der Berliner Akademie der Wissenschaften von 1700 bis 1990*, eds. Werner Hartkopf and Gert Wangermann (Berlin: Akademie-Verlag, 1991), Dokument N° 17, 216–8, here 217). See also Hans Poser, "Theoria cum praxi: Das Leibnizsche Akademiekonzept und die Technikwissenschaften," in *Wissenschaft und Weltgestaltung*, eds. Kurt Nowak and Hans Poser (Georg Olms Verlag: Hildesheim, 1999), 93–115.

3 See chapter III of M. Lærke, *Les Lumières de Leibniz. Controverses avec Huet, Bayle, Régis, et More* (Paris: Classiques Garnier, 2015).

4 Leibniz to Des Bosses, 20 September 1706, trad. in *The Leibniz—Des Bosses Correspondence*, ed. et trans. Brandon Look and Donald Rutherford (New Haven: Yale University Press, 2007), 63.

5 Leibniz, *Des controverses*, 1680, A IV, iii, 205, trad. AC, 202.

6 Leibniz to Madame de Brinon, November 1697, A I, xiv, 744.

7 Leibniz, *De rerum creatione sententiae*, 1677–1716 (?), A VI, iv, 1687.

8 See Mogens Lærke, "Apology for a *Credo Maximum*: On Three Basic Rules in Leibniz's Method for Religious Controversy," in *Leibniz: What Kind of Rationalist?* ed. Marcelo Dascal (Dordrecht: Springer, 2008), 397–407; Mogens Lærke, *Leibniz lecteur de Spinoza: La genèse d'une opposition complexe* (Paris: Champion, 2008), 169–71.

9 For a recent study that highlights important parts of Leibniz's intellectual ethics in relation to religious matters, see Claire Rösler, "I. Les négociations iréniques," in *Negotium irenicum. L'union des Églises protestantes selon G. W. Leibniz et*

D. E. Jablonski, eds. Claire Rösler-Le Van, Gottfried W. Leibniz and Daniel E. Jablonski (Paris: Classiques Garnier, 2013), 11–153.

10 Leibniz to Gabriel Wagner, 27 February 1697, GP VII 514–27, AC 386 (trans. modified).

11 Leibniz, *Praefatio ad libellum elementorum physicae*, 1678–1679 (?), A VI.iv 2009, L 289.

12 See Leibniz, *Notae ad Jacobum Thomasium*, 1663–1664, A VI.i 46–8.

13 See Leibniz, *Essais de théodicée*, § 220, GP VI 249–50.

14 See Leibniz, *De arte combinatoria*, 1666, A VI.i 230–1; *Untersuchungen*, 1670–1671 (?), A VI.i 455; and *Elementa juris naturalis*, 1670–71 (?), A VI.i 462.

15 See Mogens Lærke, "The Golden Rule: *Charitas/Prudentia*. Aspects of G. W. Leibniz's Method for Religious Controversy," in *The Practice of Reason: Leibniz and His Controversies*, ed. Marcelo Dascal (Amsterdam and Philadelphia: Andrew Benjamin Publishing, 2010), 297–319.

16 See, for example, Leibniz, *Recommandation pour instituer la science générale*, A VI.iv 695.

17 Leibniz to De Volder, 1 September 1699, GP II 195, trans. in *The Leibniz-De Volder Correspondence*, ed. and trans. Paul Lodge (New Haven: Yale University Press, 2013), 131.

18 Leibniz, *Promemoria*, November 1687, A I.v 11, AC 248–9.

19 Leibniz, *De arte dialogistica*, circa 1695, A IV.vi 71, AC 32.

20 Leibniz, *Dialogicus de religione rustici*, November 1673, A VI.iii 153, AC 27.

21 Leibniz, *Des controverses*, 1680 (?), A IV.iii 212, AC 207.

22 Leibniz to Ernst von Hessen-Rheinfels, 12 April 1686, A II.ii 23. It is worth noting also how, at the begining of the exchange, Leibniz stressed his own moderation ("*Je puis me vanter d'être un des plus dociles et des plus modérés*," Ibid. 12) and implored Arnauld to reciprocate ("*j'espère qu'il en usera avec quelque moderation*," Leibniz to Ernst von Hessen-Rheinfels, 12 April 1686, A II.ii 15).

23 Leibniz to Ernst von Hessen-Rheinfels, 12 April 1686, A II, ii, 21.

24 Leibniz to Ernst von Hessen-Rheinfels, A II.ii 12–14, see also A II.ii 23, 25. For a later, more positive, recollection of the exchange, see Leibniz to Pellisson-Fontanier, 18 (28) March 1692, A II.ii 510.

25 Leibniz to Ernst von Hessen-Rheinfals, 12 April 1686, A II.ii 25–6.

26 For this expression, see Leibniz to Ernst von Hessen-Rheinfals, 9 July 1688, A II. ii 282.

27 Leibniz to Philipp, January 1680, A II.i 790. Voir aussi Leibniz to Conring, 19 (29) March, A II.i 606, L 146 (trans. modified): "It happens, as it does for all those who wish to hold a middle ground [*medii esse volunt*], that each party considers me too inclined towards the adversary. Indeed, whenever I discuss matters with the Cartesians, I extol Aristotle where he deserves it and undertake a defense of the ancient philosophy, because I see many Cartesians read their one master only, ignoring what is held in high esteem by others, and thus unwisely impose limits on their own spirit [*ingenio*]."

28 Leibniz, "Projet pour finir les controverses en religion," in *Œuvres*, ed. Louis-Alexandre Foucher de Careil (Paris: Firmin Didot, 1858–1875), vol. I, 83, AC 202.

29 Leibniz, *Des controverses*, 1680, A IV.iii 205, AC 202. This "art of controversies" has been explored by Marcelo Dascal in texts too numerous to list here.

30 The most famous example of how Leibniz did this himself in his controversies is the *Examen religionis christianae*, also known as the *Systema theologicum*, a text where Leibniz put himself in the place of a Roman Catholic so efficiently that later editors of the text took it as proof that Leibniz had converted to Catholicism, which of course he never did (for the *Examen*, see A VI.iv 2355–455).

31 Leibniz, *Conversation du Marquis de Pianese et du Père Emery Eremite*, 1678–1680 (?), A VI.iv 2250, AC 173 (translation modified).
32 Leibniz, *Eclaircissement des difficultés que Monsieur Bayle a trouvées dans le système nouveau de l'union de l'âme et du corps*, GP IV 523–4.
33 For the notion of the "spirit of sects," see for example Leibniz to Justel, 10 (20) October 1690, A II.ii 351; Leibniz to Pellisson-Fontanier, 18 (28) March 1692, A II.ii 511; Leibniz to Nicaise, 5 June 1692, A II.ii 534. For Leibniz's critique of the Cartesian "spirit," see Mogens Lærke, "*Ignorantia inflat*. Leibniz, Huet and the Critique of the Cartesian Spirit," *The Leibniz Review* 23 (2013): 13–42.
34 On Leibniz's notion of a system, see Michel Fichant, "La notion de système dans la physique de Leibniz," in *Science et métaphysique dans Descartes et Leibniz*, ed. Michel Fichant (Paris: Presses universitaires de France, 1998), 245–66, and Nicholas Rescher, "Leibniz and the concept of a system," in *Nicholas Rescher, On Leibniz* (Pittsburgh: University of Pittsburgh Press, 2003), 106–16.
35 For this expression, see the *Nouveaux essais sur l'entendement humain*, 1704, I, i, A VI.iv 71.
36 *Ibid*. III, 10, § 4, A VI.vi 341 (emphasis mine).
37 GP III 607, L 655 (emphasis mine).
38 Leibniz, *Remarques sur la doctrine cartesienne*, 1689, A IV.iv 2046.
39 Henry More, *Divine Dialogues* (London: J. Flesher, 1668), "The publisher to the reader," A6v. For a commentary, see Alan Gabbey, "Henry More and the Limits of Mechanism," in *Henry More (1614–1687): Tercentenary Studies*, ed. Sarah Hutton (Dordrecht: Kluwer, 1990), 19–35.
40 Henry More, *The Immortality of the Soul* (London: J. Flesher 1659), III, xii, 450.
41 See Johann C. Sturm, *Collegium Experimentale sive Curiosum*, 2 vols. (Norimbergae: Wolfgang Mauritius Endterus et Johannis Andreae Endteri haeredum, 1675).
42 See Henry More, *Manual of Metaphysics*, trans. Alexander Jacob, Scholia to Chap. XIII, chap. 4, vol. II, 70–102. For a commentary, see Alexander Jacob, "The Spirit of Nature as 'hylarchic' principle of the Universe," in *Henry More, Manual of Metaphysics*, trad. A. Jacob (Hildesheim: Georg Olms Verlag, 1995), vol. II, i–xxx.
43 Leibniz, *De ipsa natura*, 1698, GP IV 504–16, AG 155–67. The text is a contribution to the controversy between Sturm and Schelhammer. In his *Natura sibi et medicis agentis . . . conceptibus dissertatio* (1697), Schelhammer had attacked Sturm's natural and experimental philosophy developed in his *Idolum natura* (1692). For commentaries on Leibniz and Sturm, see Roberto Palaia, "Naturbegriff und Kraftbegriff im Briefwechsel zwischen Leibniz und Sturm," in *Leibniz Auseinandersetzung mit Vorgängern und Zeitgenossen*, ed. Ingrid Marchlewitz and Albert Heinekamp (Stuttgart: Steiner, 1990), 157–72, and Myriam Dennehy, "Leibniz et Sturm lecteurs de Boyle," in *La Philosophie naturelle de Robert Boyle*, eds. Myriam Dennehy and Charles Ramond (Paris: Vrin, 2009), 331–59.
44 Leibniz, *De ipsa natura*, 1698, GP IV 505, AG 156 (for all three quotations).
45 Leibniz, *Deus non esse animam mundi*, 1683–1686, A VI.iv 1492; *De mundi praesenti*, 1684–1686, A VI.iv 1509; Leibniz to Des Bosses, 11 March 1706, GP II, 304–5, trans. in *The Leibniz-Des Bosses Correspondence*, 33; *Essais de théodicée*, 1710, § 195, GP VI 232. For an excellent study of Leibniz and world soul theories, see Gregory Brown, "Leibniz's Mathematical Argument Against a Soul of the World," *British Journal for the History of Philosophy* 13, no. 3 (2005): 449–88.
46 Leibniz, *Tentamen anagogicum*, 1696 (?), G VII 272, L 478.

47 Leibniz to Otto Menche [destined for Johann Christian Sturm], October–November 1694, A VI.iv 865.
48 Leibniz to Thomas Burnett, 1697, GP III 217. Emphasis added.
49 Leibniz, *Specimen dynamicum*, 1695, GM VI 241, AG 125–6. Emphasis added.
50 Leibniz to [?], May 1690, A II.ii 317–8. Emphasis added.
51 See Schultze à Leibniz, 10 (20) September 1679, A I, ii, 519.
52 Leibniz, *Praefatio ad libellum elementorum physicae*, 1678–1679 (?), A VI.iv 2008, L 288–9.
53 See Leibniz to Philipp, January 1680, A II.i 789. See also Leibniz, *G.G.L. de Principium quoddum generale non in mathematicis tantum, sed et physicus utile*, 1688, A VI.iv 2039; *Nouveaux essais sur l'entendement humain*, 1704, III.x.15, A VI.vi 344; *Leibniz to Placcius*, 18 September 1690, A II.ii 343.
54 Leibniz, *Tentamen anagogicum*, 1696 (?), G VII 272, L 478.
55 Leibniz, *Dialogus de religione rustici*, November 1673, A VI.iii 152–4, AC 27.

Bibliography

Adam, Charles. "Vie et œuvres de Descartes." In *Œuvres de Descartes*, edited by Charles Adam and Paul Tannery. Vol. 12. Paris: Léopold Cerf, 1910.

Adams, Robert M. *Leibniz: Determinist, Theist, Idealist*. New York: Oxford University Press, 1999.

Ando, Takatura. *Metaphysics. A Critical Survey of its Meaning*. 2nd enlarged ed. The Hague: Nijhoff, 1974.

Anfray, Jean-Pascal. "Le labyrinthe temporel: Simplicité, persistance et création continuée chez Leibniz." *Archives de philosophie* 77, no. 1 (2014): 43–62.

Angelini, Elisa. *Le idee e le cose: La teoria della percezione di Descartes*. Pisa: ETS, 2007.

Anstey, Peter. "Experimental Versus Speculative Natural Philosophy." In *The Science of Nature in the Seventeenth Century*, edited by Peter Anstey and John Schuster, 215–42. Dordrecht: Springer, 2005.

Antoine-Mahut, Delphine. "Cartésianisme dominant et cartésianismes subversifs: Le cas de l'infirmier de Bicêtre Jean-André Rochoux." In *La mer retentissante: Lectures de Descartes et Leibniz au XIXe siècle*, edited by Lucie Rey. *Corpus* 68 (2016): 25–56.

Antoine-Mahut, Delphine. *Descartes radical: Historicité d'un canon philosophique*. Paris: Vrin, forthcoming.

Antoine-Mahut, Delphine. "Élisabeth philosophe: un cartésianisme empirique?" In *Élisabeth face à Descartes: deux philosophes?* edited by Marie-Frédérique Pellegrin and Delphine Kolesnik-Antoine (Antoine-Mahut), 119–38. Paris: Vrin, 2014.

Antoine-Mahut, Delphine. "Experimental Method and the Spiritualist Soul: The Case of Victor Cousin." In *The Philosophy of Experience in 19th Century in France and Germany. Contexts, Methods, Problems*, edited by Delphine Antoine-Mahut and Silvia Manzo. *Perspectives on Science*, forthcoming.

Antoine-Mahut, Delphine. "Is the History of Philosophy a Family Affair? The Examples of Locke and Malebranche in the Cousinian School." In *Philosophy and Its History: New Essays on the Methods and Aims of Research in the History of Philosophy*, edited by Eric Schliesser, Justin Smith and Mogens Lærke, 159–77. Oxford and New York: Oxford University Press, 2013.

Antoine-Mahut, Delphine. "La fabrique du cartésianisme néerlandais dans les histoires de la philosophie française au XIXe siècle." In *Les Pays-Bas aux XVIIe et XVIIIe siècles*, edited by Catherine Secretan and Delphine Antoine-Mahut, 107–24. Paris: Champion, 2015.

Antoine-Mahut, Delphine. "The Leibniz-Stahl Controversy and the Renewal of French Spiritualism." In *Hegel and Schelling in Early Nineteenth-Century France*, edited by Kirill Chepurin, Adi Efal-Lautenschläger, Daniel Whistler and Ayse Yuva. Dordrecht: Springer (International Archives for the History of Ideas), forthcoming.

Antoine-Mahut, Delphine. "Les expériences physiologiques chez Henricus Regius: les pierres lydiennes du cartésianisme?" *Journal of Early Modern Studies* 2 (2013): 125–45.

Antoine-Mahut, Delphine. *L'esprit du spiritualisme français*. Paris: Vrin, forthcoming.

Antoine-Mahut, Delphine. "Maine de Biran's Places in French Spiritualism: Occultation, Reduction and Demarcation." In *Edition and English Translation of Maine de Biran, Rapport du physique et du moral chez l'homme, with various Studies of Maine de Biran's Philosophy*, edited by Darian Meacham and Joseph Spadola, 33–46. London: Bloomsbury, 2016.

Antoine-Mahut, Delphine. "Reviving Spiritualism with Monads: Francisque Bouillier's Impossible Mission (1839–1864)." In *300 Years of Leibniz's Monadology*, edited by Jeremy Dunham and Pauline Phemiser. *British Journal for the History of Philosophy* 23, no. 8 (2015): 1106–27.

Ariew, Roger. *Descartes and the First Cartesians*. Oxford: Oxford University Press, 2015.

Ariew, Roger. *Descartes and the Last Scholastics*. Ithaca: University of Cornell Press, 1999.

Ariew, Roger. "Les *Principia* et la *Summa philosophica quadripartita*." In *Descartes: Principia philosophiae (1644–1994): Atti del Convegno per il 350 anniversario della pubblicazione dell'opera (Parigi, 5–6 maggio 1994; Lecce, 10–12 novembre 1994)*, edited by Jean-Robert Armogathe and Giulia Belgioioso, 473–89. Napoli: Vivarium, 1996.

Arnauld, Antoine and Pierre Nicole. *La logique ou l'art de penser*. Critical edition by Dominique Descotes. Paris: Champion, 2011.

Arthur, Richard. "Continuous Creation, Continuous Time: A Refutation of the Alleged Discontinuity of Cartesian Time." *Journal of the History of Philosophy* 26 (1988): 349–75.

Ascione, Imma. *Il governo della prassi: L'esperienza ministeriale di Francesco D'Andrea*. Napoli: Jovene, 1994.

Azouvi, François. *Descartes et la France: Histoire d'une passion nationale*. Paris: Arthème Fayard, 2002 (Paris Hachette, 2006).

Bachelard, Gaston. *The Formation of the Scientific Mind*. Translated by Mary McAllester Jones. Manchester: Clinamen Press, 2002. First published 1938 in French.

Bachelard, Gaston. *La philosophie du non*. 5th ed. Paris: Presses universitaires de France, 1970. First published 1940.

Bachelard, Gaston. *L'engagement rationaliste*. Paris: Presses universitaires de France, 1972.

Bachelard, Gaston. *Le matérialisme rationnel*. 3rd ed. Paris: Presses universitaires de France, 1972. First published 1953.

Bachelard, Gaston. *Le rationalisme appliqué*. Paris: Presses universitaires de France, 1949.

Bachelard, Gaston. *Les intuitions atomistiques (Essai de classification)*. 2nd ed. Paris: Vrin, 1975. First published 1932.

Bachelard, Gaston. *The New Scientific Spirit.* Translated by Arthur Goldhammer. Boston: Beacon Press, 1984. First published 1934 in French.

Bacon, Francis. *The Advancement of Learning.* In *The Oxford Francis Bacon*, vol. IV, edited by M. Kiernan. Oxford: Clarendon Press, 2000.

Bacon, Francis. *The Works of Francis Bacon*, edited by James Spedding, Robert Ellis and Douglas Heath, 15 vols. London: Longman, 1857–1874.

Baier, Johann Jacob. *Biographiae professorum medicinae qui in Academia Altdorfina unquam vixerunt.* Nuremberg: Johann Daniel Tauber, 1728.

Baillet, Adrien. *La vie de Monsieur Descartes.* Paris: Chez D. Horthemels, 1691.

Balibar, Etienne. *Identité et Différence: Introduction.* Paris: Seuil, 1998.

Bardout, Jean-Christophe. *Malebranche et la métaphysique.* Paris: Vrin, 1999.

Belgioioso, Giulia. *La variata immagine di Descartes: Gli itinerari della metafisica tra Parigi e Napoli (1690–1733).* Lecce: Milella, 1999.

Bellis, Delphine. "Météores cartésiens et météores scolastiques: la lecture philosophique d'Étienne Gilson." In *Gilson et Descartes à l'occasion du centenaire de "La liberté chez Descartes et la théologie,"* edited by Dan Arbib and Francesco Marrone, 41–56. Lecce: ClioEdu Edizioni, 2015.

Bennett, Jonathan. *Learning from Six Philosophers.* Oxford: Clarendon Press, 2001, 1–98.

Bennett, Jonathan and Peter Remnant. "How Matter Must First Be Made." *Canadian Journal of Philosophy* 4 (1978): 1–11.

Beyssade, Jean-Marie. *Études sur Descartes: L'histoire d'un esprit.* Paris: Presses Universitaires de France, 2001.

Biran, Maine de. *Œuvres philosophiques*, 3 vols, edited by Victor Cousin. Paris: Ladrange, 1834–1841.

Biran, Maine de. *Exposition de la doctrine philosophique de Leibniz.* Paris: L-G. Michaud, 1819.

Blank, Andreas. "Nicolaus Taurellus on Forms and Elements," *Science in Context* 27 (2015): 659–82.

Bloch, Olivier René. "Marx, Renouvier et l'histoire du matérialisme." In *La Pensée* 191, edited by. Reprinted in *Matières à histoires*, 382–441. Paris: Vrin, 1997.

Bonaventure of Bagnoregio. "Commentaria in Quatuor libros Sententiarum." *Opera Omnia*, 10 vols. Quaracchi: Collegium S. Bonaventurae, 1882–1902.

Bontems, Vincent. "Le 'non-cartésianisme.' La méthode non-cartésienne selon Gaston Bachelard et Ferdinand Gonseth." In *Qu'est-ce qu'être cartésien?* edited by Delphine Kolesnik-Antoine, 567–80. Lyon: ENS Éditions, 2013.

Bordas-Desmoulin, Jean-Baptiste. *Le cartésianisme, ou La véritable rénovation des sciences: ouvrage couronné par l'Institut, suivi de la théorie de la substance et de celle de l'infini*, Volume 1. Précédé d'un *Discours sur la réformation de la philosophie au dix-neuvième siècle, pour servir d'introduction générale, par François Huet, Professeur à la Faculté de Philosophie et Lettres de Gand.* Paris: J. Hetzel, Libraire-Éditeur, 1843.

Bordoli, Roberto. *Dio ragione verità. Le polemiche su Descartes e su Spinoza presso l'Università di Franeker.* Macerata: Quodlibet, 2009.

Bordoli, Roberto. *Ragione e scrittura tra Descartes e Spinoza: Saggio sulla Philosophia S. Scripturae Interpres di Lodewijk Meyer e sulla sua ricezione.* Milano: Franco Angeli, 1997.

Bordoli, Roberto. *René Descartes: Henricus Regius. Il carteggio: Le polemiche.* Napoli: Cronopio, 1997.

Borghero, Carlo. *Les cartésiens face à Newton: Philosophie, science et religion dans la première moitié du XVIIIe siècle*. Turnhout: Brepols, 2011.

Borghero, Carlo and Antonella Del Prete, eds. *Immagini filosofiche e interpretazioni storiografiche del cartesianismo*. Firenze: Le Lettere, 2011.

Borghero, Carlo and Claudio Buccolini, eds. *Dal cartesianismo all'illuminismo radicale*. Firenze: Le Lettere, 2010.

Borrelli, Antonio. *D'Andrea atomista: L'"Apologia" e altri inediti nella polemica filosofica della Napoli di fine Seicento*. Napoli: Liguori, 1995.

Bos, Egbert P. and Henri A. Krop. *Franco Burgersdijk (1590–1635): Neo-Aristotelianism in Leiden*. Amsterdam and Atlanta: Rodopi, 1993.

Bos, Erik-Jan. *Correspondence Between Descartes and Henricus Regius*. Utrecht: Zeno, 2002.

Bos, Erik-Jan. "Een kleine geschiedenis van een cartesiaans versje." In *Bijzonder onderzoek: Een ontdekkingsreis door de Bijzondere Collecties van de Universiteitsbibliotheek Utrecht*, edited by Marco van Egmond, Bart Jaski and Hans Mulder, 244–51. Utrecht: Universiteitsbibliotheek Utrecht-Waanders, 2009.

Bouillier, Francisque. *Du principe vital et de l'âme pensante, ou Examen des diverses doctrines médicales et psychologiques sur les rapports de l'âme et de la vie*. Paris: J-B. Baillère et Fils, 1862.

Bouillier, Francisque. *Histoire de la philosophie cartésienne*. 2 vols. Paris: Durand and Lyon: Brun et Cie, 1854.

Bouju, Théophraste. *Corps de Toute la Philosophie divisé en Deux Parties: La premiere contient tout ce qui appartient à la Sapience, à Scavoir, la Logique, la Physique & la Metaphysique. La seconde contient tout ce qui appartient à la Prudence, à scavoir, la Morale, l'Oeconomique & la Politique. Le tout par demonstration & auctorité d'Aristote, avec esclarcissement de sa doctrine par luy-meme*. Paris: Charles Chastellain, 1614.

Boyle, Robert. *The Works of Robert Boyle*, edited by Michael Hunter and Edward E. Davies. 7 vols. London: Pickering & Chatto, 1999.

Brenner, Anastasios. *Duhem: Science, réalité et apparence*. Paris: Vrin, 1990.

Brenner, Anastasios. "Le parcours épistémologique de Milhaud." In *Science, histoire et philosophie selon Gaston Milhaud: La constitution d'un champ disciplinaire sous la Troisième République*, edited by Anastasios Brenner and Annie Petit, 19–30. Paris: Vuibert, 2009.

Brockliss, Laurence W.B. "Rapports de structure et de contenu entre les *Principia* et les cours de philosophie des collèges." In *Descartes: Principia philosophiae (1644–1994): Atti del Convegno per il 350° anniversario della pubblicazione dell'opera (Parigi, 5–6 maggio 1994; Lecce, 10–12 novembre 1994)*, edited by Jean-Robert Armogathe and Giulia Belgioioso, 491–516. Napoli: Vivarium, 1996.

Broughton, Janet and John Carriero, eds. *A Companion to Descartes*. Oxford: Blackwell Publishers, 2008.

Brown, Gregory. "Leibniz's Mathematical Argument Against a Soul of the World." *British Journal for the History of Philosophy* 13, no. 3 (2005): 449–88.

Brunschvicg, Léon. *Les Étapes de la philosophie mathématique*. Paris: Blanchard, 1993. First published 1912.

Bunge, Wiep van, Henri A. Krop, Bart Leeuwenburgh, Han van Ruler, Paul Schuurman and Michiel Wielema. *The Dictionary of Seventeenth and Eighteenth-Century Dutch Philosophers*. Bristol: Thoemmes, 2003.

Buning, Robert Otto. "Henricus Reneri (1593–1639): Descartes' Quartermaster in Aristotelian Territory." PhD thesis, University of Utrecht, 2013.

Burgersdijk, Franco. *Collegium Physicum, Disputationibus XXII absolutum . . . Editio secunda*. Lugduni Batavorum: Ex Officina Elseviriorum, 1642.

Burgersdijk, Franco. *Idea philosophiae naturalis, sive Methodus definitionum et controversiarum Physicarum: Editio novissima*. Lugduni Batavorum: Ex Officina Elseviriorum, 1645.

Burgersdijk, Franco. *Institutionum metaphysicarum libri duo: Opus Posthumum: Omni cura ac diligentia ex ipsius Authoris manuscripto collectum*. Lugduni Batavorum: Apud Hieronimum de Vogel, 1640.

Buzon, Frédéric de. *La science cartésienne et son objet*. Paris: Champion, 2013.

Buzon, Frédéric de, and Vincent Carraud. *Descartes et les Principia II. Corps et mouvement*. Paris: Presses universitaires de France, 1994.

Casmann, Otto. *Angelographia, sive Commentationum disceptationumque physicarum prodromus problematicus . . .* Frankfurt am Main: Palthen, 1605.

Cassirer, Ernst. "Das Erkenntnisproblem in der Philosophie und Wissenschaft der Neueren Zeit." 1906. In *Gesammelte Werke*, Bd 2. Hamburg: Felix Meiner Verlag, 1999.

Cassirer, Ernst. "Descartes' Kritik der mathematischen und naturwissenschaftliche Erkenntnis." 1899. In *Gesammelte Werke*, Bd 1. Hamburg: Felix Meiner Verlag, 1998.

Cassirer, Ernst. "Descartes: Lehre-Persönlichkeit-Wirkung." 1939. In *Gesammelte Werke*, Bd. 20. Hamburg: Felix Meiner Verlag, 2005.

Cassirer, Ernst. *Die Philosophie der Aufklärung*. Tübingen: Mohr, 1932.

Catalogus variorum ac rarissimorum librorum in omnibus scientiis et facultatibus, clarissimi, doctissimi viri D. Henrici Reneri . . . Utrecht: Agidius Roman, 1639.

Cavaillé, Jean-Pierre. "L'art d'écrire des philosophes." *Critique* 631 (1999): 959–80. Reprinted in *Les Déniaisés—Irréligion et libertinage au début de l'époque moderne*. Paris: Garnier, 2014.

Cavaillé, Jean-Pierre. "Libertinage et dissimulation, quelques éléments de réflexion." *Libertinage et philosophie au* XVII*e siècle* 5 (2001): 57–82.

Cellamare, Davide. "Psychology in the Age of Confessionalisation: A Case Study in the Interaction between Psychology and Theology c. 1517–c.1640." PhD thesis, Radboud University, 2015.

Charrak, André. *Empirisme et métaphysique: L'Essai sur l'origine des connaissances humaines de Condillac*. Paris: Vrin, 2003.

Charrak, André. *Empirisme et théorie de la connaissance, Réflexion et fondement des sciences au* XVIII*e siècle*. Paris: Vrin, 2009.

Clarke, Desmond M. "The Concept of Experience in Descartes' Theory of Knowledge." *Studia Leibnitiana* 8 (1976): 18–39.

Clarke, Desmond M. "The Concept of *Vis* in Part III of the *Principia*." In *Descartes: Principia philosophiae (1644–1994). Atti del Convegno per il 350° anniversario della pubblicazione dell'opera (Parigi, 5–6 maggio 1994; Lecce, 10–12 novembre 1994)*, edited by Jean-Robert Armogathe and Giulia Belgioioso, 321–39. Napoli: Vivarium, 1996.

Clarke, Desmond M. *Descartes' Philosophy of Science*. Manchester: Manchester University Press, 1982.

Clarke, Desmond M. *Descartes' Theory of Mind*. Oxford: Oxford University Press, 1992.

Clarke, Desmond M. "The Physics and Metaphysics of the Mind: Descartes and Regius." In *Mind, Method, and Morality: Essays in Honour of Anthony Kenny*, edited by John Cottingham and Peter Hacker, 187–207. Oxford: Oxford University Press, 2010.

Clatterbaugh, Kenneth. "Cartesian Causality, Explanation, and Divine Concurrence." *History of Philosophy Quarterly* 12, no. 2 (1995): 195–207.
Cohen, I. Bernard. *The Newtonian Revolution*. Cambridge, MA: Harvard University Press, 1980.
Cohen, I. Bernard and George Smith. "Introduction." In *The Cambridge Companion to Newton*, edited by I. Bernard Cohen and George Smith, 1–32. Cambridge: Cambridge University Press, 2002.
Cohen Roselfield, Leonora. *From Beast-Machine to Man-Machine*. New York: Octagon Books, 1968.
Comparato, Vittor Ivo. "Ragione e fede nelle *Discussioni istoriche, teologiche e filosofiche* di Costantino Grimaldi." In *Saggi e ricerche sul Settecento*, edited by Ernest Sestan, 48–93. Napoli: Istituto italiano per gli studi storici, 1968.
Condillac, Étienne Bonnot de. *Essai sur l'origine des connaisances humaines*, edited by Jean-Claude Pariente and Martine Pécharman Paris: Vrin, 2014. First published 1746.
Condillac, Étienne Bonnot de. *Œuvres philosophiques*, edited by Georges Le Roy. Paris: Presses universitaires de France, 1947.
Condillac, Étienne Bonnot de. *Traité des systèmes*. Paris: Fayard, 1991. First published 1749.
Cook, Monte. "Desgabets as a Cartesian Empiricist." *Journal of the History of Philosophy* 46, no. 4 (2008): 501–15.
Cornelio, Tommaso. *Discorso dell'eclissi detto nell'Accademia degli Otiosi nel dì 29 di maggio 1652, dato in luce per l'Accademico detto l'Arrestato*. Napoli: per Camillo Cavallo, 1652.
Cornelio, Tommaso. *Progymnasmata physica*. Napoli: Giacomo Raillard 1688. First published 1663.
Costabel, Pierre. *Démarches originales de Descartes savant*. Paris: Vrin, 1982.
Cottingham, John. *Descartes*. Oxford: Blackwell Publishers, 1986.
Cousin, Victor. *Fragments Philosophiques*. Paris: Charpentier, 1845. First published 1826.
Cousin, Victor. *Histoire générale de la philosophie depuis les temps les plus anciens jusqu'au XIXe siècle*. 8th ed. Paris: Didier, 1867. First published 1863.
Cousin, Victor. *Philosophie de Locke*. Paris: Didier, 1861. First published 1828.
Cover, Jan and Hartz, Glenn A. "Space and Time in the Leibnizian Metaphysic." *Noûs* 22, no. 4 (1988): 493–519.
Cover, Jan and John O' Leary Hawhtorne. *Substance and Individuation in Leibniz*. Cambridge: Cambridge University Press, 1999.
Cristofolini, Paolo. "Tommaso Cornelio et l'histoire du matérialisme." In *Gassendi et l'Europe (1592–1792)*, edited by Sylvia Murr, 335–46. Paris: Vrin, 1997.
Cross, Richard. "Eternity, Creation, and Conservation." *Religious Studies* 42, no. 4 (2006): 403–16.
Dagen, Jean. *L'ishtoire de l'esprit humain*. Paris: Klincksieck, 1977.
d'Alembert, Jean Le Rond. *Œuvres completes*. Paris: A. Blin, 1821–1822.
d'Alembert, Jean Le Rond. *Œuvres philosophiques, historiques et littéraires de d'Alembert*. Paris: J. F. Bastien, 1805.
D'Alessio, Silvana. *Masaniello: La sua vita e il mito in Europa*. Roma: Salerno Editrice, 2007.
Damiron, Jean-Philibert. *Essai sur l'histoire de la philosophie en France au XVIIe siècle*. Paris: Louis Hachette, 1846.
D'Andrea, Francesco. *Avvertimenti ai nipoti*, edited by Imma Ascione. Napoli: Jovene 1990. First published 1696.

D'Andrea, Francesco. *I ricordi di un avvocato napoletano del Seicento*, edited by Nino Cortese. Napoli: Luigi Lubrano, 1923.

D'Andrea, Francesco. *Risposta a favore del Sig: Leonardo Di Capua contro le Lettere apologetiche del P. De Benedictis Gesuita*. Napoli, 1697.

De Benedictis, Giovanni Battista. *Difesa della terza lettera apologetica*. Roma: A. De Rosi, 1705.

De Benedictis, Giovanni Battista. *Lettere apologetiche in difesa della Teologia Scolastica, e della Filosofia Peripatetica di Benedetto Aletino dedicate all'Illustriss. & Eccelentiss. Signore, il signore D. Carlo Francesco Spinelli Principe di Tarsia, &c.* Napoli: nella Stamperia di Giacomo Raillard, con licenza de' Superiori, 1694.

De La Forge, Louis. "Traité de l'esprit de l'homme (1665)." In *Œuvres philosophiques*, edited by Pierre Clair. Paris: Presses Universitaires de France, 1974.

De Liguori, Girolamo. "Teologia, filosofia e fisica di Cartesio nella Difesa della terza lettera apologetica dell'Aletino (1705)." In *L'ateo smascherato. Immagini dell'ateismo e del materialismo nell'apologetica cattolica da Cartesio a Kant*, 63–94. Firenze: Le Monnier Università, 2009.

De Raey, Johannes. *Clavis Philosophiae Naturalis Aristotelico-Cartesiana: Editio secunda* . . . Amstelodami: Apud Danielem Elsevirium, 1677.

De Raey, Johannes. *Clavis Philosophiae Naturalis, seu introductio ad naturae contemplationem, Aristotelico-Cartesiana* . . . Lugd. Batavor: Ex Officina Johannis et Danielis Elsevier, 1654.

De Raey, Johannes. *Cogitata de interpretatione*. . . Amstelaedami: Apud Henricum Wetstenium, 1692.

De Raey, Johannes. *Disputatio Philosophica de Constitutione Logicae*. . . Lugduni Batavorum: Apud Viduam et Haeredes Johannis Elsevirii, 1668.

De Renzi, Salvatore. *Napoli nell'anno 1656, ovvero documenti della pestilenza che desolò Napoli nell'anno 1656, preceduti dalla storia di quella tremenda sventura narrata da Salvatore De Renzi*. Napoli: De Pascale, 1867.

Del Prete, Antonella. "Discussioni sul metodo nel cartesianismo olandese. Il caso di Johannes de Raey." In *La ragione e le sue vie. Saperi e procedure di prova in età moderna*, edited by Carlo Borghero and Claudio Buccolini, 146–67. Firenze: Le Lettere, 2015.

Del Prete, Antonella. "Oltre Descartes: filosofia e teologia nella *Theologia pacifica* di Christoph Wittich." In *Immagini filosofiche e interpretazioni storiografiche del cartesianismo*, edited by Carlo Borghero and Antonella Del Prete, 25–45. Firenze: Le Lettere, 2011.

Dennehy, Myriam. "Leibniz et Sturm lecteurs de Boyle." In *La Philosophie naturelle de Robert Boyle*, edited by Myriam Dennehy and Charles Ramond, 331–59. Paris: Vrin, 2009.

Descartes, René. *Œuvres de Descartes*, edited by Charles Adam and Paul Tannery. 11 vols. Paris: Vrin, 1964–1978.

Descartes, René. *The Philosophical Writings of Descartes*. Translated by John Cottingham, Robert Stoothoff, and Dugald Murdoch. 2 vols. New York: Cambridge University Press, 1984.

Descartes, René. *The Philosophical Writings of Descartes*. Vol. 3. *The Correspondence*. Translated by John Cottingham, Robert Stoothoff, Dugald Murdoch and Anthony Kenny. Cambridge: Cambridge University Press, 1991.

Descartes, René. *Discourse on Method, Optics, Geometry, and Meteorology*. Translated by Paul J. Olscamp. Indianapolis: Hackett, 2001. First published 1965.

Descartes, René. *Lettres de M. Descartes, qui traitent de plusieurs belles questions concernant la morale, la physique, la médecine et les mathématiques [. . .] où l'on a joint le latin de plusieurs lettres qui n'avoient été imprimées qu'en françois, avec une traduction françoise de celles qui n'avoient paru jusqu'à présent qu'en latin, Nouvelle édition*. Paris: par la Compagnie des libraires, 1724–1725.

Descartes, René. *Principia philosophiae*. Amsterdam: Elzevier, 1644.

Des Chene, Dennis. "On Laws and Ends: Reply to Hattab and Menn." *Perspectives on Science* 8 (2000): 144–63.

Des Chene, Dennis. *Physiologia: Natural Philosophy in Late Aristotelian and Cartesian Thought*. Ithaca and London: Cornell University Press, 1996.

d'Espagnet, Jean. *La philosophie naturelle restablie en sa pureté*, edited and Translated by Jean Bachout. Paris: Edme Pepingué, 1651.

Di Capua, Leonardo. *Lezioni intorno alla natura delle mofette del signor Lionardo Di Capoa, Accademico investigante, Ultima edizione accresciuta di un'Indice copiosissimo, e delle postille nel margine*. Cologna, 1714.

Di Capua, Leonardo. *Parere del signor Lionardo Di Capoa, divisato in otto ragionamenti, ne' quali partitamente narrandosi l'origine, e'l progresso della medicina, chiaramente l'incertezza della medicina si fa manifesta*. Napoli: Antonio Bulifon, 1681.

Dibon, Paul. "Histoire des idées au XVIIe siècle: La diffusion européenne du cartésianisme." *Annuaire de l'École Pratique des Hautes Études. IVe Section: Sciences Historiques et Philologiques* 104 (1971–1972): 515–28.

Dibon, Paul. *La philosophie néerlandaise au siècle d'or. I L'enseignement philosophique dans les universités à l'époque précartésienne (1575–1650)*. Paris, Amsterdam, London and New York: Elsevier Publishing Company, 1954.

Dibon, Paul. *Regards sur la Hollande du siècle d'or*. Napoli: Vivarium, 1990.

Diderot, Denis and Jean Le Rond d'Alembert, eds. *Encyclopédie ou Dictionnaire raisonné des sciences, des arts et des métiers, par une société de Gens de Lettres: Mis en ordre et publié par M. Diderot* [. . .] *et quant à la partie mathématique par M. d'Alembert*. Paris: Briasson, David, Le Breton, Durand, 1751–65.

Dobre, Mihnea and Tammy Nyden, eds. *Cartesian Empiricisms*. Dordrecht: Springer, 2014.

Douglas, Alexander. "Spinoza and the Dutch Cartesians on Philosophy and Theology." *Journal of the History of Philosophy* 51 (2013): 567–88.

Douglas, Alexander. *Spinoza and Dutch Cartesianism*. Oxford: Oxford University Press, 2015.

Downing, Lisa. "Locke's Metaphysics and Newtonian Metaphysics." In *Newton and Empiricism*, edited by Zvi Biener and Eric Schliesser, 97–118. Oxford: Oxford University Press, 2014.

Duhem, Pierre. *The Aim and Structure of Physical Theory*. Translated by Philip P. Wiener. New York: Atheneum, 1962.

Duhem, Pierre. *Essays in History and Philosophy of Science*. Translated by Roger Ariew and Peter Barker. Indianapolis: Hackett, 1996.

Duhem, Pierre. *La théorie physique: Son objet, sa structure*. 2nd ed. (1914) rev. Paris: Vrin, 1997. First published 1906.

Duhem, Pierre. "Logical Examination of Physical Theory." *Synthese* 83, no. 2 (1990): 183–88.

Esposito, Constantino. "Suárez and the Baroque Matrix of Modern Thought." In *A Companion to Francisco Suárez*, edited by Victor M. Salas and Robert L. Fastiggi, 124–47. Leiden: Brill, 2015.

Fabiani, Jean-Louis. "Controverses scientifiques, controverses philosophiques: Figures, positions, trajets." *Enquête* 5 (1997): 11–34.

Feuerlin, Jacob Wilhelm. *Taurellus defensus: Dissertatio apologetica pro Nic. Taurello philosopho Altdorfino atheismi et deismi iniuste accusato, et ipsius Taurelli Synopsis Aristotelis Metaphysices ob raritatem recusa cum annotationibus editoris*. Nuremberg: Schmidius, 1734.

Fichant, Michel. "La notion de système dans la physique de Leibniz." In *Science et métaphysique dans Descartes et Leibniz*, 245–66. Paris: Presses universitaires de France, 1998.

Fichant, Michel. *Science et métaphysique dans Descartes et Leibniz*. Paris: Presses universitaires de France, 1998.

Fisch, Max H. "L'Accademia degli Investiganti." *De Homine* 4 (1968): 17–78.

Fontenelle, Bernard Le Bovier de. *Œuvres complètes*. Paris: Fayard, 1989–2001.

Frankfurt, Harry G. *Demons, Dreamers and Madmen: The Defense of Reason in Descartes' Meditations*. New York: The Bobbs-Merrill Company, 1970.

Freddoso, Alfred. "God's General Concurrence with Secondary Causes: Why Conservation Is Not Enough." *Philosophical Perspectives* 5 (1991): 553–85.

Frege, Gottlob. *The Basic Laws of Arithmetic*. Translated by Montgomery Furth. Berkeley: University of California Press, 1964.

Fuchs, Marko J. "Univozität und Distinktion. Metaphysische Grundstrukturen bei Duns Scotus, Suárez, Descartes und Spinoza." In *Suárez's Metaphysics and Its Historical and Systematic Context*, edited by Lukás Novák, 105–16. Berlin: Walter de Gruyter, 2014.

Futch, Michael J. *Leibniz's Metaphysics of Time and Space* (*Boston Studies in the Philosophy of Science* 258). New York: Springer, 2008.

Gabbey, Alan. "Force and Inertia in the Seventeenth Century: Descartes and Newton." In *Descartes: Philosophy, Mathematics and Physics*, edited by Steven Gaukroger, 230–319. Brighton: The Harvester Press, Barnes and Noble Books, 1980.

Gabbey, Alan. "Henry More and the Limits of Mechanism." In *Henry More (1614–1687): Tercentenary Studies*, edited by Sarah Hutton, 19–35. Dordrecht: Kluwer, 1990.

Gabbey, Alan. "Philosophia Cartesiana Triumphata: Henry More (1646–1671)." In *Problems of Cartesianism*, edited by Thomas M. Lennon, John M. Nichols and John W. Davis, 171–250. Kingston and Montreal: McGill-Queens University Press, 1982.

Galluzzi, Paolo. "L'Accademia del Cimento: 'gusti' del principe, filosofia e ideologia dell'esperimento." *Quaderni storici* 16, no. 48 (1981): 788–844.

Garber, Daniel. "Descartes and Occasionalism." In *Causation in Early Modern Philosophy*, edited by Steven Nadler, 9–26. University Park: Pennsylvania State University Press, 1993.

Garber, Daniel. *Descartes' Metaphysical Physics*. Chicago: University of Chicago Press, 1992.

Garber, Daniel. "How God Causes Motion: Descartes, Divine Sustenance and Occasionalism." *Journal of Philosophy* 84 (1987): 567–80.

Garber, Daniel. "Remarks on the Pre-history of the Mechanical Philosophy." In *The Mechanization of Natural Philosophy*, edited by Daniel Garber and Sophie Roux, 3-26. Dordrecht: Springer, 2013.

Garber, Daniel and Sophie Roux, eds. *The Mechanization of Natural Philosophy*. Dordrecht: Springer, 2012.

Garin, Eugenio. "Cartesio e l'Italia." *Giornale Critico della Filosofia Italiana* 4 (1950): 385–405.

Gasparini, Paolo. *Un viaggio al Vesuvio: Il Vesuvio visto attraverso diari, lettere e resoconti di viaggiatori*. Napoli: Liguori, 1992.

Gaukroger, Stephen W. "Bachelard and the Problem of Epistemological Analysis." *Studies in History and Philosophy of Science* 7, no. 3 (1976): 189–244.

Gaukroger, Stephen W. *Descartes: An Intellectual Biography*. Oxford: Oxford University Press, 1995.

Gaukroger, Stephen W. *Descartes' System of Natural Philosophy*. Oxford and New York: Oxford University Press, 1997.

Gay, Peter. *The Enlightenment: An Interpretation*. 2 vols. New York: W. W. Norton & Company, 1966 and 1969.

Giannone, Pietro. *Il Triregno*, edited by Alfredo Parente. Bari: Laterza 1940.

Giannone, Pietro. *Opere*, edited by Sergio Bertelli and Giuseppe Ricuperati. Napoli: Riccardo Ricciardi Editore, 1971.

Gibson, A. Boyce. *The Philosophy of Descartes*. London: Methuen & Co. Ltd, 1932.

Gilson, Étienne. *La liberté chez Descartes et la théologie*. Paris: F. Alcan, 1913.

Gilson, Étienne. *René Descartes: Discours de la méthode, texte et commentaire*. 4th ed. Paris: Vrin, 1967.

Girard, Pierre. "Giambattista Vico critique de Descartes?" In *Qu'est-ce qu'être cartésien?* edited by Delphine Kolesnik-Antoine, 503–19. Lyon: ENS Éditions, 2013.

Girard, Pierre. *Giambattista Vico. Rationalité et politique: Une lecture de la Scienza nuova*. Paris: Presses universitaires de Paris-Sorbonne, 2008.

Girard, Pierre. "Les usages de Malebranche dans la réception du cartésianisme à Naples." In *Les Malebranchismes des Lumières. Études sur les réceptions contrastées de la philosophie de Malebranche, fin XVII*[e] *et XVIII*[e] *siècles*, edited by Delphine Antoine-Mahut, 225–46. Paris: Honoré Champion, 2014.

Girard, Pierre. "Matérialisme et politique: les enjeux de la réception du cartésianisme à Naples à l'Âge classique." *Corpus* 61 (2011): 113–32.

Girard, Pierre. *Matérialisme et radicalité politique dans les premières Lumières à Naples (1649–1744)*. Paris: Honoré Champion, 2016.

Gorham, Geoffrey. "Cartesian Causation: Continuous, Instantaneous, Overdetermined." *Journal of the History of Philosophy* 42, no. 4 (2004): 389–423.

Gorham, Geoffrey. "Cartesian Temporal Atomism: A New Defence, A New Refutation." *British Journal for the History of Philosophy* 16, no. 3 (2008): 625–37.

Gorlaeus, David. *Exercitationes philosophicae post mortem auctoris editae, quibus universa fere discutitur philosophia theoretica, et plurima ac praecipua Peripateticorum dogmata evertuntur*. Leiden: Widow of Johannes Commelin, 1620.

Gorlaeus, David. *Idea physicae, cui adjuncta est Epistola cuiusdam Anonymi de terrae motu*. Utrecht: Johannes a Waesberge, 1651.

Goudriaan, Aza. *Philosophische Gotteserkenntis bei Suárez und Descartes im Zusammenhang mit der niederländischen reformierten Theologie und Philosophie des 17. Jahrhunderts*. Leiden: Brill, 1999.

Goudriaan, Aza. *Reformed Orthodoxy and Philosophy, 1625–1750: Gisbertus Voetius, Petrus van Mastricht, and Anthonius Driessen*. Leiden: Brill, 2006.

Goudriaan, Aza. "Theology and Philosophy." In *A Companion to Reformed Orthodoxy*, edited by Herman J. Selderhuis, 27–63. Leiden: Brill, 2013.

Gouhier, Henri. *La pensée religieuse de Descartes*. Paris: Vrin, 1924.

Grimaldi, Costantino. *Memorie di un anticurialista del Settecento*, edited by Vittor Ivo Comparato. Firenze: Olschki, 1964.

Grimaldi, Costantino. *Risposta alla lettera apologetica in difesa della teologia scolastica di B. Aletino. Opera nella quale si dimostra esser quanto necessaria ed utile la teologia dogmatica e metodica; tanto inutile, e vana la volgar teologia scolastica*. Colonia: S. Hecht 1699.

Grimaldi, Costantino. *Risposta alla seconda lettera apolegetica di B. Aletino. Opera utilissima a' professori della filosofia: in cui fassi vedere quanto manchevole sia la peripatetica dottrina*. Colonia: S. Hecht, 1702.

Grimaldi, Costantino. *Risposta alla terza lettera apologetica contra il Cartesio creduto da più d'Aristotele di Benedetto Aletino: Opera, in cui dimostrasi quanto salda e pia sia la filosofia di Renato dell Carte: e perchè questo si debba stimare più d'Aristotele*. Colonia: S. Hecht, 1703.

Groult, Martine. *D'Alembert et la mécanique de la vérité dans l'Encyclopédie*. Paris: Champion, 1999.

Groult, Martine. *L'Encyclopédie ou la création des disciplines*. Paris: CNRS Editions, 2003.

Guenancia, Pierre. *Descartes, chemin faisant*. Paris: Les Belles Lettres, 2010.

Guénard, Florent, Francine Markovits and Mariafranca Spallanzani, eds. "*L'ordre des renvois dans l'*Encyclopédie." *Corpus* 51 (2006).

Gueroult, Martial. "Métaphysique et physique de la force chez Descartes et chez Malebranche. Première partie: Descartes." *Revue de métaphysique et de morale* 59, no. 1 (1954): 1–37.

Guibert, Albert-Jean. *Bibliographie des Œuvres de René Descartes publiées au XVIIe siècle*. Paris: CNRS Editions, 1976.

Hamelin, Octave. *Le système de Descartes*. 2nd ed. Paris: Alcan, 1921. First published 1911.

Harris, Errol E. *The Foundations of Metaphysics in Science*. London: George Allen and Unwin, 1965.

Harris, Errol E. *The Restitution of Metaphysics*. New York: Humanity Books, 2000.

Hartkopf, Werner and Gert Wangermann, eds. *Dokumente zu Geschichte der Berliner Akademie der Wissenschaften von 1700 bis 1990*. Berlin: Akademie-Verlag, 1991.

Haslanger, Sally. "Persistence Through Time." In *The Oxford Handbook of Metaphysics*, edited by Michael J. Loux and Dean W. Zimmerman, 313–54. Oxford: Oxford University Press, 2003.

Hatfield, Gary. "Descartes' Physiology and Its Relation to His Psychology." In *The Cambridge Companion to Descartes*, edited by John Cottingham, 335–70. Cambridge and New York: Cambridge University Press, 1992.

Hatfield, Gary. "First Philosophy and Natural Philosophy in Descartes." In *Philosophy, Its History of Historiography*, edited by A. J. Holland, 149–64. Dordrecht and Boston: D. Reidel, 1985.

Hatfield, Gary. "Force (God) in Descartes Physics." *Studies in History and Philosophy of Science* 10 (1979): 113–40.

Hatfield, Gary. "Metaphysics and the New Science." In *Reappraisals of the Scientific Revolution*, edited by David C. Lindberg and Robert S. Westman, 93–166. Cambridge: Cambridge University Press, 1990.

Hattab, Helen. "Concurrence or Divergence? Reconciling Descartes' Science with His Metaphysics." *Journal of the History of Philosophy* 45 (2007): 749–88.

Hattab, Helen. *Descartes on Forms and Qualities*. Cambridge: Cambridge University Press, 2009.

Hattab, Helen. "The Problem of Secondary Causation in Descartes: A Response to Des Chene." *Perspectives on Science* 8 (2000): 93–118.

Hawley, Katherine. *How Things Persist*. Oxford: Clarendon Press, 2001.

Heereboord, Adriaan. *Meletemata philosophica, maximam partem metaphysica*. Lugduni Batavorum: Francisci Moyardi, 1654.

Helm, Paul. "Jonathan Edwards and the Doctrine of Temporal Parts." *Archiv für die Geschichte der Philosophie* 61, no. 1 (1979): 37–51.

Henry, John. "Metaphysics and the Origins of Modern Science: Descartes and the Importance of the Laws of Nature." *Early Science and Medicine* 9 (2004): 73–114.

Hoffmann, Tobias, ed. *Johannes Duns Scotus: Die Univozität des Seienden. Texte zur Metaphysik*. Göttingen: Vandenoek & Ruprecht, 2002.

Hugon, Alain. *Naples insurgée: De l'événement à la mémoire, 1647–1648*. Rennes: Presses universitaires de Rennes, 2011.

Hume, David. *An Enquiry Concerning Human Understanding and Other Writings*, edited by Stephen Buckle. Cambridge: Cambridge University Press, 2007.

Hume, David. *A Treatise of Human Nature: A Critical Edition*, edited by David F. Norton and Mary J. Norton. Oxford: Clarendon Press, 2007.

Israel, Jonathan. *Radical Enlightenment: Philosophy and the Making of Modernity 1650–1750*. Oxford: Oxford University Press, 2001.

Israel, Jonathan. *A Revolution of the Mind: Radical Enlightenment and the Intellectual Origins of Modern Democracy*. Princeton: Princeton University Press, 2010.

Jacob, Alexander. "The Spirit of Nature as 'hylarchic' Principle of the Universe." In *Manual of Metaphysics*, edited by H. More and Translated and edition Alexander Jacob, vol. II, i–xxx. Hildesheim: Georg Olms Verlag, 1995.

Jacob, Margaret. *The Newtonians and the English Revolution, 1689–1720*. Hassocks, Sussex: The Harvester Press, 1976.

Jalabert, Jacques. *La theorie leibnizienne de la substance*. Paris: Presses universitaires de France, 1947.

Janiak, Andrew. *Newton as Philosopher*. Cambridge: Cambridge University Press, 2008.

Jardine, Nicholas. "Keeping Order in the School of Padua: Jacopo Zabarella and Francesco Piccolomini on the Offices of Philosophy." In *Method and Order in Renaissance Philosophy of Nature*, edited by Daniel A. Di Liscia, Eckhardt Kessler and Charlotte Methuen, 183–209. Aldershot: Ashgate, 1997.

Jaumes, Alphonse. *De l'influence des doctrines philosophiques de Descartes et Bacon sur les progrès de la médecine*. Thèse soutenue publiquement le 22 Avril 1850, à la Faculté de Médecine de Montpellier. Montpellier: J. Martel, 1850.

Jolley, Nicholas. *Causality and Mind: Essays on Early Modern Philosophy*. Oxford: Oxford University Press, 2013.

Jolley, Nicholas. *Leibniz*. London: Routledge, 2005.

Kambouchner, Denis. *L'homme des passions: Commentaires sur Descartes*. 2 vols. Paris: Albin Michel, 1995.

Kambouchner, Denis. *Les Méditations métaphysiques de Descartes I*. Paris: Presses universitaires de France, 2005.

Keckermann, Bartholomaeus. *Scientiae metaphysicae compendiosum systema*. Hanau: Guilielmus Antonius, 1609.

Keeling, Stanley Victor. *Descartes*. 2nd ed. Oxford: Oxford University Press, 1968. First published 1934.

Kitcher, Philip. "Mill, Mathematics, and the Naturalist Tradition." In *The Cambridge Companion to Mill*, edited by John Skorupski, 56–111. Cambridge: Cambridge University Press, 1998.

Kitcher, Philip. "The Naturalists Return." *The Philosophical Review* 101 (1992): 53–114.

Knowles Middleton, Edgar William. *The Experimenters: A Study of the Accademia del Cimento*. Baltimore and London: Johns Hopkins University Press, 1971.

Kolesnik-Antoine (Antoine-Mahut), Delphine, ed. *Qu'est-ce qu'être cartésien?* Lyon: ENS Éditions, 2014.

Koyré, Alexandre. *Entretiens sur Descartes*. New York: Paris: Brentano's, 1944.

Koyré, Alexandre. *Essai sur l'idée de Dieu et les preuves de son existence chez Descartes*. Paris: Ernest Leroux, 1922.

Koyré, Alexandre. *From the Closed World to the Infinite Universe*. Baltimore: Johns Hopkins University Press, 1957.

Koyré, Alexandre. *Galileo Studies*. Translated by John Mepham. Hassocks, Sussex: The Harvester Press, 1978. First published 1939 in French.

La Grange, Jean-Baptiste. *Les principes contre les nouveaux philosophes Descartes, Rohault, Gassendi, le père Maignan, etc.* Paris: Georges Josse, 1675.

Laberthonnière, Lucien. *Études sur Descartes I*. Paris: Vrin, 1935.

Laberthonnière, Lucien. "La théorie de la foi chez Descartes." *Annales de philosophie chrétienne* 12 (1911): 382–403.

Labio, Catherine. *Origins and the Enlightenment: Aesthetic Epistemology from Descartes to Kant*. Ithaca: Cornell University Press, 2004.

Lærke, Mogens. "Apology for a *Credo Maximum*: On Three Basic Rules in Leibniz's Method for Religious Controversy." In *Leibniz: What Kind of Rationalist?* edited by Marcelo Dascal, 397–407. Dordrecht: Springer, 2008.

Lærke, Mogens. "The Golden Rule: *Charitas/Prudentia*. Aspects of G. W. Leibniz's Method for Religious Controversy." In *The Practice of Reason: Leibniz and His Controversies*, edited by Marcelo Dascal, 297–319. Amsterdam, Philadelphia: Andrew Benjamin Publishing, 2010.

Lærke, Mogens. "*Ignorantia inflat*: Leibniz, Huet and the Critique of the Cartesian Spirit." *The Leibniz Review* 23 (2013): 13–42.

Lærke, Mogens. *Leibniz lecteur de Spinoza: La genèse d'une opposition complexe*. Paris: Champion, 2008.

Lærke, Mogens. *Les Lumières de Leibniz: Controverses avec Bayle, Huet, Régis et More*. Paris: Classiques Garnier, 2015.

Lagrée, Jacqueline. "Qu'est-ce qu'être un logicien cartésien?" In *Qu'est-ce qu'être cartésien?* edited by Delphine Kolesnik-Antoine, 143–58. Lyon: ENS Éditions, 2013.

Laporte, Jean. *Le rationalisme de Descartes*. Paris: Presses universitaires de France, 1945.

Laugier, Sandra. "Science and Realism: The Legacy of Duhem and Meyerson in Contemporary American Philosophy of Science." In *French Studies in the Philosophy of Science*, edited by Anastasios Brenner and Jean Gayon, 91–112. Dordrecht: Springer, 2009.

Lecourt, Dominique. *L'épistémologie historique de Gaston Bachelard*. 5th ed. Paris: Vrin, 1978.

Lee, Sukjae. "Leibniz on Divine Concurrence." *The Philosophical Review* 113 (2004): 81–118.

Leibniz, Gottfried Wilhelm. *Œuvres*, edited by Louis-Alexandre Foucher de Careil, 7 vols. Paris: Firmin Didot, 1858–1875.

Leibniz, Gottfried Wilhelm. *The Art of Controversies*, edited and Translated by Marcelo Dascal, Quintín Racionero and Adelino Cardoso. Dordrecht: Springer, 2006.

Leibniz, Gottfried Wilhelm. *Die mathematische Schriften*, edited by Carl Immanuel Gerhardt, 7 vols. Berlin: A. Asher & Comp., 1849–1860.

Leibniz, Gottfried Wilhelm. *Die philosophischen Schriften*, edited by Carl Immanuel Gerhardt, 7 vols. Berlin: A. Asher & Comp., 1875–1890.

Leibniz, Gottfried Wilhelm. *The Leibniz—Des Bosses Correspondence*, edited and Translated by Brandon Look and Donald Rutherford. New Haven: Yale University Press, 2007.

Leibniz, Gottfried Wilhelm. *The Leibniz-De Volder Correspondence*, edited and Translated by Paul Lodge. New Haven: Yale University Press, 2013.

Leibniz, Gottfried Wilhelm. *Philosophical Essays*, edited and Translated by Roger Ariew and Daniel Garber. Indianapolis: Hackett, 1989.

Leibniz, Gottfried Wilhelm. *Philosophical Papers and Letters*, edited and Translated by Leroy E. Loemker. Dordrecht: Kluwer, 1989. First published 1956.

Leibniz, Gottfried Wilhelm. *Sämtliche Schriften und Briefe*, edited by the Brandenburgischen Akademie der Wissenschaften et al. 8 series, running volumes. Berlin: Akademie-Verlag, 1923.

Leinsle, Ulrich. *Das Ding und die Methode: Methodische Konstitution und Gegenstand der frühen protestantischen Metaphysik*, 2 vols. Augsburg: Maro, 1985.

Lennon, Thomas M. and Patricia Ann Easton. *The Cartesian Empiricism of François Bayle*. New York: Garland, 1992.

Lennon, Thomas M., John M. Nicholas and John W. Davis, eds. *Problems of Cartesianism*. Kingston and Montreal: McGill-Queen's University Press, 1982.

Levey, Samuel. "The Interval of Motion in Leibniz's *Pacidius Philalethi*." *Noûs* 37, no. 3 (2003): 371–416.

Lévy, David. "Émile Meyerson." In *L'épistémologie française, 1830–1970*, edited by Michel Bitbol and Jean Gayon, 357–74. Paris: Presses Universitaires de France, 2006.

Levy, Ken. "Is Descartes a Temporal Atomist?" *British Journal of the History of Philosophy* 13 (2005): 627–74.

Lewis, David K. *On the Plurality of Worlds*. Oxford: Blackwell Publishers, 1986.

Liard, Louis. *Descartes*. 3rd ed. Paris: Alcan, 1911. First published 1882.

Liard, Louis. *La science positive et la métaphysique*. 3rd ed. Paris: Germer Baillière et Cie, 1893. First published 1879.

Lilti, Antoine. "Querelles et controverses: Les formes du désaccord intellectuel à l'époque moderne." *Mil neuf cent* 25 (2007): 13–28.

Locke, John. *Essai sur l'entendement humain*. 2nd ed. Translated by Pierre Coste. Amsterdam, 1729.

Locke, John. *An Essay Concerning Human Understanding*, edited by Peter H. Nidditch. Oxford: Clarendon Press, 1975.

Lohr, Charles H. "Metaphysics." In *The Cambridge History of Renaissance Philosophy*, edited by Charles B. Schmitt and Quentin Skinner, 537–638. Cambridge: Cambridge University Press, 1988.

Lojacono, Ettore. "Immagini di Descartes a Napoli: da Valletta a Costantino Grimaldi (parte I)." *Nouvelles de la République des Lettres* 2 (1999): 45–65.

Lojacono, Ettore. *Immagini di René Descartes nella cultura napoletana (1644–1755)*. Lecce: Conte, 2003.

Lomonaco, Fabrizio and Maurizio Torrini, eds. *Galileo e Napoli*. Napoli: Guida, 1987.

Lugg, Andrew. "Pierre Duhem's Conception of Natural Classification." *Synthese* 83, no. 3 (1990): 409–20.

Lüthy, Christoph. *David Gorlaeus (1591–1612): An Enigmatic Figure in the History of Philosophy and Science*. Amsterdam: Amsterdam University Press, 2012.

Lüthy, Christoph. "Where Logical Necessity Turns into Visual Persuasion: Descartes' Clear and Distinct Illustrations." In *Transmitting Knowledge: Words, Images and Instruments in Early Modern Europe*, edited by Ian Maclean and Sachiko Kusukawa, 97–133. Oxford: Oxford University Press, 2006.

Lüthy, Christoph. "Entia & Sphaerae. Due aspetti dell'atomismo bruniano." In *La filosofia di Giordano Bruno. Problemi ermeneutici e storiografici*, edited by Eugenio Canone, 165-98. Florence: Leo S. Olschki, 2003.

Lüthy, Christoph and Leen Spruit. "The Doctrine, Life, and Roman Trial of the Frisian Philosopher Henricus de Veno (1574–1613)." *Renaissance Quarterly* 56 (2003): 1112–51.

Malherbe, Michel. *Introduction to the Edition of the Discours Préliminaire de l'Encyclopédie*. Paris: Vrin, 2000.

Marion, Jean-Luc. "L'instauration de la rupture: Gilson à la lecture de Descartes." In *Étienne Gilson et nous*, edited by Monique Couratier, 13–34. Paris: Vrin, 1980.

Marion, Jean-Luc. *Le prisme métaphysique de Descartes*. Paris: Presses Universitaires de France, 1986.

Markovits, Francine. *L'ordre des échanges*. Paris: Presses universitaires de France, 1986.

Markovits, Francine. *Le décalogue sceptique. L'universel en question au temps des Lumières*. Paris: Hermann, 2011.

Maull, Nancy. "Cartesian Optics and the Geometrization of Nature." In *René Descartes. Critical Assessments*, edited by Georges J. D. Moyal, vol. 4, 263–79. London and New York: Routledge, 1991.

Maupertuis, Pierre Louis Moreau de. *Discours sur les différentes figures des astres*. Paris: De l'Imprimerie Royale, 1732.

Mayer, Hans Christian. "Nikolaus Taurellus, der erste Philosoph im Luthertum: Ein Beitrag zum Problem Vernunft und Offenbarung." PhD diss. theol., University of Göttingen, 1959.

Mazauric, Simone. "Milhaud, Descartes et l'histoire des sciences." In *Science, histoire et philosophie selon Gaston Milhaud: La constitution d'un champ disciplinaire sous la Troisième République*, edited by Anastasios Brenner and Annie Petit, 117–31. Paris: Vuibert, 2009.

McDonough, Jeffrey. "Leibniz and the Foundations of Physics: The Later Years." *The Philosophical Review* 125, no. 1 (2016): 1–34.

McDonough, Jeffrey. "Leibniz: Creation and Conservation and Concurrence." *The Leibniz Review* 17 (2007): 31–60.

McGuire, James E. "Labyrinthus Continui: Leibniz on Substance, Activity, and Matter." In *Motion and Time, Space and Matter*, edited by Peter K. Machamer and Robert G. Turnbull, 290–326. Columbus: Ohio State University Press, 1976.

McGuire, James E. "Newton on Place, Time, and God: An Unpublished Source." *British Journal for the History of Science* 11 (1978): 114–29.

McMullin, Ernan. "Duhem's Middle Way." *Synthese* 83, no. 3 (1990): 421–30.

Mehl, Édouard. *Descartes en Allemagne, 1619–1620*. Strasbourg: Presses universitaires de Strasbourg, 2001.

Mengal, Paul. *La naissance de la psychologie*. Paris: L'Harmattan, 2005.

Mersenne, Marin. *Correspondance du P. Marin Mersenne, religieux minime: publiée et annotée par Cornelis de Waard; avec la collaboration de Armand Beaulieu*. 18 vols. Paris: Éditions du Centre national de la recherche scientifique, 1932–1988.

Mersenne, Marin. *L'impiété des déistes, athées et libertins de ce temps . . .* Paris: P. Bilaine, 1624.

Mersenne, Marin. *Quaestiones celeberrimae in Genesim, cum accurata textus explicatione . . .* Paris: Sebastien Cramoisy, 1623.

Meyerson, Émile. *Explanation in the Sciences*. Translated by Mary-Alice and David A. Sipfle. Dordrecht: Kluwer, 1991. First published 1921 in French.

Meyerson, Émile. *Identity and Reality*. Translated by Kate Loewenberg. New York: Gordon and Breach, 1989. First published 1908 in French.

Meyerson, Émile. *The Relativistic Deduction: Epistemological Implications of the Theory of Relativity*. Translated by David A. Sipfle and Mary-Alice Sipfle. Dordrecht: Reidel, 1985. First published 1925 in French.

Miert, Dirk van. *Humanism in an Age of Science: The Amsterdam Atheneum in the Golden Age, 1632–1704*. Leiden: Brill, 2009.

Mikkeli, Heikki. "The Foundation of an Autonomous Natural Philosophy: Zabarella on the Classification of Arts and Sciences." In *Method and Order in Renaissance Philosophy of Nature*, edited by Daniel A. Di Liscia, Eckhardt Kessler and Charlotte Methuen, 211–28. Aldershot: Ashgate, 1997.

Milhaud, Gaston. "Descartes et Newton." In *Nouvelles Études sur l'histoire de la pensée scientifique*, edited by Gaston Milhaud, 219–35. Paris: Alcan, 1911.

Milhaud, Gaston. *Descartes savant*. Paris: Alcan, 1921.

Milhaud, Gaston. "Paul Tannery." *La Revue des idées. Études de critique générale* 25 (15 janvier 1906): 3–14.

Millet, Joseph. *Histoire de Descartes avant 1637*. Paris: Librairie académique, Didier & Cie, 1867.

More, Henry. *Divine Dialogues*. London: J. Flesher, 1668.

More, Henry. *The Immortality of the Soul*. London: J. Flesher, 1659.

More, Henry. *Manual of Metaphysics*. Translated by Alexander Jacob, 2 vols. Hildesheim: Georg Olms Verlag, 1995.

Moreau, Denis. *Deux cartésiens: La polémique entre Antoine Arnauld et Nicolas Malebranche*. Paris: Vrin, 1999.

Moreau, Pierre-François. "Les enjeux de la publication en France des papiers de Leibniz sur Spinoza." *Revue de métaphysique et de morale* 2 (1998): 215–32.

Moreau, Pierre-François. "'In naturalismo'. Leibniz, Spinoza et les spiritualistes français." In *Spinoza/Leibniz: Rencontres, controverses, réceptions*, edited by Raphaële Andrault, Mogens Lærke and Pierre-François Moreau, 325–44. Paris: Presses de l'université Paris-Sorbonne, 2014.

Moreau, Pierre-François. "Spinoza et Victor Cousin." *Archivio di filosofia* 1 (1978): 327–31.

Moreau, Pierre-François. "Spinozisme et matérialisme au xixe siècle." *Raison présente* 51 (1979): 85–94.

Moreau, Pierre-François. "Trois polémiques contre Victor Cousin." *Revue de métaphysique et de morale* 88–4 (1983): 542–48.

Mozzillo, Atanasio. *La frontiera del Grand Tour: Viaggi e viaggiatori nel Mezzogiorno borbonico*. Napoli: Liguori, 1992.

Mugnai, Massimo. "Leibniz on Substance and Changing Properties." *Dialectica* 59 (2005): 503–16.

Musi, Aurelio. *La rivolta di Masaniello nella scena politica barocca*. Napoli: Guida, 2002.

Nadler, Steven, ed. *Causation in Early Modern Philosophy*. University Park: Pennsylvania State University Press, 1993.

Nadler, Steven. "Conceptions of God." In *The Oxford Handbook of Philosophy in Early Modern Europe*, edited by Desmond M. Clarke and Catherine Wilson, 525–47. Oxford: Oxford University Press, 2011.

Nadler, Steven. "Descartes and Occasional Causation." *British Journal for the History of Philosophy* 2 (1994): 35–54.

Nadler, Steven. *Occasionalism: Causation Among the Cartesians*. Oxford: Oxford University Press, 2011.

Natorp, Paul. *Descartes' Erkenntnistheorie: Eine Studie zur Vorgeschichte des Kriticismus*. Marburg: N.G. Elwert'sche Verlagsbuchhandlung, 1882.

Naudé, Gabriel. *Advis pour dresser une bibliothèque*. Paris: François Targa, 1627.

Nazzaro, Antonio. *Il rischio Vesuvio: Storia e geodiversità di un volcano*. Napoli: Guida, 2009.

Neta, Ram. "Quine, Goldman, and Two Ways of Naturalizing Epistemology." In *Epistemology: The Key Thinkers*, edited by Stephen Hetherington, 193–213. London: Continuum, 2012.

Newton, Isaac. *The Correspondence of Isaac Newton*, edited by H. W. Turnbull. Cambridge: Cambridge University Press, 1959–1977.

Newton, Isaac. *Isaac Newton's Papers and Letters on Natural Philosophy*, edited by I. Bernard Cohen and Robert E. Schofield. Cambridge, MA: Harvard University Press, 1958.

Newton, Isaac. *The Mathematical Papers of Isaac Newton*, edited by Tom Whiteside, with the assistance of Michael A. Hoskin. Cambridge: Cambridge University Press, 1967–1981.

Newton, Isaac. *Opticks: Or, a Treatise of the Reflections, Refractions, Inflections and Colours of Light*. Based on the fourth edition, London, 1730. Repr. New York: Dover, 1979.

Newton, Isaac. *The Principia, Mathematical Principles of Natural Philosophy*. Translated by I. Bernard Cohen and Anne Whitman. Berkeley: University of California Press, 1999.

Newton, Isaac. *Unpublished Scientific Papers of Isaac Newton*, edited by Alfred Rupert Hall and Marie Boas Hall. Cambridge: Cambridge University Press, 1962.

Niceron, Jean-Pierre. *Mémoires pour servir à l'Histoire des Hommes illustres dans la République des Lettres avec un catalogue raisonné de leurs Ouvrages*. Paris: Chez Briasson, 1728–1738.

Osler, Margaret. *Divine Will and the Mechanical Philosophy, Gassendi and Descartes on Contingency and Necessity in the Created World*. Cambridge: Cambridge University Press, 2005.

Palaia, Roberto. "Naturbegriff und Kraftbegriff im Briefwechsel zwischen Leibniz und Sturm." In *Leibniz Auseinandersetzung mit Vorgängern und Zeitgenossen*, edited by Ingrid Marchlewitz and Albert Heinekamp, 157–72. Stuttgart: Franz-Steiner Verlag, 1990.

Pascal, Blaise. *Œuvres complètes*, edited by Michel le Guern. 2 vols. Paris: Gallimard, 1998.

Pasnau, Robert. *Metaphysical Themes 1274–1671*. Oxford: Clarendon Press, 2011.

Paty, Michel. "D'Alembert, la science newtonienne et l'héritage cartésien." *Corpus* 38 (2001): 19–64.

Pécharman, Martine. "Signification et langage dans l'*Essai* de Condillac." *Revue de métaphysique et de morale* 1 (1999): 81–103.

Pelletier, Arnaud. "Substance, corps et phénomène dans la *Théodicée*." In *Lectures et interprétations des* Essais de Théodicée *de G. W. Leibniz*, edited by Paul Rateau, 117–38. Stuttgart: Franz-Steiner Verlag, 2011.

Pereira, Benedictus. V. *De communibous omnium Rerum naturalium principiis et affectioniusbus* [. . .]. Venetiis: apud Andrean Muschium, 1576.

Perler, Dominik and Rudolph Ulrich. *Occasionalismus: Theorien der Kausalität im arabisch-islamischen und im europäischen Denken*. Göttingen: Vandenhoeck & Ruprecht, 2000.

Perrault, Charles. *Les hommes illustres qui ont paru en France pendant ce siècle*. Paris: chez Antoine Dezallier, 1700.

Pessin, Andrew. "Descartes' Nomic Concurrentism: Finite Causation and Divine Concurrence." *Journal of the History of Philosophy* 41, no. 1 (2003): 25–50.

Pessin, Andrew. "Does Continuous Creation Entail Occasionalism? Malebranche (and Descartes)." *Canadian Journal of Philosophy* 30, no. 3 (2000): 413–39.

Piro, Francesco. "L'action des créatures et le concours de Dieu chez Leibniz. Entre transcréationnistes et durandiens." In *Lectures et interprétations des Essais de Théodicée de G. W. Leibniz*, edited by Paul Rateau, 79–96. Stuttgart: Franz-Steiner Verlag, 2011.

Pluche, Noël-Antoine. *Histoire du Ciel, où l'on recherche l'origine de l'idolâtrie, et les méprises de la Philosophie sur la formation, et sur les influences des corps célestes*. 2nd ed. Paris: Chez la Veuve Estienne, 1740.

Poirier, René. "Meyerson, Milhaud et le problème de l'épistémologie." *Conférence de la Société Française de Philosophie* (26 novembre 1960): 127–55.

Poser, Hans. "Theoria cum praxi: Das Leibnizsche Akademiekonzept und die Technikwissenschaften." In *Wissenschaft und Weltgestaltung*, edited by Kurt Nowak and Hans Poser, 93–115. Georg Olms Verlag: Hildesheim, 1999.

Quine, Willard Van Orman. "Two Dogmas of Empiricism." *Philosophical Review* 60 (1951): 20–43.

Quinn, Philip L. "Divine Conservation, Continuous Creation, and Human Action." In *The Existence and Nature of God*, edited by Alfred J. Freddoso, 55–79. Notre Dame: University of Notre Dame Press, 1983.

Ramus, Petrus. *Scholarum metaphysicarum libri quatuordecim, in totidem metaphysicos libros Aristotelis: Recens emendati per Joan. Piscatorem*. Frankfurt am Main: Heirs of A. Wechel, 1583.

Raynaud, Dominique. *Sociologie des controverses scientifiques*. Paris: Presses universitaires de France, 2003.

Regius, Henricus. *Philosophia naturalis*. 2nd ed. Amsterdam: Louis Elzevier, 1654.

Regius, Henricus. *Physiologia, sive Cognitio sanitatis, tribus disputationibus in Academia Ultrajectina publice proposita*. Utrecht: Aegidius Roman, 1641.

Reif, Patricia. "Natural Philosophy in Some Early Seventeenth-Century Scholastic Textbooks." Unpublished PhD thesis, St Louis University, 1962.

Reif, Patricia. "The Textbook Tradition in Natural Philosophy 1600–1650." *Journal for the History of Ideas* 30 (1969): 17–32.

Reiner, Hans. "Die Entstehung und ursprüngliche Bedeutung des Namens Metaphysik." *Zeitschrift für philosophische Forschung* 8 (1954), 210–37.

Renaut, Alain. "Une philosophie française de l'université allemande: Le cas de Louis Liard." *Romantisme* 88 (1995): 85–100.

Reneri, Henricus. *Disputationum physicarum quarta, De elementis* [defended by Paulus Keserui]. Utrecht, 1635.

Renouvier, Charles. *Manuel de philosophie moderne*. Paris: Paulin, 1842.

Rescher, Nicholas. "Leibniz and the concept of a system." In *On Leibniz*, edited by Nicholas Rescher, 106–16. Pittsburgh: University of Pittsburgh Press, 2003.

Rey, Lucie, ed. *La mer retentissante: Lectures de Descartes et Leibniz au XIX*[e] *siècle*. Special Issue of *Corpus* 68, 2016.

Ricuperati, Giuseppe. *La città terrena di Pietro Giannone: Un itinerario tra "crisi della coscienza europea" e l'illuminismo radicale*. Firenze: Olschki, 2001.

Ricuperati, Giuseppe. *L'esperienza civile e religiosa di Pietro Giannone*. Milano-Napoli: Riccardo Ricciardi, 1970.

Robertson, John. *The Case for the Enlightenment. Scotland and Naples 1680–1760*. Cambridge: Cambridge University Press, 2005.

Robinet, André. *Architectonique disjonctive et automates systémiques*. Paris: Vrin, 1986.

Rochoux, Jean-André. *Épicure opposé à Descartes ou examen critique du cartésianisme: mémoire envoyé au concours ouvert par l'Académie des sciences morales et politiques*. Paris: Joubert, 1843.

Rodis-Lewis, Geneviève. "Hypothèses sur l'élaboration progressive des *Méditations* de Descartes." *Archives de Philosophie* 50 (1987): 109–23.

Rodis-Lewis, Geneviève. *L'anthropologie cartésienne*. Paris: Presses universitaires de France, 1990.

Rodis-Lewis, Geneviève. "On the Complementarity of Meditations III and V: From the 'General Rule' of Evidente to 'Certain Science'." In *Essays on Descartes' Meditations*, edited by A. Oksenberg Rorty, 271–95. Berkeley: University of California Press, 1986.

Rodis-Lewis, Geneviève. "Problèmes discutés entre Descartes et Regius: L'âme et le corps." In *Descartes et Regius. Autour de l'explication de l'esprit humain*, edited by Theo Verbeek, 35–46. Amsterdam: Rodopi, 1993.

Rösler-La Van, Claire. "Les négociations iréniques." In *Negotium irenicum. L'union des Églises protestantes selon G. W. Leibniz et D. E. Jablonski*, edited by Claire Rösler-Le Van, Gottfried W. Leibniz and Daniel E. Jablonski, 11–153. Paris: Classiques Garnier, 2013.

Roux, Sophie. "Découvrir le principe d'inertie." *Recherches sur la philosophie et le langage* 24 (2006): 453–515.

Roux, Sophie. "An Empire Divided: French Natural Philosophy (1670–1690)." In *The Mechanization of Natural Philosophy*, edited by Daniel Garber and Sophie Roux, 55–98. New York, Dordrecht, Boston and London: Kluwer Academic Publishers, 2013.

Roux, Sophie. "Histoire de la physique classique et historicité des sciences chez Meyerson." In *L'histoire et la philosophie des sciences françaises à la lumière de l'œuvre d'Émile Meyerson (1859–1933)*, edited by Eva Telkes-Klein and Elhanan Yakira, 91–114. Paris: Honoré Champion, 2010.

Roux, Sophie. "Meyerson et les mathématiques." *Corpus* 58 (2010): 28–45.

Roux, Sophie. "Pour une conception polémique du cartésianisme. Ignace-Gaston Pardies et Antoine Dilly dans la querelle de l'âme des bêtes." In *Qu'est-ce qu'être cartésien?* edited by Delphine Antoine-Mahut, 315–37. Lyon: ENS Editions, 2013.

Roux, Sophie. "Was There a Cartesian Experimentalism in Seventeenth Century France?" In *Cartesian Empiricisms*, edited by Mihnea Dobre and Tammy Nyden, 47–88. Dordrecht: Springer, 2014.

Ruler, Johannes A. van. *The Crisis of Causality: Voetius and Descartes on God, Nature and Change*. Leiden: Brill, 1995.

Rutherford, Donald. "Leibniz on Infinitesimals and the Reality of Force." In *Infinitesimal Differences: Controversies Between Leibniz and His Contemporaries*, edited by Ursula Goldenbaum and Douglas Jesseph, 255–80. Berlin: De Gruyter, 2008.

Rutherford, Donald. *Leibniz and the Rational Order of Nature*. Cambridge: Cambridge University Press, 1995.

Ryan, Todd. *Pierre Bayle's Cartesian Metaphysics*. London: Routledge, 2009.

Saccenti, Mario. *Lucrezio in Toscana. Studio su Alessandro Marchetti*. Firenze: Olschki, 1966.

Savini, Massimiliano. *Johannes Clauberg: Methodus cartesiana et ontologie*. Paris: Vrin, 2013.

Schmaltz, Tad M. *Descartes on Causation*. New York: Oxford University Press, 2008.

Schmaltz, Tad M. *Descartes and the Enlightenment*. Montréal: McGill-Queen's University Press, 1989.

Schmaltz, Tad M. *Early Modern Cartesianisms: Dutch and French Constructions*. Oxford: Oxford University Press, 2016.

Schmaltz, Tad M. *Radical Cartesianism: The French Reception of Descartes*. Cambridge: Cambridge University Press, 2002.

Schmid, Xaver. *Nikolaus Taurellus, der erste deutsche Philosoph*. Erlangen: Theodor Blacsing, 1860.

Schmitt, Charles B. "Aristotle Among the Physicians." In *The Medical Renaissance of the Sixteenth-Century*, edited by Andrew Wear, Roger K. French and Iain M. Lonie, 1–15. Cambridge: Cambridge University Press, 1985.

Schouls, Peter A. *Descartes and the Enlightenment*. Edimbourgh: Edimbourgh University Press, 1989.

Scribano, Emanuela. *Angeli e beati: Modelli di conoscenza da Tommaso a Spinoza*. Roma-Bari: Laterza, 1996.

Secada, Jorge. "Descartes on Time and Causality." *Philosophical Review* 99 (1990): 45–72.

Sennert, Daniel. *Epitome naturalis scientiae*. Wittenberg: Gronenbergius, 1599–1600.

Sergeant, Rose-Mary. *The Diffident Naturalist, Robert Boyle and the Philosophy of Experiment*. Chicago: University of Chicago Press, 1995.

Séris, Jean-Pierre. *Machine et Communication*. Paris: Vrin, 1987.

Serrapica, Salvatore. *Per una teoria dell'incertezza tra filosofia e medicina: Studio su Leonardo Di Capua 1617–1695*. Napoli: Liguori Editore, 2003.

Simmons, Alison. "Descartes on the Cognitive Structure of Sensory Experience." *Philosophy and Phenomenological Research* 67 (2003): 549–79.

Sleigh, Robert C. *Leibniz and Arnauld: A Commentary on Their Correspondence*. New Haven: Yale University Press, 1990.

Smith, Norman Kemp. *Studies in the Cartesian Philosophy*. New York: Palgrave Macmillan, 1902.

Sorabji, Richard. *Time, Creation, and the Continuum*. London: Duckworth, 1983.

Sorel, Charles. *De la perfection de l'homme, où les vrays biens sont considérez, et spécialement ceux de l'âme, avec les méthodes des sciences.* Paris: Robert Le Nain, 1655.

Sorel, Charles. *La Science des choses corporelles: Première partie de la science humaine, où l'on connoist la Verité de toutes les choses du Monde par les forces de la Raison et l'on trouve la refutation de Erreurs de la Philosophie vulgaire.* Paris: Pierre Billaine, 1634.

Spink, John S. *French Free Thought from Gassendi to Voltaire.* London: Athlone Press, 1960.

Stein, Howard. "Newton's Metaphysics." In *The Cambridge Companion to Newton*, edited by I. Bernard Cohen and George E. Smith, 256–307. Cambridge: Cambridge University Press, 2002.

Stone, Harold Samuel. *Vico's Cultural History: Production and Transmission of Ideas in Naples, 1685–1750.* Leiden: E.J. Brill, 1997.

Strazzoni, Andrea. "A Logic to End Controversies: The Genesis of Clauberg's *Logica Vetus et Nova.*" *Journal of Early Modern Studies* 2 (2013): 123–49.

Strazzoni, Andrea. "La fisica aristotelico-cartesiana di Johannes de Raey." *Giornale critico della filosofia italiana* 80 (2011): 107–32.

Sturm, Johann C. *Collegium Experimentale sive Curiosum*, 2 vols. Norimbergae: Wolfgang Mauritius Endterus et Johannis Andreae Endteri haeredum, 1675.

Suárez, Francisco. *Disputationes Metaphysicae*, 2 vols. Vol. XXV of *Opera omnia*, edited by Michel André and Charles Berton, 28 vols. Paris: Ludovicus Vivès, 1856–78. Reprinted in Hildesheim: Georg Olms, 1965.

Sutton, John. *Philosophy and Memory Traces: Descartes to Connectionism.* Cambridge: Cambridge University Press, 1998.

Taurellus, Nicolaus. *Emblemata physica ethica.* Nuremberg: Lochner, 1602.

Taurellus, Nicolaus. *Medicae praedictionis methodus, hoc est recta brevisque ratio coram aegris praeterita, praesentia futuraque praedicendi, morbos scilicet morborumque causas, mortem, sanitatem, recidivam aliqua symptomata . . .* Frankfurt am Main: Johann Feierabend, 1581.

Taurellus, Nicolaus. *Philosophiae Triumphus, hoc est, Metaphysica Philosophandi Methodus, qua divinitus inditis menti notitijs, humanæ rationes eò deducuntur, ut firmissimis indè constructis demonstrationibus, apertè rei veritas elucescat, & quæ diu Philosophorum sepulta fuit Philosophia victrix erumpat: quaestionibvs enim vel sexcentis, ea quibus cum revelata nobis veritate Philosophia pugnare videnbatur, adeo vere conciliantur, ut non fidei solum servire dicenda sit, sed eius esse fundamentum.* 2nd ed. Basel: Henricpetri, 1573; Arnhem: Jansonius, 1617.

Taurellus, Nicolaus. *Synopsis Aristotelis metaphysices ad normam Christianae religionis explicate, emendatae, et completae.* Hanau: Guilielmus Antonius, 1596.

Tega, Walter. *Arbor scientiarum: Enciclopedie e sistemi in Francia da Diderot a Comte.* Bologna: Il Mulino, 1984.

Testimonium Academiae Ultrajectinae, et narratio historica quae defensae, qua exterminatae novae philosophiae. Utrecht: W. Strickius, 1643.

Thijssen-Schoute, C. Louise. *Nederlands Cartesianisme*, edited by Th. Verbeek. Utrecht: HES, 1989.

Tiles, Mary. "Technology, Science, and Inexact Knowledge: Bachelard's Non-Cartesian Epistemology." In *Continental Philosophy of Science*, edited by Gary Gutting, 157–75. Malden: Blackwell Publishers, 2005.

Torrini, Maurizio. "Il Cartesio di Giannone." In *Pietro Giannone e il suo tempo*, edited by Raffaele Ajello, vol. 1, 415–30. Napoli: Jovene, 1980.

Torrini, Maurizio. "L'Accademia degli Investiganti: Napoli 1663–1670." *Quaderni storici* 48 (1981): 845–83.

Torrini, Maurizio. "La discussione sullo statuto delle scienze tra la fine del '600 e l'inizio del '700." In *Galileo e Napoli*, edited by Fabrizio Lomonaco and Maurizio Torrini, 357–83. Napoli: Guida, 1987.

Torrini, Maurizio. *Tommaso Cornelio e la ricostruzione della scienza*. Napoli: Guida, 1977.

Torrini, Maurizio. "Un episodio della polemica tra antichi e moderni: la disputa sulla macerazione dei lini nel lago di Agnano." *Bollettino del Centro di studi vichiani* 5 (1975): 56–70.

Verbeek, Theo. "Descartes et les Pays-Bas: Philosophie et théologie." In *Il vocabolario della République des Lettres. Terminologia filosofica e storia della filosofia. Problemi di metodo*, edited by Marta Fattori, 301–6. Firenze: Leo S. Olschki, 1997.

Verbeek, Theo. *Descartes and the Dutch: Early Reactions to Cartesian Philosophy, 1637–1650*. Carbondale, Edwardsville: Southern Illinois University Press, 1992.

Verbeek, Theo. "*Ens per accidens*: Le origini della *Querelle* di Utrecht." *Giornale critico della filosofia italiana*, 6th series 71 (1992): 276–88.

Verbeek, Theo. *La querelle d'Utrecht*. Paris: Les Impressions Nouvelles, 1988.

Verbeek, Theo. "Les cartésiens face à Spinoza: l'exemple de Johannes de Raey." In *L'hérésie spinoziste: la discussion sur le Tractatus theologico-politicus, 1670–1677, et la réception immédiate du spinozisme*, edited by Paolo Cristofolini, 77–88. Amsterdam: APA Maarssen, 1995.

Verbeek, Theo. "Tradition and Novelty: Descartes and Some Cartesians." In *The Rise of Modern Philosophy*, edited by Tom Sorell, 167–96. Oxford: Oxford University Press, 1993.

Vinci, Thomas C. "Reason, Imagination, and Mechanism in Descartes' Theory of Perception." In *Oxford Studies in Early Modern Philosophy*, edited by Daniel Garber and Steven Nadler, 35–73. Oxford: Clarendon Press, 2005.

Voetius, Gijsbert. *Selectarum disputationum theologicarum pars prima*. Utrecht: Johannes a Waesberge, 1648.

Voltaire. "Lettres Philosophiques." In *Œuvres complètes de Voltaire*, edited by Louis Moland. Paris: Garnier, 1883. First published 1734.

Wade, Ira O. *The Clandestine Organization and Diffusion of Philosophical Ideas in France from 1700 to 1750*. Princeton: Princeton University Press, 1938.

Wade, Ira O. *The Intellectual Origins of the French Enlightenment*. Princeton: Princeton University Press, 1971.

Westfall, Richard S. *Force in Newton's Physics: The Science of Dynamics in the Seventeenth Century*. London: MacDonald; New York: Amercian Elsevier, 1971.

Whipple, John. "Continual Creation and Finite Substance in Leibniz's Metaphysics." *Journal of Philosophical Research* 36 (2011): 1–30.

Whipple, John. "The Structure of Leibnizian Simple Substance." *British Journal for the History of Philosophy* 18 (2010): 379–410.

Wilson, Catherine. "Descartes and the Corporeal Mind. Some Implications of the Regius Affair." In *Descartes' Natural Philosophy*, edited by Stephen Gaukroger, John Schuster and John Sutton, 659–79. London and New York: Routledge, 2000.

Wilson, Catherine. *Leibniz's Metaphysics: A Historical and Comparative Study*. Manchester: Manchester University Press, 1989.
Wilson, Margaret. "Descartes on the Perception of Primary Qualities." In *Ideas and Mechanism: Essays in Early Modern Philosophy*, 26–40. Princeton: Princeton University Press, 1999.
Wolff, Christian. *Philosophia Prima sive Ontologia* [. . .]. Francofurti and Lipsiæ: in Officina libraria Rengeriana, 1736.
Yolton, John W. "Ideas and Knowledge in Seventeenth-Century Philosophy." *Journal of the History of Philosophy* 13 (1975): 145–65.
Yolton, John W. *Locke and French Materialism*. Oxford: Clarendon Press, 1991.
Yolton, John W. *Perceptual Acquaintance from Descartes to Reid*. Minneapolis: University of Minnesota Press, 1984.
Yolton, John W. *Thinking Matter*. Oxford: Blackwell Publishers, 1984.
Zedler, Johann Heinrich. *Grosses vollständiges Universal-Lexicon aller Wissenschaften und Künste*. Leipzig and Halle: Carl Günther Ludovici, 1732–54; Repr. Graz: Akademische Druck- und Verlagsanstalt, 1961–1964.

Contributors

Jean-Pascal Anfray. Associate Professor, ENS, Paris. " 'Autant de mondes sans connexion': Leibniz et Lewis sur la compossibilité et l'unité du monde." *Les études philosophiques* 164 (2016), 537–57; "Le labyrinthe temporel. Simplicité, persistance et création continuée chez Leibniz." *Archives de philosophie* 77 (2014), 43–62; "Partes extra partes. Étendue et impénétrabilité dans la correspondance entre Descartes et More." *Les études philosophiques* 108 (2014), 37–59.

Delphine Antoine-Mahut. Professor ENS de Lyon. *Oxford Handbook on Descartes and Cartesianisms*, ed. in collaboration with Steve Nadler and Tad Schmaltz, forthcoming; *Descartes' Treatise on Man and Its Reception*, ed. in collaboration with Stephen Gaukroger, Dordrecht: Springer, 2016; "Reviving Spiritualism with Monads: Francisque Bouillier's Impossible Mission (1839–1864)." *British Journal for the History of Philosophy* 23, (2015), 1106–27.

Delphine Bellis. Associate Professor, Paul Valéry University, Montpellier. "Nos in Diem Vivimus: Gassendi's Probabilism and Academic Philosophy from Day to Day." In *Academic Scepticism in the Development of Early Modern Philosophy*, Sébastien Charles and Plinio Junqueira Smith eds., Dordrecht: Springer, 2017, 125–52; "Météores cartésiens et météores scolastiques: la lecture philosophique d'Étienne Gilson." In *Gilson et Descartes à l'occasion du centenaire de La liberté chez Descartes et la théologie*, Dan Arbib and Francesco Marrone eds., Lecce: ClioEdu Edizioni, 2015, 41–56; "Empiricism Without Metaphysics: Regius's Cartesian Natural Philosophy." In *Cartesian Empiricisms*, Mihnea Dobre and Tammy Nyden eds., Dordrecht: Springer, 2013, 151–83.

Desmond M. Clarke. Professor at Cork University, †2017. "The Physics and Metaphysics of the Mind: Descartes and Regius." In *Mind, Method, and Morality: Essays in Honour of Anthony Kenny*, John Cottingham and Peter Hacker eds., Oxford: Oxford University Press, 2010, 187–207; *Descartes' Theory of Mind*, Oxford: Oxford University Press, 1992; *Descartes' Philosophy of Science*, Manchester: Manchester University Press, 1982.

Antonella Del Prete. Associate Professor at the Universita della Tuscia. *Chemins du cartésianisme*, ed. in collaboration with Raffaele Carbone, Paris: Classiques Garnier, 2017; "Discussioni sul metodo nel cartesianismo olandese. Il caso di Johannes de Raey." In *La ragione e le sue vie. Saperi e procedure di prova in età moderna*, Carlo Borghero and Claudio Buccolini eds. Firenze: Le Lettere, 2015, 146–67; "Oltre Descartes: filosofia e teologia nella *Theologia pacifica* di Christoph Wittich." In *Immagini filosofiche e interpretazioni storiografiche del cartesianismo*, Carlo Borghero and Antonella Del Prete eds., Firenze: Le Lettere, 2011, 25–45.

Pierre Girard. Professor at the University of Lyon III. *"Comme des lumières jamais vues." Matérialisme et radicalité politique dans les premières Lumières à Naples (1647–1744)*, Paris: Honoré Champion, 2016; "Les usages de Malebranche dans la réception du cartésianisme à Naples." In *Les Malebranchismes des Lumières. Études sur les réceptions contrastées de la philosophie de Malebranche, fin XVIIe et XVIIIe siècles*, Delphine Antoine-Mahut ed., Paris: Honoré Champion, 2014, 225–46; "Matérialisme et politique: les enjeux de la réception du cartésianisme à Naples à l'Âge classique." *Corpus* 61 (2011), 113–32.

Philippe Hamou. Professor at University of Nanterre (Paris X). *Dans la chambre de l'esprit. John Locke et l'invention du mind*, Paris: Ithaque, 2018; *Locke and Cartesian Philosophy*, in collaboration with M. Pécharman, Oxford: Oxford University Press, 2018; "Experimentum crucis. Newton's Empiricism at the Crossroads," in Anne-Lise. Rey and Siegfried Bodemann eds.; *What Does It Mean to Be an Empiricist?* Dordrecht: Springer, 2018: 47–69.

Mogens Lærke. Senior Researcher at the CNRS, based at the ENS de Lyon. "Leibniz's Encounter with Spinoza's Monism, October 1675 to February 1678." In *Oxford Handbook to Spinoza*, M. Della Rocca ed., Oxford: Oxford University Press, 2018, 434–63; "Materialism, Constructivism, and Political Skepticism. Leibniz, Hobbes and the Erudite Libertines." In *The New Politics of Materialism. History, Philosophy, Science*, S. Ellenzweig and J. Zammito eds., London and New York: Routledge, 2017, 251–69; *Les Lumières de Leibniz. Controverses avec Huet, Bayle, Regis et More*, Paris: Classiques Garnier, 2015.

Christoph Lüthy. Professor at Radboud University, Nijmegen. "Conceptual and Historical Reflections on Chance (and Related Concepts)" in collaboration with Carla Rita Palmerino. In *The Challenge of Chance: A Multidisciplinary Approach from Science and the Humanities*, N.P. Landsman and E.J. van Wolde eds., Berlin : Springer, 2016: 9–47; *David Gorlaeus (1591–1612). An Enigmatic Figure in the History of Philosophy and Science*, Amsterdam: Amsterdam University Press, 2012; "Where Logical Necessity Turns into Visual Persuasion: Descartes' Clear and Distinct Illustrations." In *Transmitting Knowledge: Words, Images* *and Instruments in Early Modern Europe*, Ian Maclean and Sachiko Kusukawa eds., Oxford: Oxford University Press, 2006, 97–133.

Sophie Roux. Professor at the ENS, Paris. "The Two Comets of 1664-1665: A Dispersive Prism for French Natural Philosophy Principles." In *The Idea of Principles in Early Modern Thought*, Peter Anstey ed., London and New York: Routledge, 2017, 98–146; *The Mechanization of Natural Philosophy*, ed. in collaboration with Dan Garber, Dordrecht: Springer, 2013; L'Essai de logique *de Mariotte. Archéologie des idées d'un savant ordinaire*, Paris: Classiques Garnier, 2011.

Emanuela Scribano. Professor at Università Ca' Foscari Venezia. *A Reading Guide to Descartes' Meditations,* South Bend, Indiana: St. Augustine Press; *I volti dell'errore nel pensiero moderno. Da Bacone a Leibniz. Rivista di storia della filosofia,* in collaboration with Matteao Favaretto Camposampiero and Mariangela Priarolo, 2016; *Macchine con la mente. Fisiologia e metafisica Tra Cartesio e Spinoza*, Roma: Carocci, 2015.

Mariafranca Spallanzani. Professor at the University of Bologna. *Descartes. La règle de la raison*, Paris: Vrin, 2015; *L'arbre et le labyrinthe. Descartes selon l'ordre des Lumières*, Paris: Champion, 2009; *Immagini di Descartes nell'Encyclopédie*. Bologna: Il Mulino, 1999.

Index Rerum and Nominum[1]

1 The index omits the terms "Cartesian," "Descartes" and "physics" on purpose. Editors of collective books are not included either. Adjectival entries (for example, "atomist" or "Aristotelian") are collecting references to the corresponding adjective (thus, "atomist" or "Aristotelian"), but also to the name derived from it (in the example, "atomism" or "Aristotelianism"). Entries are in English, but they include references to terms from other languages as well: for example, the entry "system" includes references to "esprit de système".

www.ingramcontent.com/pod-product-compliance
Ingram Content Group UK Ltd.
Pitfield, Milton Keynes, MK11 3LW, UK
UKHW021300190926
475701UK00009B/168
* 9 7 8 1 0 3 2 0 9 4 8 1 6 *